Flexible Electronic Packaging and Encapsulation Technology

Flexible Electronic Packaging and Encapsulation Technology

Edited by Hong Meng and Wei Huang

WILEY-VCH

Editors

Prof. Hong Meng
Peking University
Shenzhen Graduate School
Lishui Road, Nanshan District
Shenzhen, 518055
China

Prof. Wei Huang
Northwestern Polytechnical University
127 West Youyi Road
Xi'an, 710072
China

Cover Image: © Courtesy of Hong Meng

C9783527353590_280426

Library of Congress Card No.:
applied for

British Library Cataloguing-in-Publication Data
A catalogue record for this book is available from the British Library.

Bibliographic information published by the Deutsche Nationalbibliothek
The Deutsche Nationalbibliothek lists this publication in the Deutsche Nationalbibliografie; detailed bibliographic data are available on the Internet at <http://dnb.d-nb.de>.

Print ISBN: 978-3-527-35359-0
ePDF ISBN: 978-3-527-84570-5
ePub ISBN: 978-3-527-84571-2
oBook ISBN: 978-3-527-84572-9

Typesetting Straive, Chennai, India
Printing and Binding CPI Group (UK) Ltd, Croydon, CR0 4YY

The manufacturer's authorized representative according to the EU General Product Safety Regulation is Wiley-VCH GmbH, Boschstr. 12, 69469 Weinheim, Germany, e-mail: Product_Safety@wiley.com.

Contents

Preface

With the advent of the digital and information age, flexible electronic technology has gained increasing attention from the academic and industrial worlds. The development of this technology has made electronic devices lighter, more durable, and capable of deploying electronic functions on various shapes and surfaces, opening up new possibilities for designing and manufacturing innovative products. As a cutting-edge technology with huge potential, flexible electronic technology is reshaping our everyday lives and could potentially have a profound impact on various industries. Flexible electronic packaging technology, as an important and key part of flexible electronic technology, is continually advancing with technological innovation. This powerful tool has had a profound impact on various fields, including the Internet of Things, consumer electronics, healthcare, and new energy vehicles. Given the rapid changes in this unpredictable field, we believe it is necessary to compile a professional book on flexible electronic packaging technology that systematically and straightforwardly introduces and discusses the basic theories, practical applications, and frontier developments of this technology. The rise of flexible electronic packaging technology is rooted in the pursuit of more portable, lighter, and humanized electronic devices. Today, as technology continues to advance, our demand and expectation for electronic devices are constantly increasing. What we pursue is not only functionality and powerful performance but also lightness, flexibility, and environmental friendliness. This requires us to adopt new materials and new manufacturing processes, and flexible electronic packaging technology is key to meeting these needs.

Chapters 1 and 2 will provide readers with a comprehensive background and basic concepts, including the historical development, basic components, working principles, and application areas of flexible electronic packaging technology. The goal of these two chapters is to help readers establish a comprehensive understanding and deep comprehension, providing a clear direction and solid foundation for subsequent learning and applications. In Chapters 3 and 4, we will focus on flexible substrates and related testing methods, two indispensable aspects of flexible electronic packaging. In these chapters, we provide detailed explanations and practical guides, aiming to help readers understand and master these key points. In Chapters 5 and 6, we will delve into the specific operations of flexible electronic packaging and discuss the development of flexible electronic packaging technology, including traditional packaging technology, the latest research results, and frontier technologies. Through these two chapters, we hope that readers can have a deeper understanding and knowledge of flexible electronic packaging technology and better prepare for its future development. In the following three chapters, we will explore the applications of flexible electronic packaging technology in various industries, introduce related testing standards, and reveal industry business models and success stories through case studies of flexible electronic packaging companies. Through these contents, we hope to help readers understand the practical applications and commercial value of flexible electronic packaging technology, enabling them to transform theoretical knowledge into practical skills and achieve success in their respective fields. In the final chapter, we will look forward to the development trend of flexible electronic packaging, including the development of technology, changes in the market, policy impacts, and social demands. Through this chapter, we aim to help readers have a clearer understanding and better preparation for the future.

In summary, this book offers a comprehensive overview of the development and innovative trends within the field of flexible electronic packaging technology, further underscoring its pivotal role in the advancement of future technology. Our aim is to provide a systematic understanding of this technology to our readers, regardless of whether they are newcomers to this field or have already built a certain foundation.

The chapters of this book were completed jointly by students and teachers from Northwestern Polytechnical University and Peking University Shenzhen Graduate School. Here, we would like to express our deepest gratitude to the teachers and students who contributed to this book: Chapter 1 was written by student Chen Yongji and Prof. Liu Zhenguo; Chapter 2 was written by Prof. Zong Peng'an; Chapter 3 was written by Prof. Chen Yanhui; Chapter 4 was written by student Yuan Junjie and Prof. Meng Hong; Chapter 5 was written by Prof. Yu Tao; Chapter 6 was written by student Rao Qiushi and Prof. Meng Hong; Chapter 7 was written by Prof. Zhang Yuezhou; Chapter 8 was written by student Yuan Junjie and Prof. Meng Hong; Chapter 9 was written by teacher Wei Zhenrong and Prof. Meng Hong; and Chapter 10 was written by student Liu Mingqiang and Prof. Meng Hong. The editing work of the book was completed under the personal deployment and careful guidance of academician Huang Wei. We thank them for their selfless contributions and for investing their valuable time. We also thank our families for their support and encouragement, and the contributions of students Chen Yongji and Yuan Junjie in the editing process of the book. We also appreciate our readers for choosing this book as part of their learning journey, and we look forward to your achievements and breakthroughs in this field. We are willing to accompany you on this path full of challenges and opportunities, witnessing the development and transformation of flexible electronic packaging technology together.

23 July 2023

Hong Meng
Wei Huang

1

Overview of Flexible Electronic Encapsulating Technology

Zhenguo Liu and Yongji Chen

Northwestern Polytechnical University Ningbo Research Institute, Qingyi Road 218, 315821, Ningbo, China

1.1 Flexible Electronics Overview

Flexible electronics, with its unique flexibility, ductility and high efficiency and low cost manufacturing process, have wide application prospects in information, energy, medical, national defense and other fields. As with traditional integrated circuit (IC) technology, manufacturing processes and equipment are also the main drivers for the development of flexible electronics technology. Flexible electronics manufacturing technology level indicators include chip feature size and substrate area size; the key is how to create a smaller feature size of flexible electronic devices on a larger format substrate at a lower cost.

Compared to traditional electronics, flexible electronics have greater flexibility and can adapt to a certain extent to different working environments to meet the requirements of the device's deformation. Flexible electronics covers organic electronics, plastic electronics, bioelectronics, nanoelectronics, and printed electronics, including radio frequency identification (RFID), flexible display, organic electroluminescent (Organic Light-Emitting Diode, OLED) display and lighting, chemical and biological sensors, flexible photovoltaics (PVs), flexible memory or storage, flexible batteries, wearable devices, and many other applications. With its rapid development, the involved fields have been further expanded, and now it has become one of the research hotspots in cross-disciplinary research (Figure 1.1) [1].

In recent years, with the further improvement of flexible electronic technology, we have seen some unimaginable products. For example, the current attention is on folding-screen (Figure 1.2) and wrap-around-screen cell phones. In fact, whether it is a folding screen or a wrap-around screen, the essence of the use of flexible screen technology is that it is a form of flexible electronics technology. The flexible screen, flexible chip, and flexible electrode are only the tip of the iceberg of flexible electronics technology. In fact, information technology involves a variety of sensing, information transmission, information processing, energy storage, and other links that are expected to achieve flexibility [2].

Flexible Electronic Packaging and Encapsulation Technology, First Edition.
Edited by Hong Meng and Wei Huang.

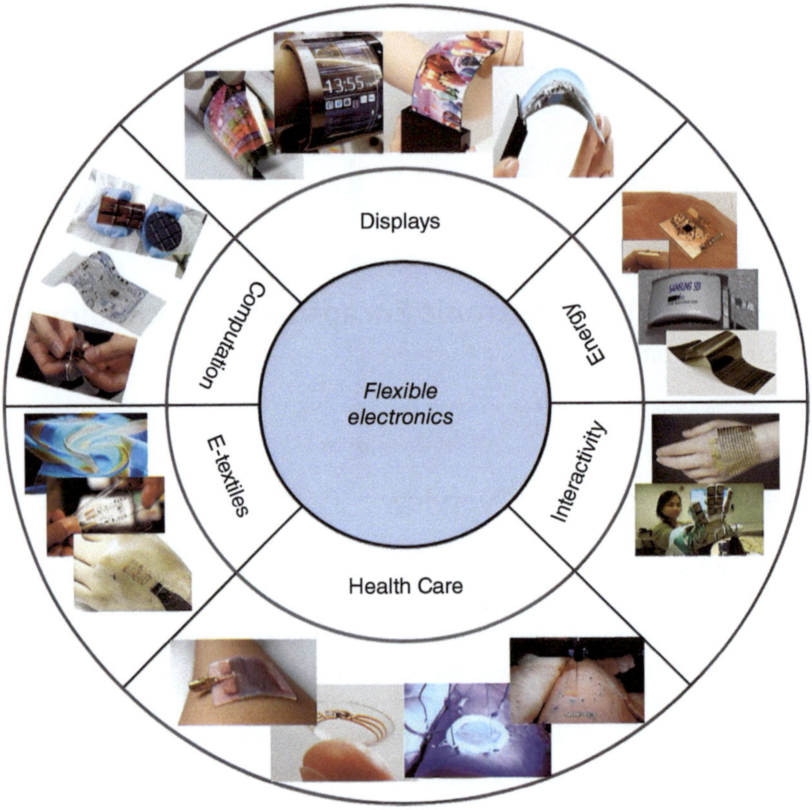

Figure 1.1 The fields of flexible electronics.

So, how exactly is flexible electronics achieved?

First, let us understand the materials used in flexible electronics. Common materials for flexible electronics include flexible substrates, metallic materials, organic materials, inorganic semiconductor materials, and carbon materials represented by graphene (Figure 1.3).

After the raw materials are available, let us look at how flexible electronic devices are manufactured. There are three common flexible electronics fabrication methods: transfer printing, inkjet printing, and fiber structure formation. Among them, transfer printing is a series of functional arrangement techniques used to deterministically assemble micromaterials and nanomaterials into spatially organized structures with two- and three-dimensional layouts [4]. Inkjet printing, on the other hand, as the name implies, allows the direct deposition of functional materials to form patterns on substrates [5]. Flexible electronics fabrication methods based on fiber structures are well suited for wearable electronics that are lightweight, durable, flexible, and comfortable [6].

Since, as mentioned above, flexible electronics have so many advantages and broad application prospects, why has their development been slow to open up?

Figure 1.2 Folding-screen phone.

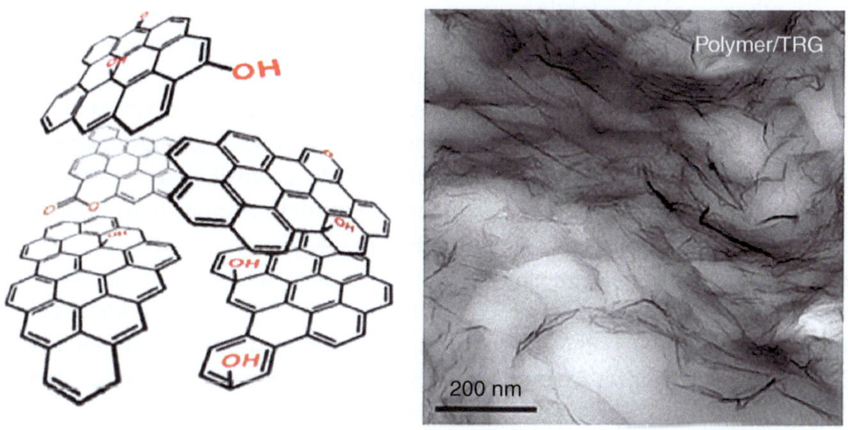

Figure 1.3 Thermally reduced graphite oxide (TRG). Source: Kim et.al. [3]; © 2010, Reproduced with permission from American Chemical Society.

Two major obstacles impede the development of flexible electronics: mechanics and encapsulation. Shen Yang, vice president of the School of Materials at Tsinghua University, has said that the first challenge in the development of flexible electronics is the mechanics of the problem: flexible electronic components in repeated folding and bending will be constantly subjected to alternating stress over time, making them easy to crack. This problem can be overcome mainly through structural design. The second challenge is the problem of electronic encapsulating, which is to integrate the components on the flexible substrate tightly encapsulated together and achieve the desired function.

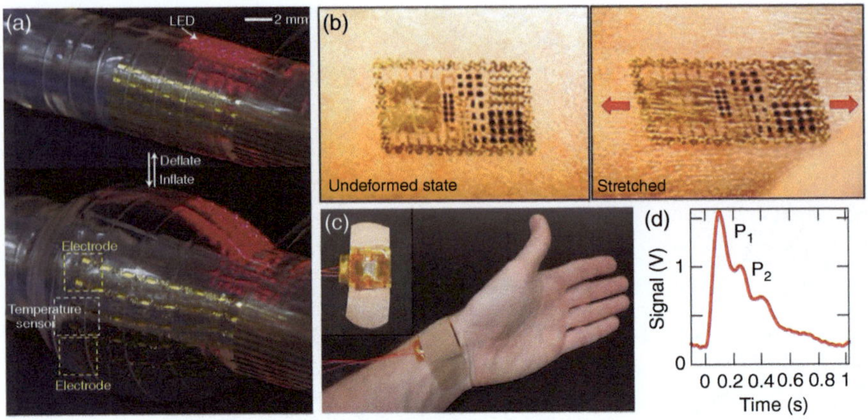

Figure 1.4 Flexible sensors. Source: Hammock et al. [7]; 2013, Reproduced with permission from John Wiley and Sons.

Furthermore, the slow progress in the development of flexible electronics can be attributed to the absence of a significant "viral effect" in terms of application scenes. In other words, the folding-screen cell phone is not an industry pain point. However, another application of flexible electronics – flexible sensors – may be the real revolutionary change in the industry application scene.

Using flexible sensors and conductors, scientists can convert the external force or heat into electrical signals, which are transmitted to the robot's computer for signal processing, so that it can be made transparent, flexible, extensible, freely bendable, foldable, and wearable electronic skin in order to monitor the human body indicators in real time and accurately [7], as shown in Figure 1.4.

Recently, the Institute of Mechanics of the Chinese Academy of Sciences, in cooperation with researchers from Dalian University of Technology and Beijing University of Aeronautics and Astronautics, has developed a thin-film patch-type flexible curvature sensor for wearable devices from the mechanical structure design (see Figure 1.5). This sensor can accurately measure the dynamic bending curvature and bending angle of the measured surface, and its bending measurement results are not affected by tensile deformation. So, in practical application, it does not require the sensor to be perfectly bonded to the measured surface but simply fit, so it is no problem at all even with gloves or tights on. Also, this sensor is very suitable for integration with wearable apparel and can be applied to flexible smart wearable devices such as joint flexion monitoring, gesture recognition, and sitting posture monitoring [8].

Currently, there are two main approaches to the selection of flexible electronic materials internationally. One approach is to shift from traditional inorganic materials to organic materials, such as polymer materials and organic semiconductors, for flexible electronic applications. Another approach involves the combination of organic and inorganic materials, utilizing composite materials to develop flexible electronic devices.

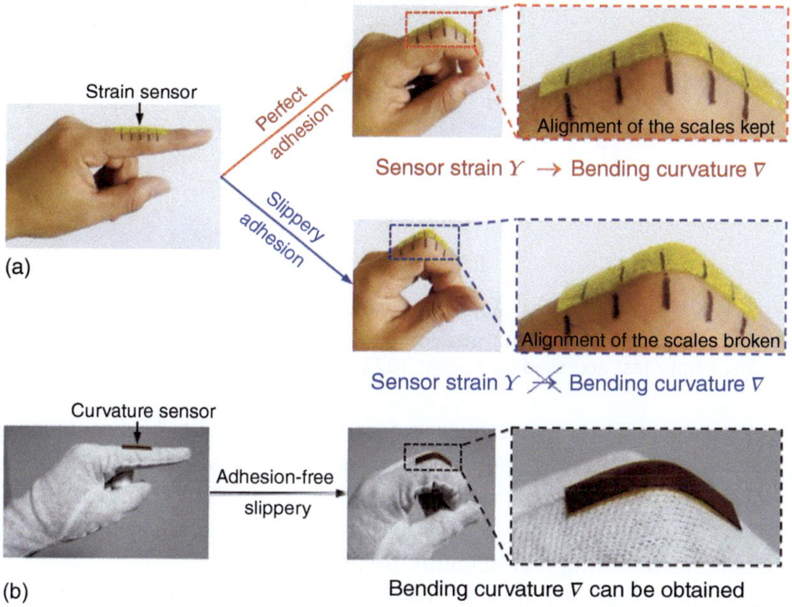

Figure 1.5 Curvature sensors for joint flexion deformation, gesture recognition, and sitting posture monitoring. (a) Strain sensor. (b) Curvature sensor. Source: Liu et al. [8]; © 2018, Reproduced with permission from John Wiley and Sons.

Since the discovery of graphene, two-dimensional materials consisting of single layers of atoms, such as boron nitride, molybdenum disulfide, and black phosphorus, have garnered attention from the semiconductor industry. Research related to these two-dimensional materials holds promise for the advancement of flexible electronics.

The successful application of flexible displays, flexible sensors, and other flexible electronic components signifies the transition of flexible electronics from theory to practice. This advancement may herald a new era of electronic device revolution, bridging the gap between humans and machines and fostering closer interactions.

1.2 Development of Flexible Electronic Encapsulating Technology

One generation of encapsulating, one generation of products. One generation of encapsulating, one generation of products. After the flexible electronic device is manufactured, before it is brought to market as a product, it needs to be packaged to isolate water vapor to ensure its stable operation and complete function. The encapsulating of flexible electronic devices is the same as the encapsulating of traditional electronic devices and is a branch of the encapsulating of electronic devices. When it comes to the word encapsulating, it first originated from the encapsulating of ICs,

(a) (b)

Figure 1.6 TO-type encapsulation (a) and double inline encapsulation (b). Source: Reproduced with permission from huangye88.com / http://yiqiyibiao.huangye88.com/xinxi/5529un9e0ef8b2.html / last accessed 02 August 2023.

and the development of IC encapsulating was developed along with the development of IC chips. The history of the development of encapsulating is also the history of the continuous improvement of chip performance and the continuous miniaturization of the system. As the size of IC devices shrinks and the operating speed increases, new and higher encapsulating requirements are placed on ICs.

Therefore, before introducing flexible electronic device encapsulating, we review the development history of the IC encapsulating industry, which will inspire us to learn as well as develop flexible electronic device encapsulating. The history of the IC encapsulation industry can be divided into two stages: traditional encapsulation and advanced encapsulation, with the year 2000 serving as a crucial boundary [9].

Traditional encapsulating: The development of traditional encapsulating technology can be further subdivided into three phases.

Phase 1 (before 1980): This phase corresponds to the through hole (TH) era, characterized by jack mounting to printed circuit boards (PCBs), pin counts below 64, fixed pitch, a maximum mounting density of 10 pins/cm^2, and encapsulation types such as transistor outline (Transistor Outline Encapsulation, TO) and dual inline (Dual Inline Encapsulation, DIP), as represented in Figure 1.6.

Phase 2 (1980–1990): This phase marks the surface mount technology (SMT) era, characterized by the use of leads instead of pins, wing-shaped or ding-shaped leads, two-sided or four-sided leads, pitch ranging from 1.27 to 0.44 mm, suitable for 3–300 leads, mounting density of 10–50 pins/cm^2, and encapsulation types such as small outline encapsulation (SOP) and quad-edge flat Packaging (QFP), as depicted in Figures 1.7 and 1.8.

Phase 3 (1990–2000): This phase corresponds to the area array encapsulation era, represented by ball grid array (BGA) and chip scale encapsulation (CSP). In this phase, solder balls are used for encapsulation, greatly reducing the distance between the chip and the system. Moreover, the multi-chip module (MCM) represents the integration of multiple chips on a high-density multilayer interconnected substrate using SMT, enabling the assembly of various electronic components and subsystems, as shown in Figure 1.9.

(a) (b) (c)

Figure 1.7 Narrow pitch small outline encapsulation (SSOP) (a), thin small outline encapsulation (TSOP) (b), and thin shrink small outline packaging (TSSOP) (c). Source: Reproduced with permission from Guangzhou Nuode Electronics Co., Ltd / https://www.nodpcba.com/news/2769-cn.html / last accessed 02 August 2023.

(a) (b)

Figure 1.8 Quad-edge flat encapsulation (QFP) (a), Quad (dual) edge pinless flat encapsulation QFN/DFN (b). Source: Reproduced with permission from Guangzhou Nuode Electronics Co., Ltd / https://www.nodpcba.com/news/2769-cn.html / last accessed 02 August 02 2023.

(a) (b)

Figure 1.9 Solder ball array encapsulation BGA (a), chip level encapsulation (CSP) (b). Source: Reproduced with permission from biyuzg / http://www.biyuzg.com/ / last accessed 02 August 02 2023.

Advanced encapsulation: Since the mid-1990s, driven by the demand for multifunctional system products and advancements in CSP encapsulation and multilayer substrate technology, the IC encapsulation industry has entered the era of three-dimensional (3D) stacked encapsulation. The key features of advanced encapsulation include: (i) the evolution from encapsulating components to encapsulating systems; (ii) the shift from single-chip to multi-chip development; (iii) the transition from multi-chip module (MCM) flat encapsulation to 3D encapsulation; and (iv) the adoption of flip chip (FC) connection and

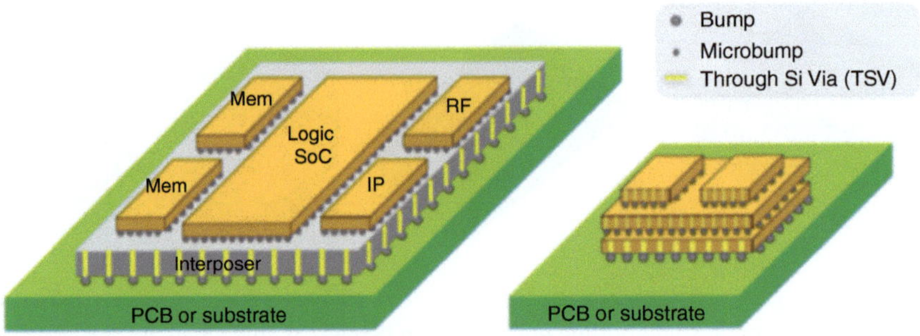

Figure 1.10 2.5D IC encapsulation (a), 3D IC encapsulation (b). Source: Reproduce with permission fromANSYScorporation.

silicon through-hole connection (Through Silicon Vias, TSV) as the main bonding methods. Advanced encapsulation technologies include FC, wafer-level encapsulation, and encapsulation-on-encapsulation (PoP)/system-in-a-encapsulation (Sip)/TSV, which exhibit the following characteristics.

3D encapsulating technology: Multi-chip component planar encapsulating technology integrates multiple IC chips to achieve the integration of encapsulated products in terms of area, then allows chip integration to achieve vertical integration, which is the main efficacy of 3D encapsulating technology. 3D encapsulating can be achieved in two ways: die stacking and encapsulation stacking within the encapsulation. 3D encapsulating is a combination of FC, wafer-level, and POP/Sip/TSV encapsulating technologies, and its development is divided into three stages: the first stage uses lead and FC bonding technology to stack chips; the second stage uses encapsulation body stacking (POP); and the third stage uses through-silicon-via technology to achieve chip stacking (Figure 1.10).

FC technology is not a specific encapsulation type but a circuit interconnection technology that facilitates direct connection between the chip's bare face and the substrate. Unlike wire bond (WB) and tape automated bonding (TAB) technologies that restrict the chip pads to the periphery of the chip, FC technology allows for better electrical performance by utilizing the entire chip area, eliminating the need for interconnection leads. The general structure of FC encapsulation is illustrated in Figure 1.11.

Wafer-level encapsulation (WLP) technology is a product of the combination of FC technology with surfacemount technology and solder ball array encapsulating technology under the continuous pursuit of miniaturization in the market and is an improved and enhanced chip-level encapsulation. WLP is very different from the traditional encapsulating method (first cut and then sealed, after the encapsulation area is at least greater than 20% of the original chip area). WLP technology is first applied to the whole wafer at the same time for many chip encapsulations, testing, and finally cutting into a single device and directly mounted on the substrate or PCB, so the volume after the encapsulation is equal to the original size of the chip, and production costs are also significantly reduced. WLP technology can also be called standard WLP (fan-in WLP) and then evolved into diffusion WLP (fan-out

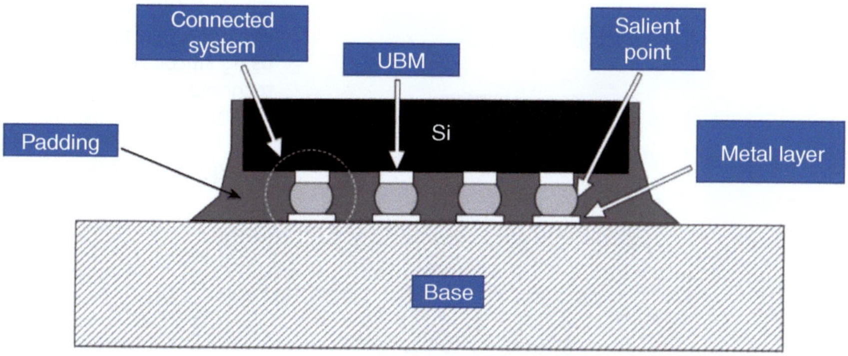

Figure 1.11 General structure of flip chip encapsulation body. Source: Reproduced with permission from www.dymek.cn.

WLP), which is based on wafer reconfiguration technology, rearranging the chip on an artificial wafer and then following similar steps to the standard WLP process for encapsulating, as depicted in Figure 1.12.

PoP is an out-of-encapsulation encapsulation, which refers to the longitudinal arrangement of logic and storage components of the IC encapsulation form. It uses two or more BGA stacks with generally strong resistance under the logic operation located at the bottom and storage components located in the upper, using solder balls to combine the two encapsulations. It is mainly used in the manufacture of advanced portable devices and smartphones used in advanced mobile communication platforms, as shown in Figure 1.13.

Figure 1.12 Wafer-level encapsulation devices. Source: Reproduced with permission from Daimei Instrument Technology Service (Shanghai) Co., Ltd / https://www.dymek.cn/ / last accessed 02 August 2023.

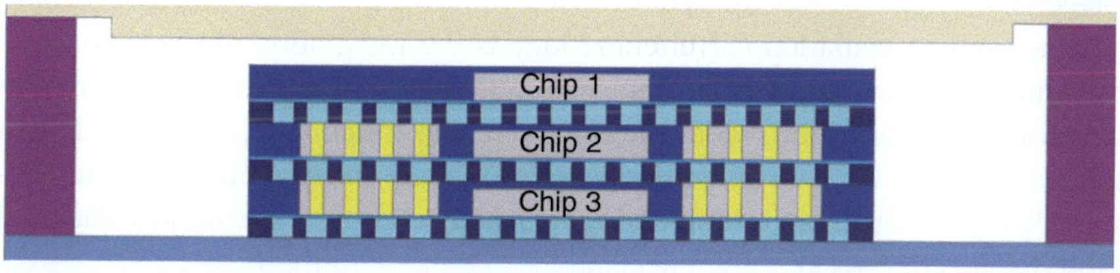

Figure 1.13 Three-dimensional chip stacking structure. Source: Reproduced with permission from www.dymek.cn.

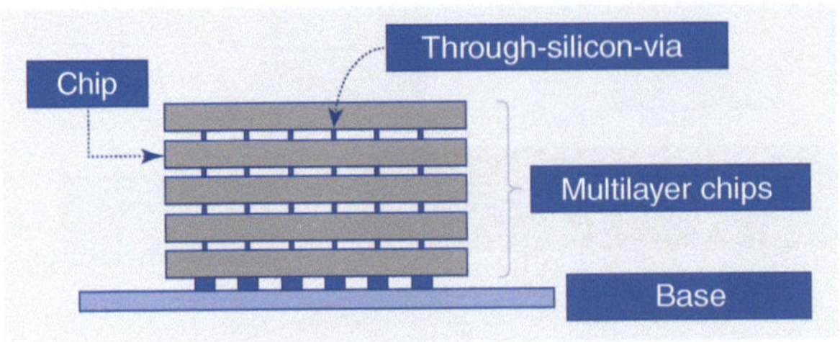

Figure 1.14 3D encapsulation structure using silicon through-hole technology. Source: Reproduced with permission from https://zhuanlan.zhihu.com/p/396703245.

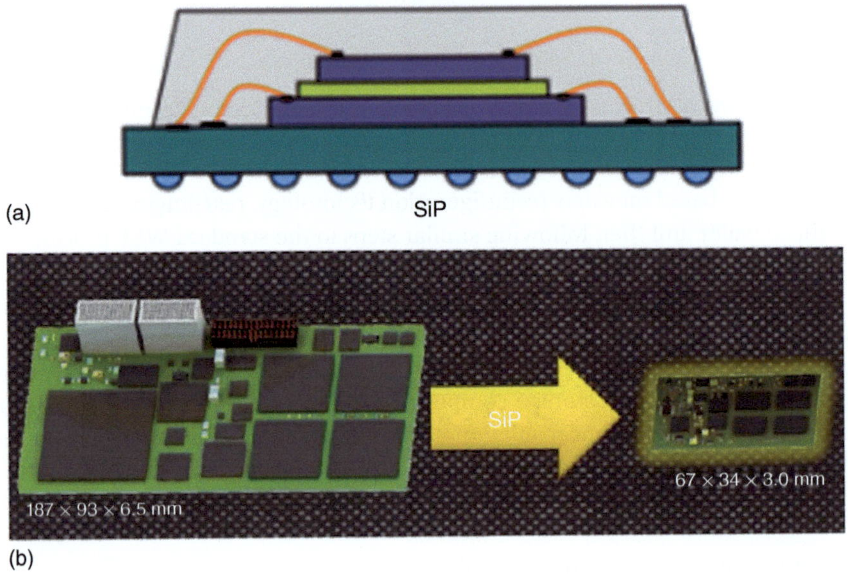

Figure 1.15 System-level encapsulation structure schematic (a) and system-level encapsulation devices (b). Source: Reproduced with permission from https://smartwear.zol.com.cn/458/4582055.html.

TSV technology is also a circuit interconnection technology that interconnects chips by creating vertical conduction between chip and chip and wafer to wafer. Unlike previous IC encapsulation bonding and stacking techniques that use bumps, TSV enables the highest density of chip stacking in the three-dimensional direction, the smallest form factor, and greatly improves chip speed and low power performance. TSV is a key technology for 2.5D and 3D encapsulations, as illustrated in Figure 1.14.

System in a encapsulation (SIP) technology is the integration of multiple functional chips, including processors, memories, and other functional chips, into a single encapsulation to achieve an essentially complete function. It corresponds to a system on a chip (SOC). The difference is that SOC is an encapsulation in which different chips are encapsulated side by side or stacked, while SOC is a highly integrated chip product (Figure 1.15).

Overall, encapsulating technology has evolved from traditional encapsulations (DIP, SOP, QFP, PGA, etc.) to advanced encapsulations (BGA, CSP, FC, WLP, TSV,

3D stacking, SIP, etc.). In terms of structure, encapsulations have evolved from the early transistor outline encapsulation (TO) to DIP and to SMT in the 1980s. The SOP was derived from SMT in the 1980s, and gradually the small outline J-lead (SOJ), Thin Small Outline Encapsulation (TSOP), and Very Small Outline Encapsulation (VSOP) were derived. SOP, Shrink Small Outline Encapsulation (SSOP), Thin Shrink Small Outline Encapsulation (TSSOP), Small Outline Transistor (SOT), SOJ, TSOP, VSOP, SSOP, SOT, small outline integrated circuit (SOIC), etc., and the later plastic leaded chip carrier (PLCC), QFP. After the IC function and the number of I/O pins gradually increased, Intel took the lead in updating the QFP package to the ball grid array package (BGA) in 1997. The recent mainstream packaging methods are chip scale package (CSP) and overlay package or flip chip package (Flip Chip). From the encapsulating materials, including metal, ceramic, plastic, and other materials used for encapsulating, many high-intensity working conditions of the circuit such as military and aerospace level still have a large number of metal encapsulating. Therefore, the general development process of IC encapsulation can be summarized as follows:

Structure: TO-DIP (Dual In-Line Encapsulation) – PLCC – QFP – BGA – CSP.
Material aspects: (metal, ceramic) – (ceramic, plastic) – plastic.
Pin shape: long-lead direct insertion – short-lead or leadless placement – ball-and-bump soldering.
Assembly method: TH cartridge – surface assembly – direct mounting.

The current global mainstream encapsulating technology of ICs is the third-generation encapsulating technology, namely BGA encapsulation, chip-level encapsulation (CSP), and FC. Among them, FC encapsulating technology, which is considered necessary to promote low-cost, high-density portable electronic device manufacturing process, has been widely used in the consumer electronics industry. The fourth-generation encapsulating technologies, such as WLP, silicon through-hole technology (TSV), and system-level encapsulating (SIP), are still being promoted on a small scale, but they will become the mainstream of future encapsulating methods under the technology upgrade.

In the encapsulating of flexible electronics, the above encapsulating technologies are borrowed and improved, and now the following two processes are mainly used in flexible electronics encapsulating.

1.2.1 Flip Chip Process

FC technology is widely used in traditional microelectronic encapsulating, which has the following characteristics [10]: (i) provides a good point connection for the signal, which can make more effective use of the chip area and has an ultrahigh encapsulating density; (ii) has light mass and small form factor; and (iii) the lead inductance becomes smaller, crosstalk becomes weaker, the signal transmission time is shortened, and thus has excellent high-frequency performance. In flexible electronic encapsulating, FC technology also has these characteristics, and because of the flexible substrate as a carrier, making the product bendable and flexible can

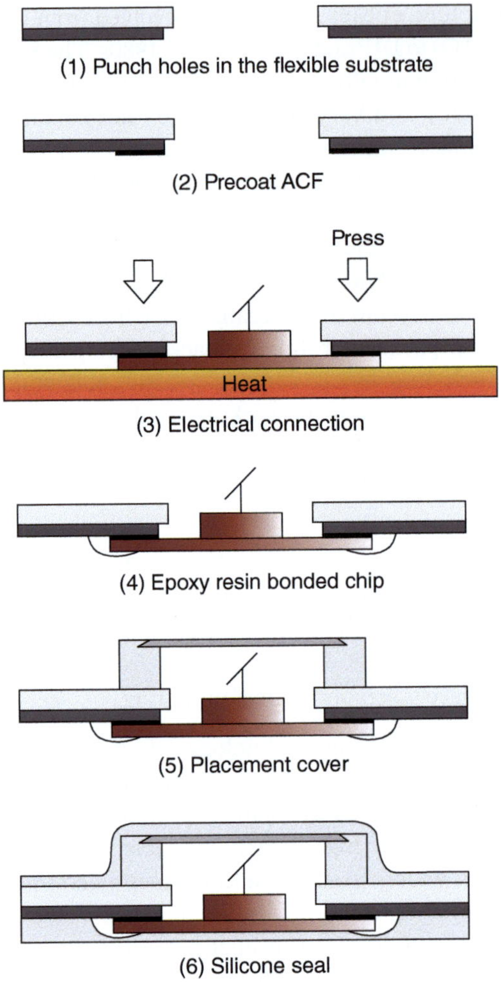

Figure 1.16 The encapsulating process of flip chip. Source: Reproduced with permission from Zhang and Pan. [11]; © 2010, Springer-Verlag.

make full use of space to reduce the size. FC encapsulating methods are solder bump method and conductive adhesive bonding. The former uses the reflow process between the chip and the substrate to form a solder ball, but the reflow requires a high temperature, and commonly used flexible substrate materials including polyimide and polyester (PET) are only suitable for low-temperature (less than 200 °C) bonding technology, so flexible electronic encapsulating is often used in the conductive adhesive connection. Figure 1.16 shows an application of FC technology for flexible encapsulating, and its main process includes [11]: (i) cutting a hole in the flexible substrate that has been made to place the chip; (ii) pasting the anisotropic conductive film (ACF) to the substrate pad for subsequent electrical connection; (iii) adjusting the chip pad so that it is aligned with the pads on the substrate and realizing the physical and electrical connection of the assembly through heating and compression; (iv) filling the side of the chip with epoxy resin to strengthen the structural strength of the assembly to prevent cracks when bending;

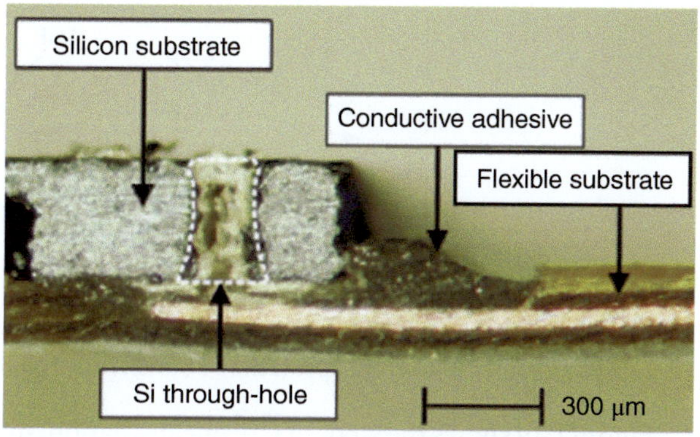

Figure 1.17 Cross-sectional micrograph of MEMS connected to FPCB. Source: Reprinted with permission from Hocheng et al. [12]; © 2012 The Institute of Physics.

(v) placing the mask above the chip, and the gap between the mask and the chip provides space for the chip to move, which is suitable for certain moving parts. The mask is made by injection molding process and its materials are epoxy resin and polydimethylsiloxane; (vi) polydimethylsiloxane is coated around the whole device to seal it, completing the whole encapsulating process. Benfield et al. [12] used FC technology to realize micro-electro-mechanical system (MEMS) and the electrical and structural connection between the MEMS and the flexible printed circuit board (FPCB) is shown in Figure 1.17. The combination of the silicon through-hole design on the chip can effectively reduce the encapsulation size by up to 33%.

1.2.2 Progress of CIF-Based Flexible Electronic Encapsulating Technology

Chip in Flex (CIF) is a kind of structure that fixes the driver IC on the flexible circuit board with die flexible film, which uses a flexible additional circuit board as a carrier to encapsulate the chip and the flexible substrate circuit together. The chip can be encapsulated in a flexible substrate with a thickness of less than 100 μm chip. Flex-on-flex assembly technology, making it more versatile and reliable. At the same time, its cost will be further reduced, both in terms of raw materials and process, which has great significance for its mass production. These advantages make it gradually become a new trend in the current flexible electronic encapsulating.

Compared with embedded encapsulating and CIP (Chip In Plastic) encapsulating, especially when combined with wafer-level encapsulating technology, the main advantages of CIF include [13]: (i) the process temperature is lower and can reach below 200 °C, which is more favorable for the curing of ACFs; (ii) the process steps are also greatly reduced, mainly in the form of certain ACF process steps can be omitted, such as ACF cutting and pre-lamination; (iii) the mechanical and electrical connection between the chip and the flexible substrate is more reliable; and (iv) a certain degree of integration and flexibility can be achieved.

As shown in Figure 1.18, the CIF encapsulation consists of an ultrathin silicon wafer (40 μm thick), a FPCB, and an ACF. PI in the figure is polyimide. The CIF structure has excellent bending performance, and Kim et al. [14] demonstrated that no chip fracture was observed for a minimum bending radius of 5 mm.

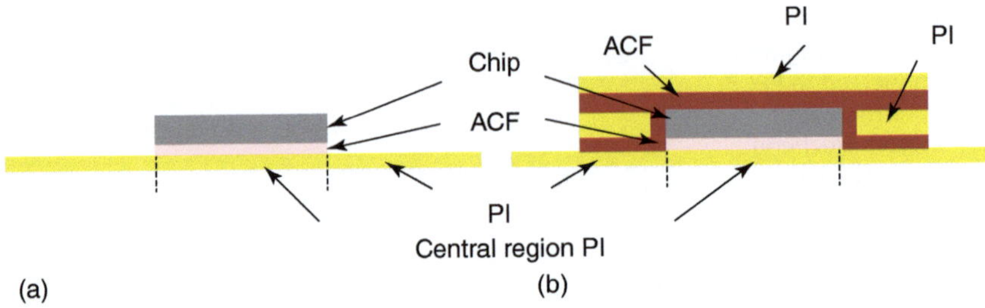

Figure 1.18 Schematic diagram of COF encapsulation and CIF encapsulation. (a) COF encapsulation and (b) CIF encapsulation. Source: Adapted from Suk et al. [14].

Figure 1.19 shows the steps of CIF encapsulation, mainly including: (i) pre-coating a layer of ACF on a 50 μm thick wafer; (ii) cutting the pre-coated thin ACF wafer into small units; (iii) placing the pre-coated thin ACF wafer on the first layer of flexible substrate for COF encapsulation; (iv) bonding the second layer of substrate with ACF; (iv) bonding the third layer of substrate with ACF; and (iv) finishing the encapsulation with FOF process CIF encapsulation. It should be noted that parameters such as encapsulation temperature, time, and force must be taken into account in the FOF encapsulation process. The optical image and cross-section of the finished CIF encapsulation are shown in Figures 1.20 and 1.21, respectively.

1.3 Encapsulating Technology of Several Important Flexible Electronic Devices

1.3.1 Organic Light-Emitting Diode

Organic light-emitting diode (OLED) is favored by the market for its advantages of surface light source, cold light, energy savings, fast response, flexibility, ultrathinness, and low cost, etc. OLED technology is also known as the third-generation display and lighting technology. At present, global manufacturers continue to invest in technology research and development. OLED flat panel display technology is tending toward mass production technology and is increasingly mature with market demand for high-speed growth stage, but because of its poor stability, water, oxygen, and heat are extremely sensitive, and encapsulating technology is particularly critical. For flexible OLED devices, rigorous encapsulating to achieve longer life and improved stability is a must for their mass production (Figure 1.22) [16].

The key core material of flexible OLED devices is an ultrathin organic electroluminescent layer, which is extremely sensitive to water, oxygen, and heat and is responsible for its poor stability [17]. Oxygen quenches the trilinear excitons, which directly leads to a significant decrease in the luminescence quantum efficiency. It also oxidizes the organic materials of the hole transport and light-emitting layers, leading to the opening of unsaturated double bonds and a decrease in the device's luminescence efficiency and electron–hole transport capacity. Water vapor tends to make the organic semiconductor in the light-emitting layer through hydrolysis reactions. The metals used in the electrodes are very reactive and are susceptible to both

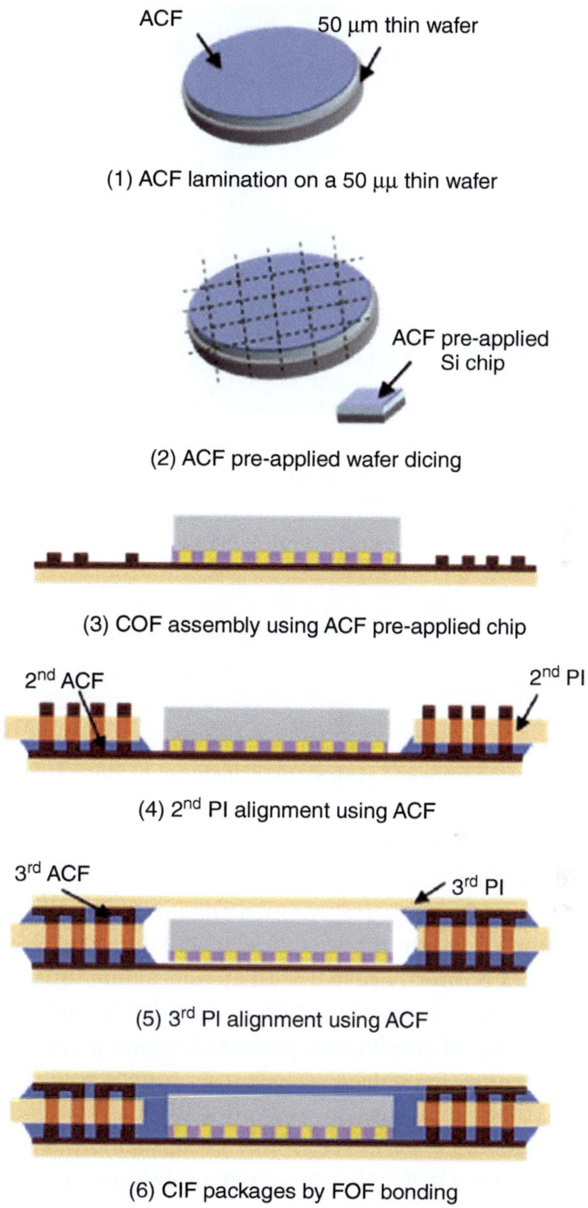

Figure 1.19 CIF encapsulating process steps. Source: Adapted from Kim et al. [14].

oxidation and hydrolysis by reaction with the water vapor that penetrates [18]. Due to the special characteristics of the substrate material, the water and oxygen barrier ability of flexible devices is worse than that of rigid materials, and the requirements for encapsulating are higher. Strict encapsulation of flexible OLED devices is necessary to extend device life and improve device stability, which is the key to mass production of flexible OLED devices.

OLED devices have extremely demanding requirements for encapsulating, and practical OLED devices usually require a water vapor transmission rate (WVTR) of

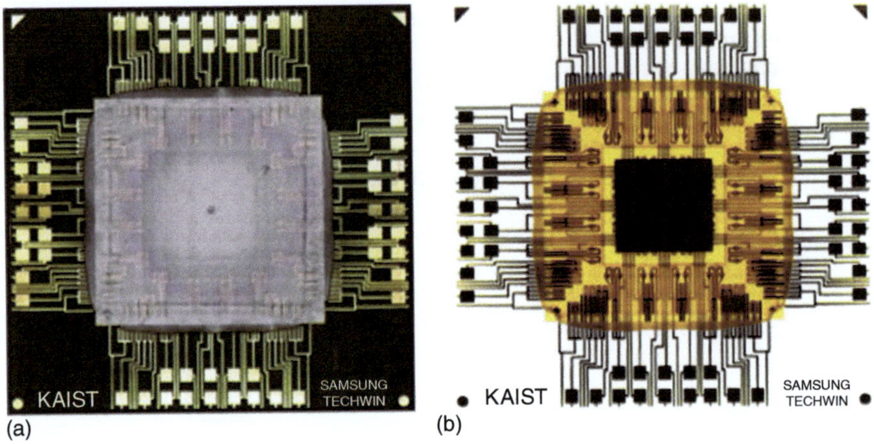

Figure 1.20 Optical image of CIF. Source: Suk et al. [15]; © 2015, Reproduced with permission from Elsevier.

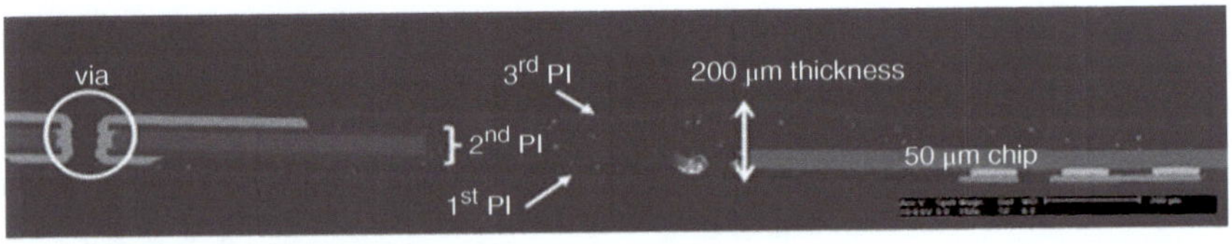

Figure 1.21 Cross-sectional view of the CIF encapsulation. Source: Reprinted with permission from Ref. [15]; © 2015, Institute of Electrical and Electronics Engineers.

less than 10^{-6} g/m^2/day and an oxygen vapor transmission rate (OVTR) of less than 10^{-5} cm^3/m^2/day [19, 20].

Currently, there are two methods for encapsulating flexible OLED devices. One method is in-line encapsulation, where the metal electrode surface of flexible OLED devices is directly coated with an alternating inorganic–organic barrier layer [21], as shown in Figure 1.23. This encapsulation method is widely used and offers the advantage of avoiding the process of breaking the vacuum seal of the OLED device, preventing the metal electrode from coming into contact with water, oxygen, dust, and other impurities, thus ensuring the long-term performance requirements of the OLED device. However, in-line encapsulation is a more complex process overall, making it challenging to achieve roll-to-roll production. The in-line processing temperature (≥90 °C) during the preparation of the barrier layer can also cause damage to the OLED device, affecting its lifespan. Currently, most domestic and foreign companies employ in-line encapsulation technology.

Another method is embossed encapsulation, where the metal electrode surface of OLED devices is laminated with a high barrier film through an adhesive/film layer, as shown in Figure 1.24. The embossed encapsulation process is relatively simple, allowing for separate optimization of the OLED device and the high barrier layer preparation process. However, during encapsulation, the OLED device may be exposed to water, oxygen, and dust, and the encapsulation material may also be

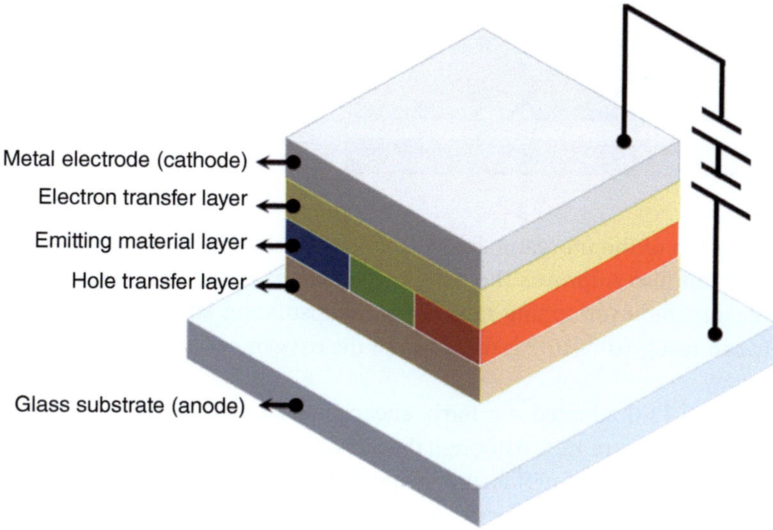

Figure 1.22 OLED light-emitting element structure diagram. Source: Reproduced with permission from http://bushu.swdkcc.com/home/result/info/id/444/catId/56.html.

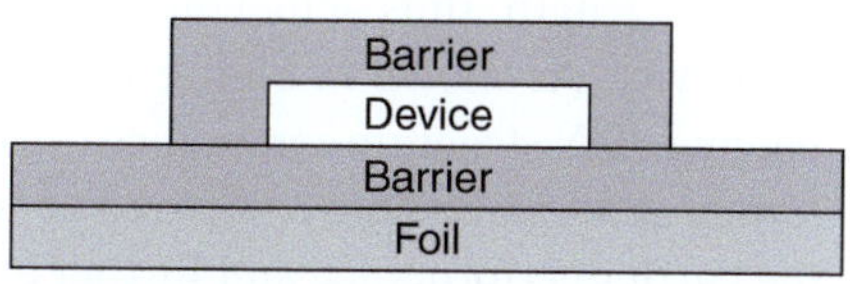

Figure 1.23 In-line encapsulating structure of flexible OLED devices. Source: Adapted from Park et al. [21].

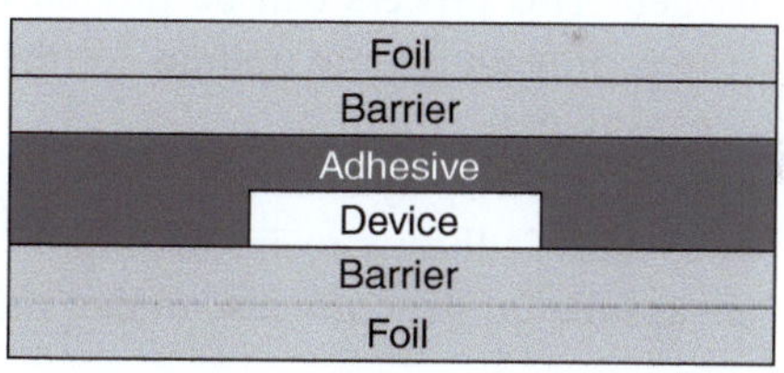

Figure 1.24 Structure of embossed encapsulation for flexible OLED devices. Source: Adapted from Park et al. [21].

destructive to the metal electrode. Additionally, the use of adhesive films that are not highly resistant can result in side leakage issues, allowing water and oxygen to penetrate the device through the adhesive layer and cause damage. Furthermore, under-reacted adhesives can react with OLED devices, impacting their lifespan.

Representative in-line encapsulation technologies include:

Barix encapsulation technology was first developed in Vitex System, United States of America, which is an inorganic–organic overlapping multilayer encapsulation structure [22, 23]. In the process of preparing large-size flexible OLED devices, some unavoidable small raised particles have appeared on the surface of the film layer before cathode plating [24]. Because the cathode metal layer is very thin, these small raised particles are enough to pierce the cathode, causing pinholes in the cathode metal layer, which eventually lead to hydroxide entering the device along the pinholes and causing damage to the device. The Barix encapsulation is coated with an

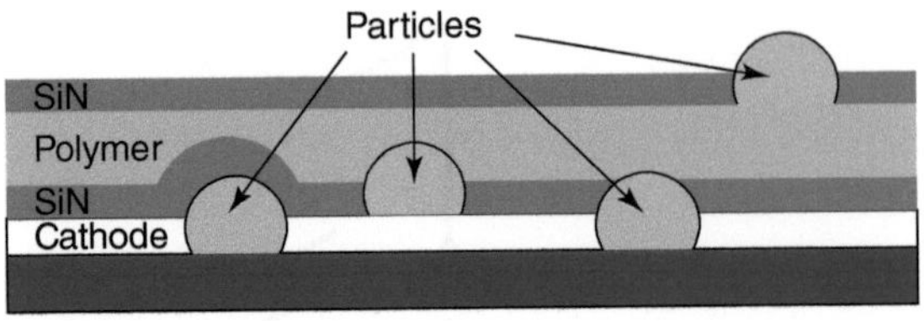

Figure 1.25 Schematic diagram of cross section of multilayer structure. Source: Reproduced with permission from Brand et al. [26]; © 2014, Elsevier.

organic layer on top of the inorganic layer, which not only fills the pinholes in the inorganic layer but also improves the bending resistance of the device [25]. With this inorganic–organic overlapping multilayer encapsulation structure, the water permeability can reach 10^{-4}–10^{-6} g/m^2/day and the oxygen permeability can reach 0.005 cc/cm^2/day.

The Dutch Holst [26] adopted the Barix encapsulation method, and the device structure is shown in Figure 1.25. Although the SiN film can cover the pinholes of the metal layer, the SiN film itself will form new pinholes. Coating an organic polymer layer on the SiN to flatten the interface can effectively fill the defects of the SiN layer and extend the path of water and oxygen into the OLED device, and the organic polymer can also effectively improve the bending resistance of the device. Then a layer of SiN film is plated to reduce the impact of pinholes on the device by increasing the channel length, thus achieving the purpose of improving the barrier properties of the device. As shown in Figure 1.26, the device encapsulated with the SiN/organic polymer/SiN structure can reduce the number of black spots by three orders of magnitude at the same time compared to one layer of SiN alone, significantly improving the barrier performance and further enhancing the device's resistance to water and oxygen. This process can be produced by roll-to-roll and strip-seam coating processes, providing more options for the industrialization of large-size OLED device encapsulation.

Samsung Display uses an inorganic/organic/inorganic Barix structure, followed by a layer of adhesive resin, and finally a coated hybrid encapsulation. This five-layer encapsulation structure also makes the encapsulation effect much better.

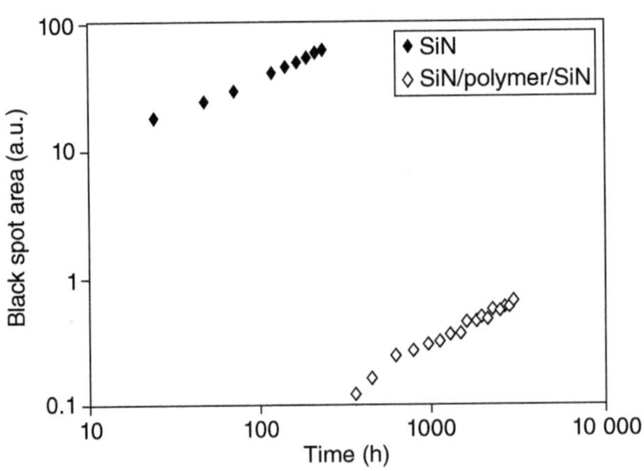

Figure 1.26 Relationship between the number of black spots and time. Source: Reproduced with permission from Brand et al. [26]; © 2014, Elsevier.

Atomic layer deposition (ALD) is considered an extension of chemical vapor deposition (CVD) technology, which is a controlled chemical reaction [27, 28]. Unlike CVD, the ALD technique breaks down the chemical reaction into two steps, allowing the parent material to separate after a certain level of reaction with a sudden charge of a large amount of inert gas, which removes the excess parent material from the chamber and prevents the process from proceeding further. Most ALD techniques are binary reactions, and the above method enables controlled and quantitative deposition of both reactants, allowing the preparation of smooth, continuous, and defect-free films. Lee [29] proposed a hybrid barrier layer in which a layer of graphene was first deposited by chemical vapor deposition and then a layer of Al_2O_3 was deposited by ALD. Compared with the Al_2O_3 layer alone, the water resistance and bending resistance of the hybrid barrier layer were superior. Lee's group pioneered the use of electrode materials and inorganic materials together as a barrier layer, providing a broader idea for OLED device encapsulation. The group has pioneered the use of electrode materials and inorganic materials as a barrier layer, providing a broader idea for OLED device encapsulation. In the literature [30], it is described that the water permeability of OLED devices was reduced to 0.058 g/m^2/day by using ALD to deposit 20 nm Al_2O_3 followed by 200 nm SiN_x as the barrier layer, and the lifetime of OLED devices reached 260 hours. In the literature [31], the water permeability of OLED devices could be as low as 4.75×10^{-5} g/m^2/day by using ALD to deposit Al_2O_3 and ZrO_2 as the barrier layers, and the OLED device lifetime could be extended to 10,000 hours.

Vacuum thermal deposition is one of the traditional vacuum coating techniques that requires equipment to reach at least 10^{-3} Pa. Wei Bin [32] proposed the use of MgF_2 and ZnS films grown by vacuum thermal deposition as a hybrid barrier layer, and the device's water barrier capability was greatly enhanced. The yellow OLED devices encapsulated with three pairs of MgF_2/ZnS had a lifetime of more than 500 hours at 2000 cd/m^2 initial luminance with a 56% luminance drop without UV resin sealing. With UV resin sealing, the device's half-life could reach 245 hours. Using MgF_2/ZnS as the passivation layer with good light transmission, this encapsulation method is simple to prepare and inexpensive for the encapsulation of OLED devices with top light-emitting structures. In general, the in-line encapsulation process is developing toward simplicity, controllability, and low cost.

The off-line encapsulating technology is a high-barrier film laminated to the metal electrode surface of flexible OLED devices using encapsulant/film, which is overall less difficult to prepare than the in-line process. The advantage of off-line encapsulating technology is that the performance of the encapsulant/film and high-barrier film can be optimized separately, but in the process of encapsulating flexible OLED devices may be exposed to water and oxygen or dust, which may cause damage to the flexible devices and is not conducive to achieving the goal of high efficiency and long life.

The first-generation flexible barrier film materials used for flexible electronic encapsulating are mainly organic films, such as polyethylene (PE), PET, and polyvinylidene chloride (PVDC). The structural properties of these materials themselves determine their limited water barrier capacity. The second-generation flexible barrier materials uses aluminum foil and aluminized film as a barrier layer because of its excellent barrier performance and simple preparation process, but the aluminum foil needs to have a certain thickness to achieve the water barrier requirements, and the cost of barrier materials rises. The aluminum foil material is brittle and easy to crack at the folds, which affects its barrier properties. In addition, aluminum foil and aluminized films are opaque, which limits their applications [33]. The development of third-generation high-barrier films started in the mid-1980s when Mitsubishi Plastics in Japan prepared barrier films by vaporizing inorganic silica onto polyvinyl alcohol, PET, polyamide, and other substrates. In 2001, with the help of a 100 kW high-power electron gun developed in Germany, a technological breakthrough was achieved in the silicon oxide barrier layer, which led to the rapid development of oxide barrier film applications. From 2007 onward, high-barrier film research began to be applied to solar cell encapsulating, OLED, and quantum dot display industries.

The barrier film layer has undergone the evolution of single-layer, double-layer, and multilayer structures. Single-layer barrier film has a simple structure, mature technology, and high light transmission rate, but poor barrier properties; double-layer barrier film has simple technology, high light transmission rate, but the barrier properties still cannot meet the requirements of the encapsulating OLED index; multilayer barrier film structure and process is complex, the polymer layer and organic layer overlapping structure makes its barrier properties high, generally on the surface of the flexible substrate PET, PEN or PI first flattening treatment, i.e. coating a layer of organic flattening material, and then making silicon nitride, silicon oxide, or alumina inorganic barrier layer by PECVD or ALD process; followed by cycling the organic flattening layer/inorganic barrier layer, which can be compared to the water and oxygen barrier performance of glass and can meet the index requirements of OLED encapsulating. Cho et al. [34] deposited AlN_x/UVR/AlN_x on the PEN substrate as a barrier layer and used it as a flexible substrate for OLED encapsulating. Aging tests were performed at 85 °C and 85% RH, and it was found that the brightness of the encapsulated OLED devices decreased by only 9% after 750 hours compared with those encapsulated on standard glass substrates.

The encapsulant used for flexible electronic encapsulating generally refers to liquid encapsulation adhesive, and the encapsulation film is mainly encapsulation adhesive film. Liquid encapsulation adhesives are usually divided into solvent-based and solvent-free types. The solvent is easy to damage the metal electrode, the light-emitting layer and the carrier transport material in the flexible OLED device, so the packaging adhesive is solvent-free, UV-curable, neutral and cannot corrodes the metal electrode. Encapsulation adhesive film is divided into hot-melt adhesive film and room-temperature pressure-sensitive adhesive film. To use hot-melt adhesive film for encapsulation, the adhesive film needs to be heated to 140 °C or more, so that it can be heated to a molten state to bond the battery to the high-barrier film. However, the 140 °C process temperature will cause thermal deformation of the substrate and will also damage the OLED light-emitting layer film and other functional layer materials, resulting in significant degradation of the device's optoelectronic performance. The pressure-sensitive adhesive film has a protective film on both sides and can be bonded to the barrier film and the OLED device after removing the protective film. Compared with hot-melt adhesive film, the process temperature of pressure-sensitive adhesive film is very low, and the lamination can be done at room temperature to avoid deformation and decomposition of the substrate and functional layers by heat. To prevent water vapor leakage from the edges of OLED devices in hot and humid environments, high-barrier pressure-sensitive films can be used. Therefore, the use of pressure-sensitive adhesive film requires that the encapsulation film can be laminated at low or normal temperatures, the lamination process is free of air bubbles, and the film should have barrier properties.

1.3.2 Flexible Solar Cell Encapsulating

Flexible solar cells will be one of the most important mainstays of future PV applications due to their lightweight, bendability, and low installation cost. It has a wide application space in building integrated photovoltaic (BIPV), building attached photovoltaic (BAPV), mobile objects, portable devices, and aerospace. Flexible solar cells generally have the following types: flexible inorganic solar cells, flexible organic solar cells, flexible calcium titanite solar cells, and flexible dye-sensitized solar cells. Flexible inorganic solar cells include flexible amorphous silicon solar cells and flexible copper indium gallium selenide solar cells. This book mainly introduces the encapsulation of flexible amorphous silicon solar cells and flexible chalcogenide solar cells.

1.3.3 Flexible Amorphous Silicon Solar Cells

In the early days of solar cells, the main raw material was crystalline silicon, but the cost was high, so it was only used in space exploration. Due to the high cost of crystalline silicon, crystalline silicon solar cells consume more silicon materials, and due to the development of technology, finally in 1974 Carlson developed the earliest amorphous silicon solar cells in the laboratory. An amorphous silicon cell (a-Si) is a solar cell made of amorphous silicon film deposited on a conductive glass (non-flexible) or stainless steel and special plastic substrate.

In 1977, Carlson and his team successfully developed an amorphous silicon Schottky barrier cell with an energy conversion efficiency of 5.5%. Following this, in 1978, Osaka University in Japan achieved a conversion efficiency of 4.5% with the development of an amorphous silicon PIN cell. In the fall of 1981, Osaka University further improved the technology by preparing an a-SiC:H/a-Si:H PIN heterojunction solar cell, which surpassed an energy conversion efficiency of 8%. This cell utilized a P-type broadband a-SiC:H as the window material. By 1982, the efficiency of this a-SiC:H/a-Si:H PIN heterojunction solar cell exceeded 10%. Subsequently, in 1987, the conversion efficiency of amorphous silicon cells reached 12%. Sanyo in Japan achieved a milestone in 1990 by developing an amorphous silicon cell with a conversion efficiency of 15.8%. In 1994, amorphous silicon cells with a back surface field structure, prepared using the PECVD method, were introduced in Japan with a conversion efficiency of 18.9% [35].

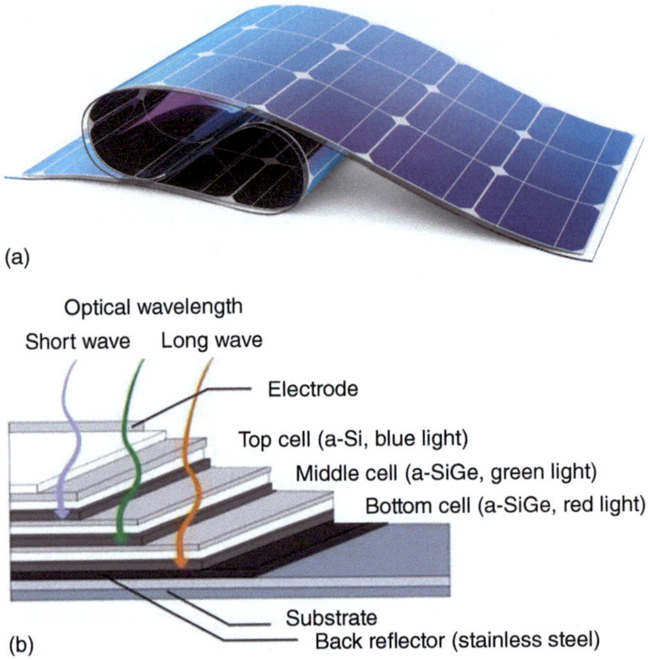

Figure 1.27 Flexible amorphous silicon solar cell (a) Source: iaremenko/Adobe Stock and its general structure (b). Source: Reprinted with permission from Yin and Jiang. [37]; © 2012, Guangzhou Chemical Industry.

Flexible amorphous silicon solar cells are mainly made of stainless steel and special plastics as substrates [36]. As shown in Figure 1.27, flexible amorphous silicon solar cells are mostly designed with sandwich structure, divided into three light-absorbing layers: the top layer using amorphous silicon material with an optical energy gap of 1.8 eV, which facilitates the absorption of blue light; the middle layer using amorphous silicon and 10–15% germanium alloy material with an optical energy gap of 1.6 eV, which facilitates the absorption of green light; and the bottom layer using amorphous silicon and 40–50% germanium alloy material with an optical energy gap of 1.4 eV, which facilitates the absorption of green light. The optical energy gap is 1.4 eV, which can absorb red light and far-red light and absorb different solar spectrum bands in different layers. The high PV conversion rate makes up for the low conversion rate of amorphous silicon compared to polycrystalline silicon and monocrystalline silicon panels in sunny weather. The light that is not absorbed when the light source enters is reflected back by the back reflection layer (Ag/ZnO) of the substrate and is absorbed on the way out, thus converting and outputting more electricity than normal solar products. The cell is covered with a protective layer of ethylene vinyl acetate (EVA) polymer and encapsulated with a transparent, highly inert fluoroplastic film. The surface of the film is patterned to absorb light from all angles, effectively preventing sunlight loss. The flexible amorphous silicon solar cell can work normally in an environment from −40 to 80 °C, and even if a part is pierced by a bullet, the rest of the cell can still generate electricity without being affected by it [37].

1.3.4 Flexible Perovskite Solar Cells

Perovskite solar cells, which are solar cells using chalcogenide-type organometallic halide semiconductors as light-absorbing materials, belong to the third-generation solar cells. Among them, the most commonly used light-absorbing layer is a hybrid organic–inorganic lead or tin halide-type material, such as methylammonium lead halide, which is cheap and easy to produce.

The term "chalcogenide solar cell" is derived from the chalcogenide crystal structure of the light-absorbing layer material ABX_3. Chalcogenide crystals exhibit an ABX_3 structure, typically cubic or octahedral. As illustrated in Figure 1.28, in a chalcogenide crystal, B ions occupy the central position of the cubic cell and are surrounded by six X ions, forming a coordination cuboctahedron with a coordination number of 6. A ions are located at the top corner of the cubic cell and are surrounded by 12 X ions, forming a coordination octahedron with a coordination number of 12. A ions and X ions have similar radii, creating a dense cubic stack when combined [38].

Chalcogenide solar cells have rapidly gained attention as a research hotspot for new solar cell technologies, owing to their high light absorption properties and excellent carrier separation and transport characteristics. In just a decade, their efficiency has increased from 3.8% in 2009 to the current level of 25.8%. These solar cells utilize common and inexpensive materials with high abundance and high tolerance for raw material purity and defects.

One of the significant advantages of chalcogenide solar cells is that they do not require the vacuum and high-temperature processes associated with traditional crystalline silicon cells. Instead, they can be manufactured using solution-based methods. This feature makes them highly promising for mass production and the application of low-cost, high-efficiency flexible solar cells. They have the potential to rapidly supplement the market and address applications that cannot be covered by traditional crystalline silicon solar cells.

However, it should be noted that the materials used in each functional layer of chalcogenide solar cells are sensitive to water vapor, oxygen, ultraviolet light, and pressure in the air. These environmental factors can significantly reduce the service life of the cells. Encapsulation technology offers an effective solution to isolate the working components from the external environment, preventing pollution and corrosion from various impurities. It is a method employed to improve the service life of precision electronic components (Figure 1.29).

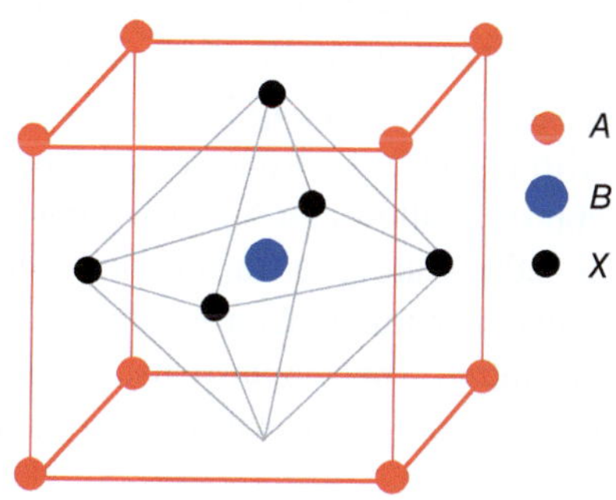

Figure 1.28 Crystal structure of chalcogenide. Source: Reproduced with permission from Rongrong et al. [38]; © 2019, Chinese Physical Society.

Figure 1.29 Calcium titanite solar cell device. Source: Reproduced with permission from Guo et al. [39]; © 2018, Royal Society of Chemistry.

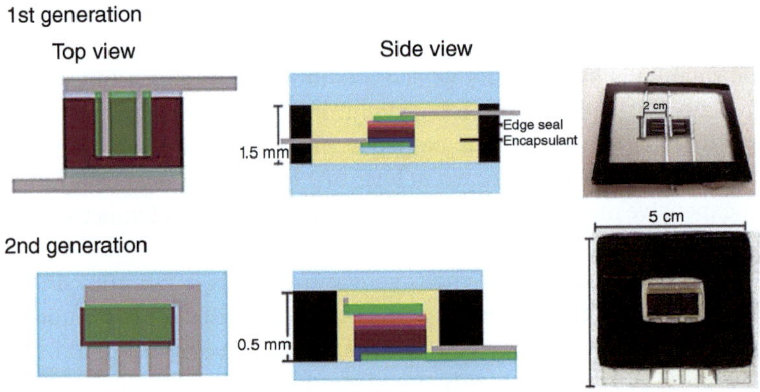

Figure 1.30 Top view, side view, and physical view of the first-generation encapsulating technology and the second-generation encapsulating technology. Source: Reproduced with permission from Juarez-Perez and Haro [40]; © 2016, Elsevier.

Currently, there are two common cell encapsulation techniques for chalcogenide solar cells [40], as illustrated in Figure 1.30.

The first-generation encapsulation technique involves the use of evaporative metal injectors and welded metal strips to conduct current from the cell to the outside. The edges of the strips are sealed, with the device positioned in the center of the closed cavity.

The second-generation encapsulation technique utilizes transparent indium tin oxide (ITO) electrodes to separate the chalcogenide from the metal electrodes. This ensures a lateral gap between the electrodes and the calcium titanite solar cells, with the ITO electrodes positioned directly on one side of the encapsulation. This approach allows for better sealing of the entire device. Both techniques are known as "edge sealing" encapsulation techniques.

Polymer materials are commonly used for encapsulation due to their excellent insulation, thermoplasticity, and mechanical strength. The dense encapsulation layer effectively isolates water and oxygen from the air, enabling cost-effective

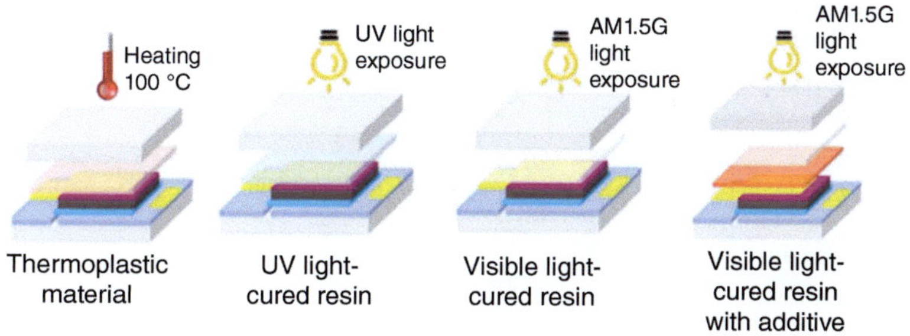

Figure 1.31 Several operating conditions for the preparation of polymeric encapsulation layers. Source: Reproduced with permission from Shi et al. [41]; © 2020, American Association for the Advancement of Science.

large-area encapsulation. Common materials used for encapsulation include polyisobutylene, PE, thermoplastic polyurethane, EVA, and cyclized perfluoropolymers. Various preparation methods for encapsulation layers are available, such as thermal curing, UV curing, and visible light curing, as shown in Figure 1.31 [41].

Recently, Dr. Lei Shi from the Australian Centre for Advanced Photovoltaics and Prof. Anita Ho-Baillie from the University of Sydney [40] investigated the encapsulated chalcogenide solar cell system using gas chromatography–mass spectrometry (GC–MS) and found that the polymer-glass "blanket-cover" encapsulation technique can form an absolutely hermetic system and greatly improve the operating life of chalcogenide solar cells. The polymer-glass encapsulation technique allows the efficiency of chalcogenide solar cells to remain above 95% after 4000 hours of operation under IEC 61215:2016 standard test conditions. The (GC–MS) technique reveals that the absolute hermetic encapsulation effectively prevents the diffusion of various decomposition gas molecules and maintains the equilibrium of the system, and the residual vapor can promote the regeneration of chalcogenide at night, which increases the cycle life of the cell. The "blanket" encapsulation technique, in which the entire device is encapsulated with a polymer so that the device is in close contact with the polymer and there are no cavities inside. Interestingly, the authors added a layer of Cover Glass to the encapsulated cell, which effectively prevents the escape of CH_3Br and NH_3 from the decomposition of chalcogenide, an effect that cannot be achieved with polymer encapsulation alone. The polymer acts not only as a water vapor barrier but also as a strong binder between the chalcogenide solar cell and the cover glass, further enhancing the hermetic performance of the system (Figure 1.32).

For the encapsulation of calcium titanite solar cells, the researchers employed polyisobutylene and polyolefin. They subjected the cells to hygrothermal and humidity freeze tests within a temperature range of −40 to 80 °C. Remarkably, the cells encapsulated with this technique showed no degradation after 1800 hours of operation, surpassing the 1000-hour requirement stipulated in the test standard. The short-circuit current density and open-circuit voltage of the cells remained stable even after the test, a characteristic commonly observed in calcium titanite solar cells containing methylamine ions.

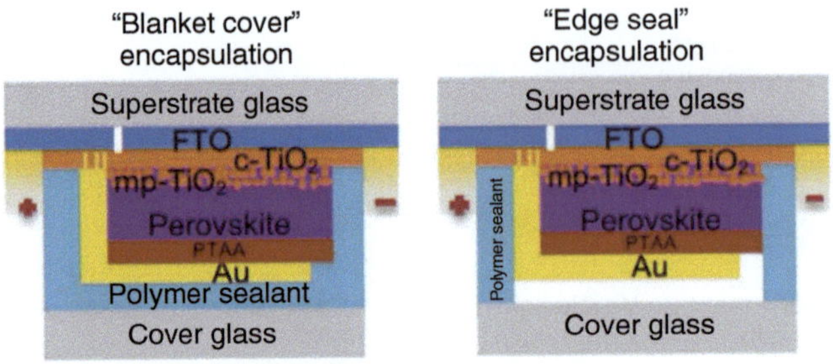

Figure 1.32 Cross-sectional diagram of "blanket cover" and "edge seal" encapsulations. Source: Reproduced with permission from Juarez-Perez and Haro [40]; © 2020, American Association for the Advancement of Science.

This study demonstrates that the use of simple and cost-effective polymer-glass combination encapsulation techniques, such as polyisobutylene or polyolefin-based encapsulation, can confer exceptional durability to organic–inorganic hybrid chalcogenide solar cells. Notably, these cells exhibited remarkable resilience against severe humidity freeze tests, despite the known low thermal stability of methylammonium ions, which has typically been a challenge in achieving highly stable chalcogenide solar cells. The polymer-glass "blanket-cap" encapsulation technology ensures absolute hermeticity, creating a stable operating environment that prevents decomposition products from escaping the system and significantly extending the lifespan of chalcogenide solar cells. This work offers new insights for designing more stable chalcogenide solar cells and greatly advances the commercialization of this technology [40].

1.4 Flexible Electronic Encapsulating Materials

1.4.1 Selection Principle of Flexible Electronic Encapsulating Materials

Flexible electronic encapsulating materials are utilized to support and protect flexible electronic devices, as well as establish interconnections between devices and external circuits. They play multiple roles, including mechanical support, environmental sealing, signal transmission, heat dissipation, and shielding. These materials consist of substrates, wiring, frames, interlayer media, sealing materials, and more. Among them, flexible electronic encapsulating substrate materials primarily provide mechanical support and airtight protection and facilitate heat dissipation for flexible electronic devices and their interconnections.

The selection of flexible electronic encapsulating materials should consider the following aspects:

(1) **Thermal expansion coefficient**: Matching the thermal expansion characteristics of the substrate and flexible electronic devices
(2) **Dielectric properties (permittivity and loss)**: Ensuring fast response and minimal delay in electrical signal transmission within the circuit

(3) **Thermal conductivity**: Facilitating heat dissipation from the working circuit
(4) **Mechanical properties**: Possessing adequate strength, hardness, and toughness.

Flexible electronic encapsulating substrate materials are commonly composed of plastic encapsulating materials, typically thermosetting materials such as epoxy, caseinate, polyester, and silicone. Among these, epoxy resin is the most widely used. Plastic encapsulating materials are popular in the electronic encapsulation industry due to their low cost, maturity, and simple production processes. However, they also have notable drawbacks. Most plastic materials lack sufficient density, exhibit poor thermal conductivity, possess thermal expansion coefficients that do not match flexible electronic devices, have high dielectric loss, and tend to be brittle. Plastic encapsulating materials are susceptible to moisture absorption, which leads to expansion and the potential for device failure, particularly with epoxy materials being highly affected by moisture. This significantly impacts the reliability of encapsulation, making them unsuitable for industries with higher reliability requirements, such as military and aerospace. Through ongoing research on encapsulating materials, these issues can be effectively mitigated by adjusting the ratio of plastic encapsulating materials and optimizing processing methods. Ideally, plastic encapsulating materials should possess the following characteristics: high raw material purity, low viscosity, minimal impurities, low water absorption, good heat resistance, high thermal conductivity, matching thermal expansion coefficients, ease of processing and molding, minimal raw material waste, good flame retardant properties, and excellent environmental performance without toxicity or pollution.

1.4.2 Desirable Properties of Flexible Electronic Encapsulating Materials

To minimize the impact of environmental factors such as water, oxygen, and dust on flexible electronic devices, flexible electronic encapsulating materials must meet the performance requirements specified in national standards. General performance indicators include insulation, breakdown strength, heat resistance, mechanical strength, and WVTR.

(1) **Insulation**: Flexible electronic encapsulating materials should exhibit high insulation resistance, typically exceeding 30 MΩ under normal conditions.
(2) **Breakdown strength**: When the electric field strength exceeds a certain threshold, flexible electronic encapsulating materials may experience a breakdown, leading to a loss of insulation properties. Generally, the breakdown strength requirement for flexible electronic encapsulating materials is above 10 KV/mm.
(3) **Heat resistance**: With increasing temperature, the resistance, breakdown strength, and mechanical strength of flexible electronic encapsulating materials tend to decrease. Therefore, flexible electronic encapsulating materials should maintain stable and reliable insulation and sealing performance while operating within specified temperature ranges for extended periods.

(4) **Mechanical strength**: Depending on specific requirements, different mechanical strength indicators such as tensile strength, compressive strength, bending strength, shear strength, tear strength, and impact resistance may be specified for different flexible electronic encapsulating materials.

(5) **Water and oxygen transmission rates**: Water and oxygen transmission rates, including WVTR and oxygen transmission rate, are measured in grams per square meter per day (g/m^2/day) and cubic centimeters per square meter per day (cm^3/m^2/day), respectively. As water vapor molecules are smaller than oxygen molecules and are more difficult to block, the WVTR is used to evaluate the encapsulation's effectiveness. The WVTR indicates the weight of water vapor that passes through a unit area of material under specific time, temperature, and humidity conditions. The oxygen transmission rate, on the other hand, measures the volume of oxygen that permeates through a unit area of material within a given time under constant temperature and pressure differences. Typically, flexible electronic encapsulating devices require a WVTR below 10^{-6} g/m^2/day and an oxygen transmission rate below 10^{-5} cm^3/m^2/day.

(6) **Other characteristics**: Certain flexible electronic encapsulating materials may exist in liquid form, such as various resins. Their properties include viscosity, fixed content, acid value, drying time, curing time, and more. Additional characteristics of flexible electronic encapsulating materials may encompass permeability, oil resistance, elongation, shrinkage, solvent resistance, arc resistance, and others.

1.5 Overview of the Development of Flexible Electronic Packaging at Home and Abroad

The development of flexible electronic packaging has seen significant progress both domestically and internationally. Currently, Japanese companies such as Sumitomo Chemical, Shin Kong Electric, Toppan, Tanaka, and Mitsui High-tec hold a substantial market share of around 25% globally. DuPont, a global leader in fluorine material development and production, is also a key player in the future competitive landscape of electronic encapsulating materials, with its market share steadily expanding. DuPont's involvement in the field has provided new development ideas for other fluorine material companies. Given the numerous advantages of fluorine materials and their experience in fluorine encapsulating technology, it is anticipated that fluorine materials will play a significant role in the field of flexible electronic encapsulation.

China represents the main application market for flexible electronic encapsulating materials, accounting for approximately 40% of the global market demand, followed by the United States with a share of around 15%. However, China currently lacks well-known enterprises focused on the development and production of functional materials in the field of flexible electronic encapsulating, unlike DuPont and Sumitomo Chemical. Although domestic fluorine chemical giants such as Juhua and San Aifu have shown interest in this area, they are still in the initial stages and cannot yet compete with international giants like DuPont. Therefore, flexible electronic encapsulating materials could be another area where foreign companies have a competitive advantage over domestic ones.

The global encapsulating industry has experienced single-digit growth, with statistics indicating that Taiwan accounts for 53% of sales, occupying half of the global encapsulating industry. Mainland China and the United States follow with shares of 21% and 15%, respectively, ranking second and third. Malaysia, South Korea, Singapore, and Japan each hold shares of 4%, 3%, 2%, and 2%, respectively. In terms of market share, domestic encapsulating companies have entered the global top tier and possess a certain level of international competitiveness.

Over the next five years, the competition pattern in the electronic encapsulating materials market is expected to be established. Leading global companies in this field include DuPont, Evonik, EPM, Mitsubishi Chemical, Sumitomo Chemical, Mitsui High-tech, Tanaka, Shin Kong Electric, Panasonic, Hitachi Chemical, Kyocera Chemical, Gore, BASF, Henkel, AMETEK Electronics, Toray, Maruwa, Lida Fine Ceramics, NCI, Chaozhou Sanhuan, Nippon Micrometallic, Toppan, Dainippon Printing, Germany Posel, and Ningbo Kangqiang, among others. Various authorities predict that the global electronics encapsulating market will experience rapid growth with a stunning CAGR from 2023 to 2028.

References

1 Nathan, A., Ahnood, A., Cole, M.T. et al. (2012). Flexible electronics: the next ubiquitous platform. *Proceedings of the IEEE* 100: 1486–1517.

2 Stoppa, M. and Chiolerio, A. (2014). Wearable electronics and smart textiles: a critical review. *Sensors* 14: 11957–11992. https://doi.org/10.3390/s140711957.

3 Kim, H., Abdala, A.A., and Macosko, C.W. (2010). Graphene/polymer nanocomposites. *Macromolecules* 43: 6515–6530.

4 Carlson, A., Bowen, A.M., Huang, Y. et al. (2012). Transfer printing techniques for materials assembly and micro/nanodevice fabrication. *Advanced Materials* 24: 5284–5318. https://doi.org/10.1002/adma.201201386.

5 Yin, Z.P., Huang, Y.A., Bu, N.B. et al. (2010). Inkjet printing for flexible electronics: materials, processes and equipments. *Science Bulletin* 55: 3383–3407.

6 Zeng, W., Shu, L., Li, Q. et al. (2014). Fiber-based wearable electronics: a review of materials, fabrication, devices, and applications. *Advanced Materials* 26: 5310–5336. https://doi.org/10.1002/adma.201400633.

7 Hammock, M.L., Chortos, A., Tee, B.C. et al. (2013). 25th Anniversary Article: the evolution of electronic skin (e-skin): a brief history, design considerations, and recent progress. *Advanced Materials* 25: 5997–6038. https://doi.org/10.1002/adma.201302240.

8 Liu, H., Zhao, H., Li, S. et al. (2018). Adhesion-free thin-film-like curvature sensors integrated on flexible and wearable electronics for monitoring bending of joints and various body gestures. *Advanced Materials Technologies* 4: 1800327. https://doi.org/10.1002/admt.201800327.

9 Wang, G., He, X., Tang, H. et al. (2016). Research progress and prospect of flexible electronic packaging technology. *Electronic Process Technology* 37 (5): 253–256. https://doi.org/10.14176/j.issn.1001-3474.2016.05.002.

10 O'Laughlin, D. (2022). Semiconductor encapsulating history and primer. *Semiconductor Services*. https://semiwiki.com/semiconductor-services/308968-semiconductor-encapsulating-history-primer (accessed 18 October 2023).

11 Zhang, Y. and Pan, W. (2005). MEMS packaging technology. *Nanotechnology and Precision Engineering* 3 (3): 194–198.

12 Hocheng, H., Chen, C., Chou, Y. et al. (2010). Study of novel electrical routing and integrated encapsulating on bio-compatible flexible substrates. *Microsystem Technologies* 16 (3): 423–430.

13 Benfield, D., Lou, E., and Moussa, W. (2012). A encapsulating solution utilizing adhesive-filled TSVs and flip chip methods. *Journal of Micromechanics and Microengineering* 22 (22): 65009–65017.

14 Suk, K., Son, H., Chung, C. et al. (2012). Flexible Chip-on-Flex (COF) and embedded Chip-in-Flex (CIF) encapsulations by applying wafer level encapsulation (WLP) technology using anisotropic conductive films (ACF). *Microelectronics Reliability* 52: 225–234.

15 Kim J., Lee T., Shin J., et al. (2015). Ultra-Thin Chip-in-Flex (CIF) technology using Anisotropic Conductive Films (ACF) for Wearable Electronics Applications. *2015 IEEE 65th Electronic Components and Technology Conference (ECTC)*: 714–718.

16 Fukagawa, H., Sasaki, T., Tsuzuki, T. et al. (2018). Long-lived flexible displays employing efficient and stable inverted organic light-emitting diodes. *Advanced Materials* 30 (28): 1706768. (1–7).

17 Riedel, D., Dlugosch, J., Wehlus, T. et al. (2015). Extracting and shaping the light of OLED devices. In: *Organic Light Emitting Materials and Devices XIX*. International Society for Optics and Photonics, 1364. SPIE.

18 Nam, T., Park, Y.J., Lee, H. et al. (2017). A composite layer of atomic-layer-deposited Al_2O_3 and graphene for flexible moisture barrier. *Carbon* 116: 553–561.

19 Sun, F.B., Duan, Y., Yang, Y.Q. et al. (2014). Fabrication of tunable Al_2O_3: alucone thin-film encapsulations for top-emitting organic light-emitting diodes with high performance optical and barrier properties. *Organic Electronics* 15 (10): 2546–2552.

20 Han, P., Lai, T.C., Wan, M. et al. (2019). Outstanding memory characteristics with atomic layer deposited $Ta_2O_5/Al_2O_3/TiO_2/Al_2O_3/Ta_2O_5$ nanocomposite structures as the charge trapping layer. *Applied Surface Science* 467–648: 423–427.

21 Park, S.H.K., Oh, J., Hwang, C.S. et al. (2005). Ultra-thin film encapsulation of organic light emitting diode on a plastic substrate. *ETRI Journal* 27 (5): 545–550.

22 Yang, Y. (2014). Packaging and integration characteristics of OLED devices based on flexible polymer substrate. *Jilin University Daily* 20 (4): 136–139.

23 Kim, B.J., Park, H., Seong, H. et al. (2017). A single-chamber system of initiated chemical vapor deposition and atomic layer deposition for fabrication of organic/inorganic multilayer films. *Advanced Engineering Materials* 19 (6): 1600819.

24 Dameron, A.A., Davidson, S.D., Burton, B.B. et al. (2008). Gas diffusion barriers on polymers using multilayers fabricated by Al_2O_3 and rapid SiO_2 atomic layer deposition. *Journal of Physical Chemistry C* 112 (12): 4573–4580.

25 Yamashita, K., Mori, T., and Mizutani, T. (2001). Encapsulation of organic light-emitting diode using thermal chemical-vapor-deposition polymer film. *Journal of Physics D: Applied Physics* 34 (5): 740.

26 Brand, J.D., Baets, J.D., Mol, T.V. et al. (2014). Systems-in-foil devices, fabrication processes and reliability issues. *Microelectronics Reliability* 48: 1123–1128.

27 Tanskanen, A. and Karppinen, M. (2015). Iron-based inorganic-organic hybrid and superlattice thin films by ALD/MLD. *Dalton Transactions (Cambridge, England: 2003)* 44 (44): 19194–19199.

28 Karttunen, A.J., Tynell, T., Karppinen, M. et al. (2016). Layer-by-layer design of nanostructured thermoelectrics: first-principles study of ZnO: organic superlattices fabricated by ALD/MLD. *Nano Energy* 22: 338–348.

29 Lee, L., Yoon, K.H., Jung, J.W. et al. (2018). Ultra gas-proof polymer hybrid thin layer. *Nano Letters* 18 (9): 5461–5466.

30 Sang, H.K.P., Ji, Y.O., Hwang, C.S. et al. (2005). Ultra thin film encapsulation of organic light emitting diode on a plastic substrate. *ETRI Journal* 27: 5.

31 Meyer, J., Schneidenbach, D., Winkler, T. et al. (2009). Reliable thin film encapsulation for organic light emitting diodes grown by low-temperature atomic layer deposition. *Applied Physics Letters* 94: 233305.

32 Liao, Y., Yu, F., Long, L. et al. (2011). Low-cost and reliable thin film encapsulation for organic light emitting diodes using magnesium fluoride and zinc sulfide. *Thin Solid Films* 519: 2344–2348.

33 Park, J.S., Chae, H., Chung, H.K. et al. (2011). Thin film encapsulation for flexible AMOLED: a review. *Semiconductor Science and Technology* 26 (3): 034001.

34 Cho, A.R., Kim, E.H., Park, S.Y. et al. (2014). Flexible OLED encapsulated with gas barrier film and adhesive gasket. *Synthetic Metals* 193: 77–80.

35 Wu, P., Wang, H., and Chen, L. (2009). Research progress and development trend of laminated solar cells. *Science and Technology Review* 27 (3): 95–98.

36 Kim Wiley, L. and Jiang. (2010). Application analysis of amorphous silicon thin film solar cells. *Solar Energy Technology and Products* 7: 34–38.

37 Yin, B. and Jiang, F. (2012). Research progress of amorphous silicon thin film solar cells. *Guangzhou Chemical Industry* 40 (8): 31–33.

38 Rongrong, C., Boyd, C.C., Burkhard, G.F. et al. (2018). Encapsulating perovskite solar cells to withstand damp heat and thermal cycling. *Sustainable Energy & Fuels* 2: 2398–2406.

39 Guo, L.X., Xin, Z., Jiao, S.Z. et al. (2019). Research progress of interface passivation of n-i-p perovskite solar cells. *Acta Physica Sinica* 68 (15): 158803. https://doi.org/10.7498/aps.68.20190468.

40 Juarez-Perez, E.J. and Haro, M. (2020). Perovskite solar cells take a step forward. *Science* 368 (6497): 1309–1309. https://doi.org/10.1126/science.abc5401.

41 Shi, L. et al. (2020). Gas chromatography-mass spectrometry analyses of encapsulated stable perovskite solar cell. *Science* 368 (6497): eaba2412. https://doi.org/10.1126/science.aba241.

2

Basic Concepts Related to Flexible Electronic Packaging

Peng-an Zong and Mengran Chen

College of Materials Science and Engineering, Nanjing Tech University, Nanjing City, China

Flexible electronic packaging primarily involves the incorporation or alteration of conductive materials within flexible structures using nonintegrated mechanical techniques such as embedding or stitching. These components are subsequently packaged and secured to achieve an integrated functionality. The primary objective of flexible electronic packaging is to safeguard chips while enhancing the overall stability and safety performance of the electronic product. As technology advances, traditional electronic packaging methods utilizing materials like metal, glass, and ceramics face susceptibility to environmental influences, thereby posing challenges to electronic packaging. The advent of novel electronic packaging technologies enables the enhancement of overall device quality. Moreover, the progress in materials science, mechanics, and related fields has presented new demands and challenges concerning the extensibility of electronic packaging materials.

Extensibility, in the context of electronic packaging, pertains to the capacity of electronic devices to uphold their exceptional performance even when subjected to various deformations, distortions, and environmental conditions. This attribute significantly enhances the portability and adaptability of electronic devices. The objective of this chapter is to elucidate fundamental concepts associated with flexible electronic packaging, aiming to offer valuable insights and guidance for the design and comprehension of flexible electronic devices.

2.1 Composition of Flexible Electronic Packaging

Due to its distinctive flexibility and stretchability, flexible electronic packaging technology has emerged as a prominent trend in electronic manufacturing, offering vast application prospects across various industries. It encompasses essential elements such as flexible substrates, adhesive layers, crosslinked conductive materials, electronic components, and cover layers (illustrated in Figure 2.1) [1]. The primary objective of this section is to delve into the fundamental aspects of flexible

Flexible Electronic Packaging and Encapsulation Technology, First Edition.
Edited by Hong Meng and Wei Huang.

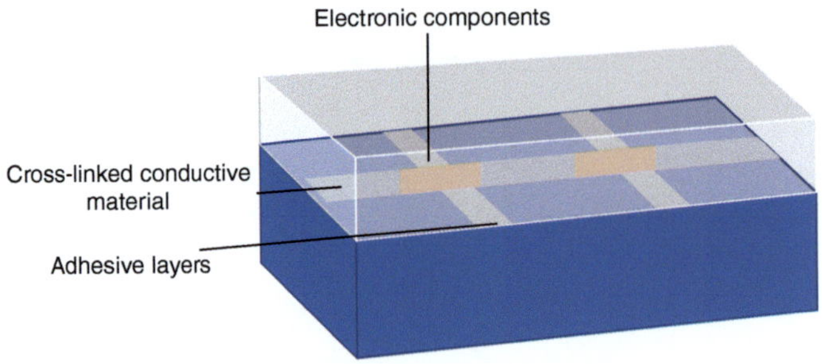

Figure 2.1 Fundamental building block of flexible electronics.

packaging technology, providing a comprehensive understanding of its principles and mechanisms.

2.1.1 Flexible Substrate

In contrast to traditional substrates, which are predominantly fabricated from rigid materials like quartz, glass, silicon wafers, and metal electrodes, flexible substrates exhibit a unique set of characteristics and are gaining considerable attention in various applications. These flexible substrates not only inherit the advantageous features of their rigid counterparts, such as cost-effectiveness and electromagnetic stability, but also offer additional benefits such as flexibility, lightweight construction, thin film properties, and excellent light transmittance. Furthermore, their inherent flexibility enables them to conform to complex device surfaces while maintaining consistent performance. They are easily manipulable, transportable, and seamlessly integrable with other structures or devices, thus attracting significant research interest in the field. Figure 2.2 showcases several commonly employed flexible substrates, including polyvinyl alcohol (PVA) film, polyimide (PI) film, polyethylene terephthalate (PET) film, polydimethylsiloxane (PDMS) film, polyethylene naphthalate (PEN), as well as textile materials [2].

PDMS film, known for its convenience, availability, chemical stability, transparency, and excellent thermal stability, serves as a frequently employed flexible substrate. One notable advantage of PDMS lies in its distinguishable adhesive and nonadhesive regions, which become evident under UV radiation, facilitating effective integration with electronic materials. Consequently, PDMS finds widespread application in this regard. However, the prevailing approach to attain flexibility in electronic devices primarily revolves around reducing substrate thickness, a method predominantly suitable for relatively flat substrate surfaces. This approach falls short when it comes to fulfilling the requisites of stretchable electronic devices that necessitate seamless adhesion to diverse irregular surfaces [3]. To achieve the desired stretchability in wearable sensors, two primary methodologies are presently employed: directly bonding thin conductive materials with low Young's modulus onto flexible substrates or assembling devices employing conductors possessing intrinsic stretchability.

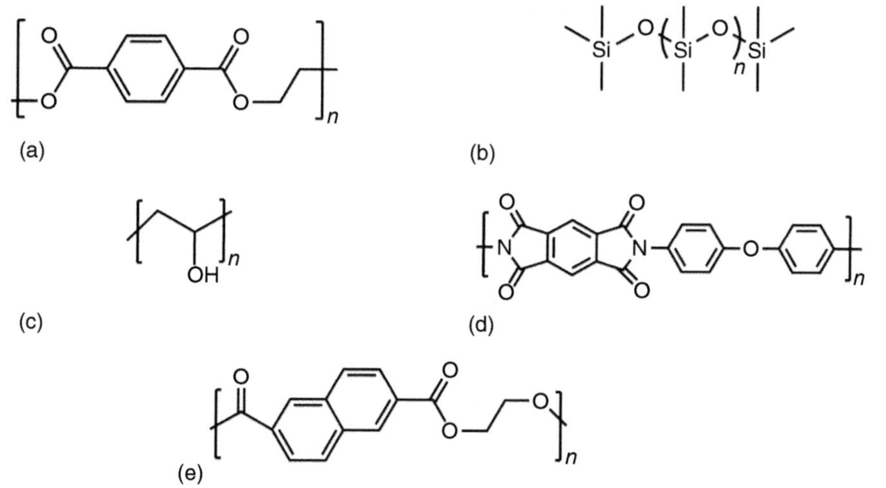

Figure 2.2 Molecular structures of common flexible substrates: (a) polyethylene terephthalate (PET); (b) polydimethylsiloxane (PDMS); (c) polyvinyl alcohol (PVA); (d) polyimide (PI); (e) polyethylene naphthalate (PEN).

2.1.2 Electronic Components

Electronic components serve as the fundamental building blocks of flexible electronic products, encompassing essential functionalities such as resistance, capacitance, transistors, information storage, and computing units. These components constitute the core units for microsystem functions and circuit execution. Within the realm of electronic components, the term "electronic devices" primarily pertains to products that retain the molecular structure of their raw materials throughout the processing and manufacturing stages. These products exhibit a distinct structural pattern and performance characteristics, representing devices that do not require energy generation, such as resistors, capacitors, and inductors. Due to their lack of electron generation, voltage or current control, and transformation capabilities, they are often referred to as passive devices. Components can be further categorized into two types: circuit components and connecting components. Circuit components encompass diodes, resistors, and other similar elements, while connecting components comprise connectors, sockets, printed circuit boards (PCBs), connecting cables, and other associated parts.

Electronic devices primarily encompass products that undergo alterations in the molecular structure of raw materials during processing and manufacturing. These products are comprised of two or more components, enabling the creation of products with distinct performance characteristics compared to single-component ones. Notable examples include transistors, electronic tubes, and integrated circuits. By virtue of their capacity to generate and control electrons as well as transform current and voltage through functions such as amplification, switching, rectification, oscillation, and modulation, electronic devices that rely on external energy sources are categorized as active devices. Devices can be further classified into two main types: active devices and discrete devices. Active devices possess the characteristic of consuming their own electrical energy and necessitate external power sources for operation. On the other hand, discrete devices encompass components such as bipolar junction transistors, field-effect transistors, thyristors, semiconductor resistors, and capacitors.

As researchers delve deeper into the exploration of electronic components, the clear distinction between components and devices has become increasingly challenging to define. Consequently, it is now common to collectively refer to them as "electronic components" or simply as "components." In terms of essential characteristics, there is no fundamental difference between electronic components and components in traditional electronic technology. Currently, certain components employ inorganic semiconductor materials such as silicon. However, these materials exhibit rigidity and susceptibility to damage during deformation. To mitigate such risks, these components can be initially placed on rigid microcell islands, which are subsequently integrated with a flexible substrate carrying the microcell islands. This approach helps prevent damage to the components. In the realm of flexible electronics, the prominent distinction from traditional electronics lies in the utilization of organic electronic components. Organic materials provide favorable conditions for reducing the weight and thickness of components while enhancing their flexibility and stretchability [4, 5].

2.1.3 Crosslinked Conductive Materials

The term "crosslinked conductor" denotes a metal film or graphene material that is affixed to a flexible substrate, exhibiting exceptional conductivity. This material plays a crucial role in connecting electronic components dispersed across different locations on a flexible substrate, thereby establishing a complete flexible circuit that functions as a wire within flexible electronic systems [2]. When the substrate experiences substantial stretching, the structural configuration of the crosslinked conductive material ensures the utmost stability of the electromagnetic properties of the components. In the realm of flexible electronic products, commonly employed metal conductors in crosslinked conductive materials include copper foil, aluminum foil, and copper beryllium alloy films. Rolled copper foil demonstrates remarkable ductility and resistance to bending, making it a cost-effective option. Conversely, electrolytic copper foil possesses a relatively rough surface and exhibits ease of integration with flexible substrates. Consequently, rolled copper foil and electrolytic copper foil represent prevalent choices as crosslinked conductive materials.

2.1.4 Adhesive Layer

The adhesive layer possesses essential characteristics such as heat resistance, superior bonding strength, and flexibility, enabling it to effectively bond diverse functional layers of electronic components in flexible electronic packaging. Currently, widely used materials for the adhesive layer include acrylic resin and epoxy resin. Acrylic resin exhibits versatility, as its properties and applications can be tailored by selecting distinct resin structures, formulations, production processes, and solvent compositions. On the other hand, epoxy resin showcases strong adhesive properties to various substances owing to the presence of polar hydroxyl groups and ether bonds in its molecular chain. Additionally, the low shrinkage and minimal internal stress of epoxy resin contribute to its enhanced adhesion strength. Furthermore, cured epoxy resin demonstrates favorable physical and chemical properties, such as exceptional adhesion strength to both metal and nonmetal surfaces, favorable dielectric properties, and negligible deformation shrinkage. Moreover, epoxy resin maintains a desirable balance between flexibility and stability.

In addition to the utilization of acrylic and epoxy resin materials, recent advancements in research have led to the development of an innovative adhesive inspired by the remarkable suction capabilities of octopuses. Drawing inspiration from the working mechanism of octopus suction cups, researchers have chosen dimethylsiloxane as the foundational material to fabricate a nanosheet featuring an array of suction cup-shaped structures with porous surfaces that mimic those found on octopus suction cups. These pores remain open under normal room temperature conditions; however, upon heating the nanosheet to 32 °C, they undergo automatic contraction, generating a suction effect [6]. This intelligent adhesive exhibits versatile applications, including the manufacturing of flexible electronic products, compound semiconductor devices, and even serving as a substrate for wearable health sensors, obviating the need for additional adhesives.

2.1.5 Coating Layer

In the realm of flexible electronics packaging technology, the cover layer, also referred to as the packaging layer, typically constitutes a composite structure comprising a variety of materials. Its primary function is to safeguard the flexible circuit from the detrimental effects of electromagnetic fields, water vapor, oxygen, dust, chemical reagents, and other impurities, thus mitigating strain on the flexible circuit during bending processes. As the composite layer undergoes bending or flexing, it is crucial for it to offer effective isolation or treatment against the complex environmental conditions encompassing multiple elements. This capability serves to minimize the occurrence of torsional stress, which can lead to the separation of rigid microcell islands from the flexible substrate [7]. Consequently, the material employed in the encapsulation layer must possess desirable attributes such as robust heat resistance, bending endurance, fatigue resistance, and conformability. Commonly utilized cover layer materials encompass acrylic resin, epoxy resin, and PI.

2.2 Flexible Electronic Packaging Structure

There are two primary methodologies for achieving flexibility in electronic devices and systems: introducing flexibility through materials and introducing flexibility through design. The first approach, introducing flexibility through materials, primarily involves utilizing flexible semiconductor materials that are sensitive and deformable, enabling the realization of flexible electronic devices. In the early stages, flexible materials such as glass foil, PET film, and stainless steel substrates had limited dielectric constants and coupling coefficients, leading to compromised performance due to their susceptibility to deformation. Nevertheless, in recent years, novel flexible materials like graphene, black phosphorus, and liquid metals (e.g. GaIn–Si) have emerged, exhibiting promising performance and demonstrating potential applications in the domains of electronics, optics, and mechanics. These advanced materials have opened up new possibilities for the development of flexible electronic devices [8].

The subsequent phase involves the integration of flexibility into the structure, which necessitates a harmonious interplay between the material and the structure to attain flexibility in electronic devices and systems. From a mechanical perspective, interconnect structures characterized by wrinkles and folds can bolster the overall tensile strength of the materials, while fine-tuning their deformation capacity can further augment flexibility. Currently, a wide range of conductive materials can be manipulated into diverse configurations, encompassing wavy, mesh, strip, serpentine, and even multidimensional shapes, employing advanced techniques such as photolithography, transfer printing, and electron-beam deposition, thereby facilitating the introduction of flexibility into the structure. Furthermore, the introduction of flexibility into the structure can be classified into the following three types:

2.2.1 Curved Structures of Hard Thin Films

Curved structures of rigid thin films pertain to quasi-periodic or periodic curvatures attained by transferring silicon and other rigid thin film strips onto a flexible elastic substrate via transfer printing techniques. Researchers have also explored the utilization of transfer technology based on soft lithography to facilitate the packaging of flexible electronic components [9]. Through numerous iterative experiments, it has been demonstrated that this technology can induce bending waves in silicon ribbons on flexible substrates. By manipulating the amplitude and wavelength of the silicon ribbon bending wave, this technology enables the integration of various electronic materials onto curved surfaces, while mitigating tensile fracture and enhancing compressive performance during subsequent deformation. This mechanism primarily relies on the dimensional deformation within the plane of the substrate and longitudinal displacement to alleviate internal stresses. Distinguished scholars, namely Professors Rogers from the University of Illinois and Huang from Northwestern University in the United States [10, 11], have successfully achieved direct integration of flexible structures on curved surfaces using diverse materials. The process of fabricating the curved silicon strip on a PDMS substrate is illustrated in Figure 2.3. In 2017, they jointly proposed a non-bent structure with a wire width equivalent to its thickness, which exhibited sixfold greater extensibility compared to rigid thin-film materials [12]. By increasing the wire's thickness, this structure not only significantly reduces material resistance but also enhances electrical and thermal properties. This innovative structure has found applications in the development of stretchable devices, including LED arrays, solar cells, and flexible antennas.

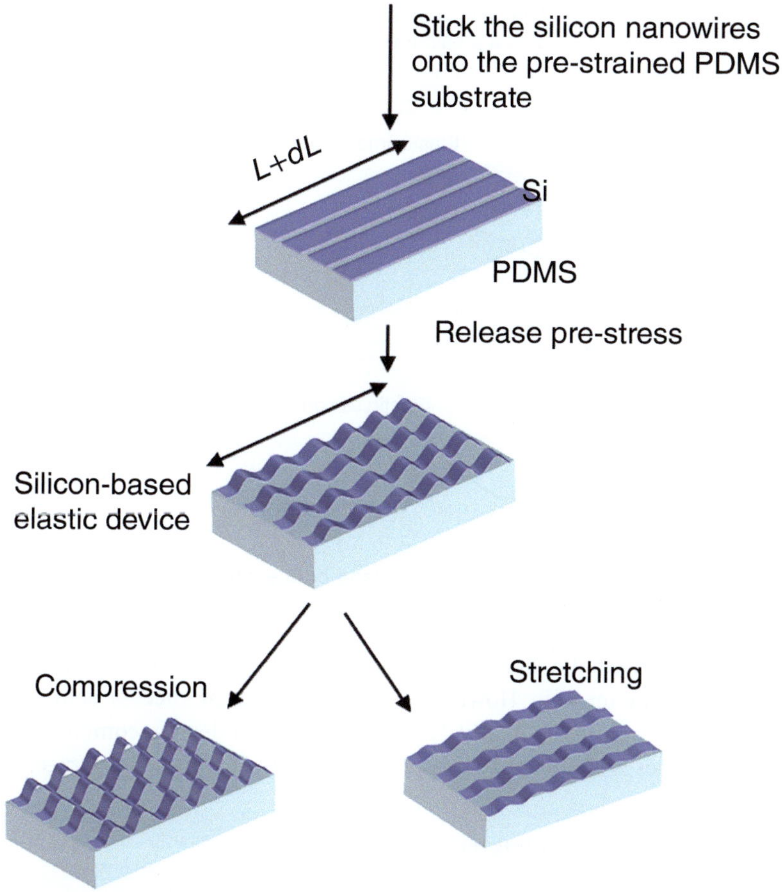

Figure 2.3 The process for fabricating the PDMS-based curved silicon strip structure.

2.2.2 Island-Bridge Structure

The advent of interconnecting structures aims to overcome the limitations of tensile fracture in curved structures. The fundamental principle involves interconnecting multiple microelectronic devices in series using flexible wires, thereby establishing an island bridge structure. By leveraging the bendability of these wires, the island-bridge structure formed by interconnecting microelectronic components enhances the extensibility of electronic devices and augments the overall flexibility of the structure. Figure 2.4 illustrates a direct-connected island-bridge structure, where various microelectronic components on the PDMS substrate are interconnected through bent wires. In the lead bridge structure, when external forces vary along the x or y direction of the substrate, the noncoplanar serpentine interconnect structure effectively compensates for the impact of external strain [13].

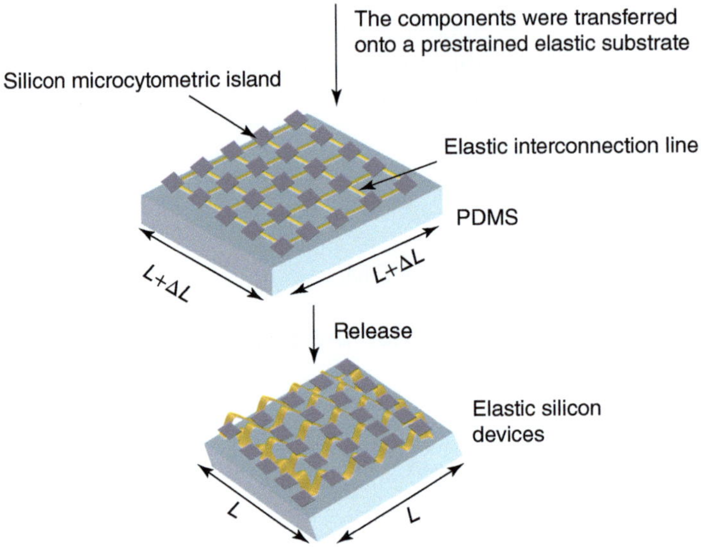

Figure 2.4 Construction of PDMS substrate-based island-bridge interconnect structure.

This mechanism improves the ductility of the structure while endeavoring to maintain stability under tension. However, despite the partial success achieved by the island-bridge structure, its relatively lower integration density compared to other configurations poses challenges for applications requiring higher coverage.

2.2.3 Pre-strained Super-Soft Interconnect Structure

The implementation of the pre-strained structure involves affixing functional components onto a pre-stretched substrate while the wires are deliberately bent and arched out of the plane to achieve device ductility. By combining pre-strained direct and bent interconnections, electronic systems that fulfill a wide range of flexible requirements can be achieved. The pre-strain and serpentine interconnect structure is illustrated in Figure 2.5 [14]. In this approach, researchers subjected the serpentine connecting lines and substrate to pre-stretching up to their deformation limit and then transferred them onto the base using transfer printing technology to release strain on the substrate. Experimental findings, along with numerical simulations, substantiate that the pre-strained substrate exhibits twice the elastic modulus and elongation compared to the non-pre-strained structure.

2.2.4 Open Grid Structure

The Open Grid Structure represents an open mesh configuration developed to enhance the properties of silicon-based thin-film electronic device materials. Within this structure, the thin film material is capable of in-plane rotation during

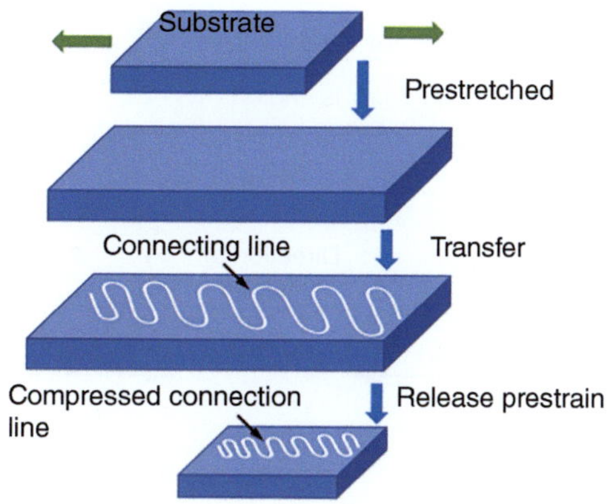

Figure 2.5 Pre-strain combined with serpentine interconnect scheme.

deformation, thereby enhancing its flexibility and stretchability. However, it is worth noting that the unique shape of the open mesh structure limits its suitability for numerous applications.

2.3 Encapsulation Principle

Packaging technology assumes paramount significance in the design and manufacturing of electronic devices, spanning from individual semiconductor components to intricate systems such as large-scale computers. Its fundamental objective is to furnish protective enclosures that secure, seal, and shield the devices from environmental influences. Furthermore, packaging technology serves as a vital link connecting the internal components of the device with external circuits. Consequently, the packaging stands as a pivotal constituent of integrated circuit products. This section elucidates the principles governing packaging, encompassing aspects such as permeation, the preparatory procedures involved in packaging films, the barrier characteristics of films, and the mechanical evaluation of films.

2.3.1 Basic Principle of Penetration

In the realm of flexible electronic packaging, penetration primarily pertains to the phenomenon wherein contaminants, such as water vapor and oxygen, infiltrate electronic devices via barrier films, driven by factors such as pressure and concentration. By regulating the permeation rate through barrier films, the exposure of electronic devices to water vapor and oxygen can be limited, thereby enhancing their lifespan and performance. The penetration process can be delineated into two sequential steps: firstly, impurities such as water vapor permeate the barrier film and get absorbed into the solid, and secondly, the impurities diffuse within the material [15]. Consequently, under steady-state conditions characterized by uniform thin films, the permeation rate (P) equals the product of the two inherent characteristics of materials, solubility coefficient (S) and the diffusion coefficient (D) [16]:

$$P = S \cdot D \tag{2.1}$$

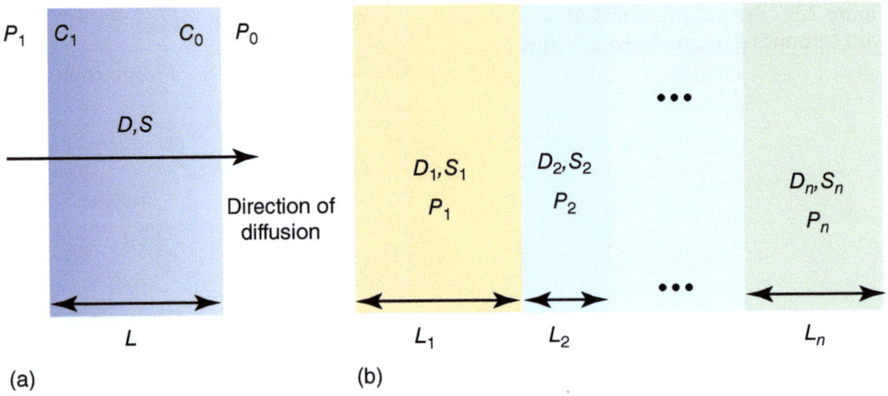

Figure 2.6 Gas or vapor permeation mechanisms in (a) mono-layer and (b) multilayer composite material.

When it comes to gas permeation, the P of barrier films is typically quantified in units of cm^3/m^2/day, taking into account of specific temperature and pressure parameters. On the other hand, in the context of water vapor permeation under specific temperature and concentration conditions (often expressed in terms of relative humidity, RH), the conventional unit for measuring P is in grams per square meter per day (g/m^2/day).

Fick's first law and Henry's law can illustrate this process well. Figure 2.6a schematically illustrate the permeation model with given boundary conditions [17, 18]. The diffusion flux J exhibits a direct proportionality to the concentration c:

$$J = -D \cdot \nabla(c) \tag{2.2}$$

Assuming that Henry's law establishes a linear relationship between the concentration c and the partial pressure ΔP through $c = S\Delta P$, then the concentration in Eq. (2.2) can be replaced with the S and ΔP. In steady state, the equation can be written as:

$$J = \frac{DS(P_0 - P_1)}{L} \tag{2.3}$$

To determine the total mass that passes through the layer, the diffusion flux is multiplied by time [19]. Nonetheless, the investigation of transient permeation has garnered significant interest owing to the utilization of barrier layers with extended lag times. In the case of time-varying gas permeation, Fick's second law can be applied:

$$\frac{\partial c}{\partial t} = \nabla(D \cdot \nabla c) = D \cdot \left(\frac{\partial^2 c}{\partial x^2} + \frac{\partial^2 c}{\partial y^2} + \frac{\partial^2 c}{\partial z^2}\right) \tag{2.4}$$

Under the assumption that the diffusion coefficient D remains constant regardless of concentration and exhibits isotropy, this scenario is applicable only when there is no chemical interaction between the permeating substance and the solid material. In the presence of an infinite source and sink, the calculation can be performed using the following equation [20]:

$$c(x,t) = C_1\left(1 - \frac{x}{l}\right) - \frac{2C_1}{\pi}\sum_{n=1}^{\infty}\frac{1}{n}\sin\frac{n\pi x}{l}e^{-Dn^2\pi^2 t/l^2} \tag{2.5}$$

The aforementioned equation is generally applicable to the diffusion process occurring in a single-layer thin film, where C_1 represents the constant c on one end, and l denotes the film thickness.

The permeation of H_2O vapor and O_2 through multilayer barrier films has emerged as a significant concern in the context of widely utilized multilayer packaging configurations. The equation representing the overall permeation coefficient P_T of a structure of n layers, featuring uniform thickness L_n and permeation coefficient P_T, as depicted in Figure 2.6b, can be expressed as follows:

$$\frac{1}{P_T} = \frac{L_{\text{total}}}{\frac{L_1}{P_1} + \frac{L_2}{P_2} + \frac{L_3}{P_3} + \cdots + \frac{L_n}{P_n}} \tag{2.6}$$

In Eq. (2.6), L_{total} is the cumulative depth of the multilayer structure. In steady-state scenarios, presuming a uniform J across all barrier layers, the P_T of the multilayer film can be determined. Recent research findings suggest that steady state, as well as transient permeation due to delay, can influence the P_T of multilayer structures, comprising both organic and inorganic layers. The impact of transient permeation within multilayer structures holds considerable significance, emphasizing its crucial role in the assessing P_T.

The delay of single and multilayer film are described by Eqs. (2.7) and (2.8), respectively [20]:

$$L = \frac{l^2}{6D} \tag{2.7}$$

$$L = \frac{\sum_{i=1}^{n}\left\{\frac{l_i^2}{2D_i}\sum_{i=1}^{n}\left[\frac{l_m^2}{D_m}\prod_{j=1}^{i-1}k_j\right] - \frac{l_i^3}{3D_i^2}\prod_{j=1}^{i-1}k_j\right\} + \sum_{i=1}^{n}\left\{\frac{l_i}{D_i}\prod_{j=1}^{i-1}k_j\sum_{\beta=i+1}^{n}\left[\frac{l_\beta}{\prod_{j=1}^{\beta-1}k_j}\sum_{m=\beta}^{n}\left(\frac{l_m}{D_m}\prod_{j=1}^{m-1}k_j\right) - \frac{l_\beta^2}{2D_\beta}\right]\right\}}{\sum_{i=1}^{n}\frac{l_i}{D_i}\prod_{j=1}^{i-1}k_j} \tag{2.8}$$

In Eq. (2.8), $k_j = S_j/S_{j+1}$. In Eq. (2.7), the lag time of a single layer film is contingent upon its l and D. In contrast, the delay of a multilayer film not only relies on the l and D, but also incorporates the additional factor of the solubility of the individual layers.

2.3.2 Permeation Mechanism of Water Vapor and Gas

The permeation modes of gases and moisture within polymers and inorganic films display distinctive characteristics [17]. Polymers feature a dynamic internal lattice structure, where interlaced polymer chains form interstitial spaces characterized by continual variations in size and geometry. While moisture and gases can penetrate through these spaces, their presence is transient and unpredictable [18]. Direct utilization of individual polymer layers in organic device packaging is inadequate, as they fail to meet the rigorous requirements for barrier performance. Hence, achieving high barrier properties often entails employing polymers that have been coated with inorganic or multilayer films.

The advancement of ultrahigh barrier films primarily revolves around the utilization of inorganic films, aiming to address the limitations inherent in polymerics. The permeation mechanism of inorganic films, deposited through techniques like physical magnetron sputtering, thermal evaporation, as well as PECVD, remains a subject of ongoing investigation. Currently, the diffusion process within numerous silicon oxides, collectively known as "glasses," occurs via narrow interstices within the Si–O lattice [18]. Penetration within deposited films is likely to transpire through both macroscopic and nanoscale defects, including interstices [19]. The extent of permeation is influenced by various factors, such as the dimensions of the permeating molecules, and the defect size and density within the barrier films, which stem from the structural incompleteness.

The SiO_x films deposited on PET substrates exhibit a grainy structure, as evidenced by the AFM images depicted in Figure 2.7a [18]. Analysis of transmission electron microscopy (TEM) images in Figure 2.7b reveals the presence of nanoscale defects between adjacent SiO_x grains, which can be mitigated through optimization of process conditions [22]. The surface morphology of SiO_x deposited by PECVD on glass substrates with varying resolutions is depicted in Figure 2.8a–c. The images reveal significant voids and a relatively rough surface marked by a root mean square (RMS) surface roughness of 3.8 nm, attributed to grain formation. The specific structure of SiO_x, as depicted in Figure 2.8d, serves as a pathway for the permeation of water vapor and oxygen [18, 19, 21, 22, 24]. Infiltration can be categorized as hindered or unhindered depending on the size of permeating entities, like water

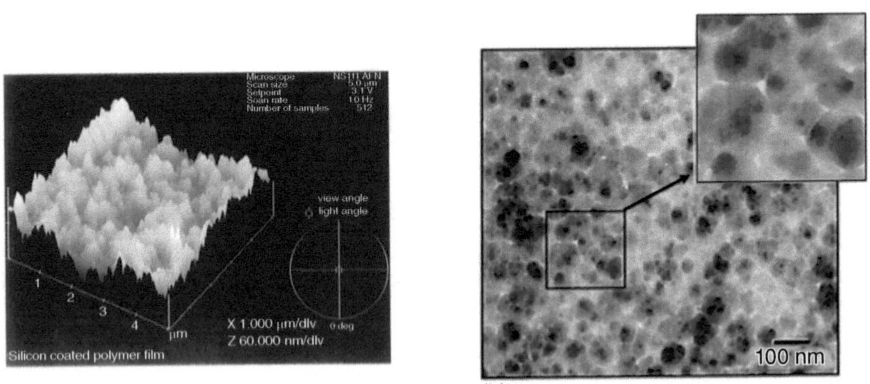

(a) (b)

Figure 2.7 Surface morphology and defect analysis of SiO_x barrier layer on PET substrate via PECVD: (a) 45° AFM image of SiO_x surface morphology on PET substrate; (b) TEM image with zero-loss energy filtering of SiO_x barrier layer. Source: Reproduce with permission from Tropsha and Harvey [17]; © 1997, Journal of Physical Chemistry B; Reproduce with permission from Erlat et al. [18]; © 1997, Journal of Physical Chemistry B.

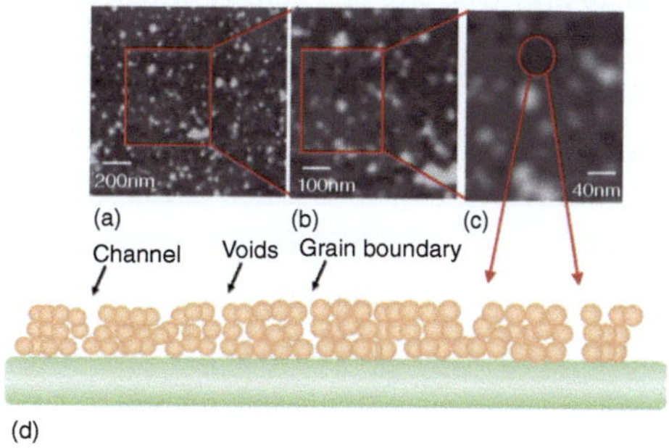

Figure 2.8 SiO_x deposition and cross-section analysis by PECVD. (a)–(c) AFM images of SiO_x on glass substrates at various resolutions; (d) schematic cross-section illustration of SiO_x deposited by PECVD based on AFM images, highlighting defects including voids, channels, and grain boundaries. Source: Reproduce with permission from Erlat et al. [24]; © 2000, Journal of Materials Research.

molecules with a diameter of 0.33 nm and oxygen defects with a diameter of 0.32 nm [19, 21]. The resistance to penetration is influenced by the dimensions of the permeating molecules, the size and distribution density of defects within the barrier films, which arise due to the structural incompleteness of the deposited inorganic films. In most deposition techniques for thin film, small islands initiate on the substrate surface and undergo growth by agglomerating with neighboring islands, eventually forming a continuous film [25]. As this growth mechanism does not adhere to a thermodynamic equilibrium process, it gives rise to defective structures, facilitate the permeation of water vapor and gases when compared to closely stacked films [18, 19, 21, 23, 24].

Apart from the inherent defects resulting from the growth mechanism, controlled defect penetration can be attributed to the geometric shadowing of micro-sized defects and the structural morphology induced by particles on the substrate [25, 26]. Sobrinho et al. have proposed a defect-driven permeation model, asserting that these micro-sized defects are pivotal in determining the barrier properties. According to their model, while inorganic coatings are considered impermeable, the geometry of the defects governs the permeation process. In this context, the size and density of defects within the coating are the primary factors influencing permeation [26–28]. Data pertaining to the defect size and density of inorganic films deposited on organic substrates are presented in Table 2.1.

The activation rate theory is a commonly employed framework in numerous studies to investigate the diffusion mechanisms of water vapor and oxygen [16, 17, 24, 25]. Specifically, the activation energy linked to water permeation represents the energy required to elevate the permeant to an activated state, surmounting the potential barrier matrix. This process enables the permeant to advance to the next available position. The movement of an ideal gas through silicon oxide and a glassy polymer below the glass transition temperature (T_g) can be accurately elucidated by applying the Arrhenius equation [28]:

$$P = P_0 e^{-\Delta E_p/RT} \tag{2.9}$$

The variables P, P_0, and ΔE_p represent the permeability, pre-exponential factor, and activation energy of permeation, respectively, measured in units of kJ/mol. R denotes the ideal gas constant, measured in J/mol/K, while T represents the temperature in Kelvin. By conducting permeability measurements at various

Table 2.1 The size and density of defects in deposited inorganic films on organic substrates.

Materials	Methods	Radius (μm)	Density (1/mm^2)	Substrate
SiO_2	PECVD	0.6	11–1100	PET
Si_3N_4	PECVD	0.6	5–1000	PET
Al	CVD	1.0	25–400	PET
Al	Sputtering method	2–3	200	PET
AlO_xN_y	Sputtering method	0.5–1.4	600	PET
Al	CVD	0.5	100–1000	BOPP

temperatures, one can calculate the activation energy of permeation [29, 30]. Sobrinho et al. employed the activation rate theory to investigate the diffusion of oxygen and water vapor via SiO_x and SiN_x-coated PET substrates. Their results indicated that, within the experimental margin of error, the activation energy for oxygen permeation remained consistent across pure PET, SiO_x/PET, SiO_x/PET/SiO_x, and SiN_x-coated samples, while SiO_x exhibited a distinct activation energy for oxygen permeation. These observations imply that inorganic barrier layers, like SiO_x and SiN_x, have imperfect structures that facilitate gas permeation through defects, with PET acting as the primary rate-limiting barrier. On the contrary, the activation energy for water vapor permeation differs from that of the bare PET substrate within the same barrier layer structure, as illustrated in Table 2.2 [26]. These outcomes indicate that the diffusion of water vapor deviates from that of ideal gases and cannot be solely explained by a defect-dominated permeation mechanism. Moreover, due to the strong interaction between water vapor and PET as well as silica, more intricate diffusion mechanisms arise compared to those involving oxygen [17, 31]. Currently, there remains controversy among researchers concerning the observed activation energy that has been reported and its interpretation. Nonetheless, it is widely accepted to characterize barrier performance by considering the interaction between water vapor and the barrier film, thereby facilitating the design of suitable encapsulation strategies for organic electronic films.

Table 2.2 Activation energy required for oxygen (OTR) and water vapor (WVTR) permeating PET before and after being coated with inorganic materials.

Materials	Activation energy of oxygen (kJ/mol)	Activation energy of water vapor (kJ/mol)
PET	29.3	47.2
SiO_2/PET	30.1	60.5
SiO_2/PET/SiO_2	31.4	62.7
SiN_x/PET	32.8	50.9
SiN_x/PET/SiN_x	31.2	64.5

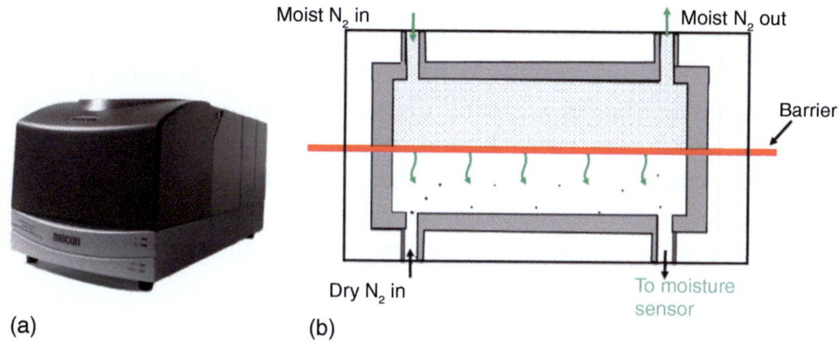

Figure 2.9 Permeation rate measurement setup: (a) The MOCON permatran-W for permeation rate measurement; (b) experimental setup for ASTM1249 permeation rate measurement.

2.3.3 Barrier Performance Measurement

The evaluation of barrier layer performance relies on permeability, measuring the barrier layer's resistance to water vapor and oxygen. Achieving ultrahigh-performance barrier layers in organic electronic packaging necessitates the development of precise and sensitive water vapor permeation testing methods. Accurately measuring and evaluating the remarkably low permeability of water vapor ($\sim 10^{-6}$ g/m^2/day) and oxygen poses a major challenge in assessing the exceptional performance of packaging films. The experimental setup for both methods exhibits fundamental similarities, as depicted in Figure 2.9b. In this setup, the thin film sample is positioned between a wet chamber and a dry chamber under standard atmospheric pressure, enabling water vapor to diffuse through the thin film from the wet chamber to the dry chamber. The transmitted water vapor is directed toward an infrared moisture sensor, which generates an electric signal proportional to the amount of water vapor that enters the sensor. By analysis and calculating of the electric signal, the water vapor transmission rate (WVTR) and other parameters can be determined. In contrast, the Mocon Aquatran permeation analyzer utilizes Coulometric detection as the primary measurement technique, distinguishing it from the infrared sensor employed in the standard ASTM F1249. This Coulometric detection approach enhances testing sensitivity by an order of magnitude compared to the conventional ASTM F1249 standard.

The Mocon Aquatran typically has a detection limit of approximately 10^{-4} g/m^2/day, which proves effective in analyzing most single- and multi-layer films. However, its sensitivity falls short of detecting the ultralow transmission rates exhibited by advanced packaging materials. Presently, the calcium corrosion test, also known as the calcium bond test, serves as the prevailing method for measuring the ultralow transmission rates of barrier films. This test primarily relies on the corrosion level of a calcium film to gauge the transmission rates of barrier films. The principle behind this method lies in the fact that at room temperature, the Gibbs free energy (ΔG) of calcium oxide approximates −1200 kJ/mol, enabling it to swiftly oxidize in the presence of air. Consequently, calcium exhibits remarkable sensitivity to the detection of oxygen and water vapor. Moreover, calcium itself is a conductive, opaque metal, while its oxide is dielectric and transparent in appearance. Thus, changes in calcium conductivity or optical transparency can be indirectly measured to monitor the oxidation and corrosion rate. Through optical microscopy or visual inspection, this test method enables the differentiation between diffusion permeation and defect permeation by observing alterations in transparency. The potential chemical reaction that may occur between calcium and air can be represented as follows:

$$2Ca + O_2 \rightarrow 2CaO \tag{2.10}$$

$$Ca + 2H_2O \rightarrow Ca(OH)_2 + H_2 \tag{2.11}$$

According to reports, when both H_2O vapor and O_2 exist, the reaction of Ca with H_2O dominates (as shown in Eq. (2.11)), while the degradation of Ca caused by the reaction with oxygen constitutes less than 5% [32]. During the testing process, it becomes challenging to distinguish the corrosive influence of H_2O vapor and O_2 on Ca separately [33]. Therefore, the variation in Ca conductivity over time also needs to take into account the permeability of H_2O vapor and O_2. This combined effect is commonly referred to as the effective WVTR. However, in most literature reports employing Ca for permeation measurement, the difference between WVTR and effective WVTR has not been explicitly addressed. To address this concern, the measurement values mentioned in the subsequent text, along with those cited in the references for the Ca corrosion test, are expressed using "effective WVTR" to encompass the comprehensive influence of H_2O vapor and O_2 permeation. Figure 2.10a illustrates an example of an experimental setup for monitoring Ca conductivity, which facilitates the measurement of Ca corrosion. In this setup, a $4.5 \times 7\,\text{mm}^2$ calcium sensor is connected to an aluminum sensor for conductivity measurement. An encapsulation layer is deposited atop the Ca sensor, and the changes in Ca conductivity are monitored through a information collection system. The calculation of the WVTR involves applying formula (2.12), which determines the ratio of the conductivity change of Ca over time. By considering the peneration rate of both H_2O vapor and O_2, the effective WVTR provides a more accurate representation of the corrosive effects experienced by Ca.

$$\text{WVTR}\,(\text{g/m}^2/\text{day}) = -n\delta_{\text{Ca}}\rho_{\text{Ca}}\frac{\text{d}(G)}{\text{d}t}\frac{l}{w}\frac{\text{M}(\text{H}_2\text{O})}{\text{M}(\text{Ca})}\frac{\text{Area}\,(\text{Ca})}{\text{Area}\,(\text{Window})} \tag{2.12}$$

The molar equivalent of the reaction is denoted by n in Eq. (2.12). Considering the reaction with H_2O prevails, it is assumed that $n = 2$. The resistivity (δ_{Ca}) and density (ρ_{Ca}) of Ca are $3.4 \times 10^{-8}\,\Omega\text{m}$ and $1.6\,\text{g/cm}^3$, respectively. G denotes the conductivity of Ca, and $\text{d}(G)/\text{d}t$ represents the linear fit of the conductivity change over time (Figure 2.10b). l is the length and w is the width of the Ca sensor, while $\text{M}(H_2O)$ and M(Ca) represent molar weight of H_2O and Ca, respectively. The experimental setup maintains a consistent area ratio and length ratio between the Ca sensor and the water vapor permeation window due to the fixed geometry.

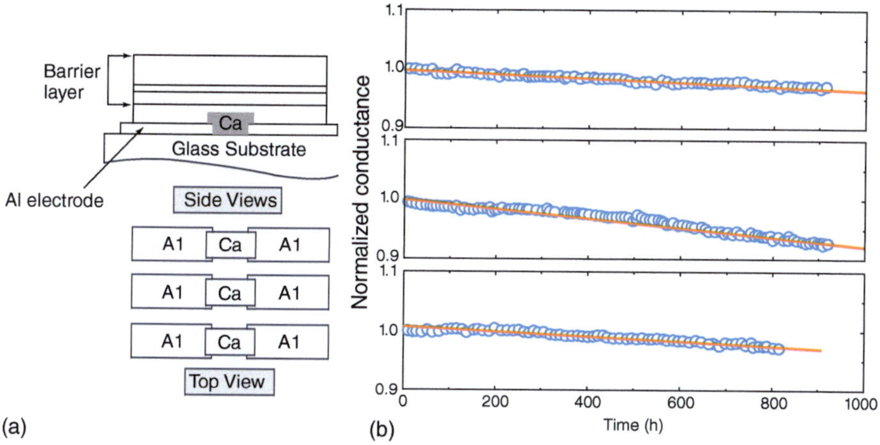

Figure 2.10 Ca corrosion test and conductivity change. (a) Side view of the test setup with the barrier layer (300 nm) on Ca and glass substrate and top view with aluminum contacts; (b) linear fitting function of conductivity change over time.

In Ca corrosion experiments, both conductivity and optical transmittance can be simultaneously measured. Conductivity measurement does not require electrical contact, whereas transmittance measurement necessitates an optical microscope or photographic camera [33]. Although both H_2O vapor and O_2 can spoil the Ca sensor and induce alterations in light transmission, this test is unable to differentiate between permeation due to H_2O vapor and O_2. Presently, Ca corrosion experiments can accurately measure effective WVTRs below 1×10^{-7} g/m^2/day [30, 33, 34]. Consequently, the Ca corrosion experiment stands as one of the most reliable methods for assessing the ultrahigh-performance barrier films demanded by organic electronic devices.

To ensure the accurate measurement of effective WVTR applying conductivity monitoring of Ca corrosion, it is assumed that oxygen uniformly permeates through the encapsulation barrier. However, it is important to acknowledge that errors in the effective WVTR value can arise from defects at the sample edges (lateral permeation) or localized defects which lead to localized Ca oxidation.

2.3.4 Thin-Film Barrier Technology for Organic Devices

Organic devices exhibit a heightened vulnerability to degradation stemming from the presence of water vapor and oxygen, highlighting the crucial need for high-performance encapsulation films with excellent barrier properties. Typically, the WVTR of barrier materials should be less than 10^{-6} g/m^2/day, whereas the OTR should range from 10^{-3} to 10^{-5} cm^3/m^2/day/atm [35, 36], to ensure the proper functioning of OLED devices. However, the specific packaging requirements for many OPVs and OLED devices are still uncertain, with stability and lifetime being the primary areas of investigation. Currently, only a limited number of studies

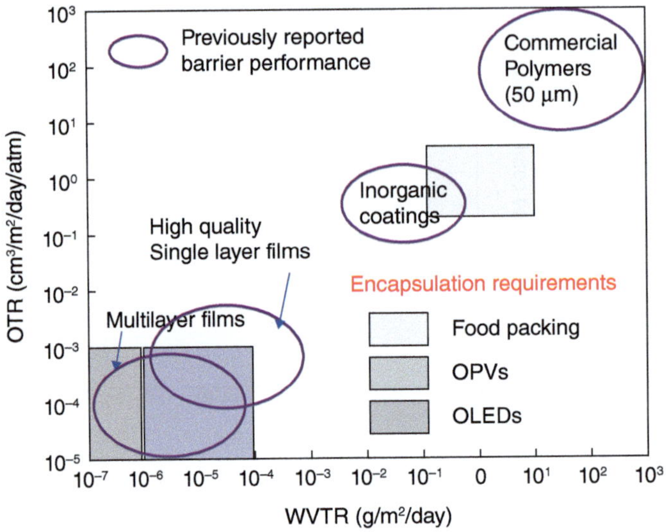

Figure 2.11 Barrier performance requirements and reported performance in technical applications.

have established a connection between WVTR and device lifetime [23, 37, 38]. Consequently, establishing a direct correlation between these parameters remains challenging for numerous organics and devices. In summary, the standard of barrier for a specific organic device is determined by the structure of the device, the materials employed, and the specific functional requirements it entails.

Many of the thin film package technologies used for organic electronics are now widely used for packaging food as well as drug. As shown in Figure 2.11, polymer films for food and pharmaceutical packaging have WVTRs ranging from 1 to 40 g/m^2/day [35, 39, 40]. Depositing inorganic films in vacuum is commonly utilized for enhancing the barrier properties of polymers and currently has demonstrated a 10–100 fold increase in the whole barrier property (Figure 2.11). Nonetheless, the majority of the deposited barrier materials for food and drug are insufficient to protect organic electronics. Therefore, it is urgent to develop films with high barrier property to facilitate the commercialization of organic electronics.

2.3.4.1 Single-Layer Film Package

Due to excellent barrier properties, the utilization of inorganics in the fabrication of encapsulation films holds significant potential for enhancing the lifetime and stability of organic electronic devices. Among the various approaches, the single-layer film encapsulation method offers relative simplicity compared to the multilayer film encapsulation technique. However, single-layer encapsulated films may exhibit certain defects, leading to subpar barrier performance. Nonetheless, the adoption of physical vapor deposition (PVD), PECVD, atomic layer deposition (ALD), and plasma-enhanced ALD (PEALD) techniques for the preparation of single-layer encapsulation films can effectively mitigate such defects, thereby enhancing the barrier effect. Currently, the primary inorganic materials employed in barrier film preparation involve the oxides of Si–O–N and Al–O–N systems. Among all methods for fabricating single-layer encapsulation films, PECVD and ALD are the mostly used. PECVD, characterized by its relatively high deposition rate, is approximately two orders of magnitude faster than ALD. Consequently, at low temperatures, monolayer SiO_x and SiN_x films for protection of organic electronics are frequently manufactured by PECVD [41, 42]. These films possess sufficient density, showing high WVTR values exceeding 0.01 g/m^2/day. Consequently, organic electronic devices encapsulated using PECVD surpass the packaging target requirements by several orders of magnitude [39, 42, 43].

As previously mentioned, the correlation between the lifetime of a device and the transparency of the packaging layer remains relatively unexplored. Instead, the viability of the packaging layer is often demonstrated by integrating it with actual devices and observing an enhancement over their lifetime. For instance, in a study conducted by W. Huang et al., the lifespan of an unpackaged OLED was approximately 6 hours. However, when the OLED was encapsulated with a single layer of PECVD SiN_x, the device exhibited a lifespan of 600 hours following testing, although the article does not provide information on the WVTR value [43]. This observation underscores the exceptional protective capability of SiO_x:H thin films prepared through the PECVD process when applied to OLED devices [44]. Moreover, researchers have discovered that utilizing hexamethyldisiloxane (HDMSO) and oxygen as deposition precursors enables the formation of a single-layer composite film comprising SiO_2 and siloxane. It was indicated that under storage conditions of 65 °C and 85% RH, the film can safeguard OLED devices for up to 7500 hours.

Furthermore, the use of ALD to deposit Al_2O_3 films has gained attention for single-layer encapsulation films because of its high density, relatively low defect density, and ability to shape-coat surfaces at comparatively low temperatures [45–47]. ALD, deposition involves introducing gas-phase precursors into a reaction chamber in an alternating pulse manner, leading to chemical adsorption reactions on the surface of the deposition substrate. A study demonstrated that depositing a 25 nm Al_2O_3 barrier layer on a PEN substrate using ALD achieved optimal performance, resulting in an effective WVTR of 1.7×10^{-5} g/m^2/day. However, integration of this film into practical devices to evaluate its influence on device lifespan is lacking [48]. Additionally, layered structures composed of Al_2O_3 and ZrO_2, as well as ALD-made double-layer structures of SiO_x and Al_2O_3, exhibit remarkable barrier properties. Although these structures can be considered multilayer structures due to the use of different materials, they are included here because they are all prepared using ALD. In one study, Meyer et al. deposited a 100 nm-thick film at 80 °C using an alternating nanolayer structure consisting of Al_2O_3 and ZrO_2 layers of 2–4 nm thick [30]. The barrier performance was compared to a monolayer Al_2O_3 of the identical thickness. The results revealed that the alternating nanolayer film structure exhibited superior barrier properties compared to the single Al_2O_3 layer.

This improvement can be attributed to the ZrO_2 layer, which not only hinders water vapor corrosion of the Al_2O_3 layer but also prevents the gathering of the Al–OH that has not been reacted, thus inhibiting the void formation in the Al_2O_3 layer and improving its barrier properties. Moreover, reports indicate that depositing a composite layer of Al_2O_3 and SiO_2 by ALD at 175 °C can further enhance the performance of the barrier layer. This improvement cannot be solely explained by the layering theory, as it is attributed to the SiO_2 layer filling the pinhole defects in the Al_2O_3 layer, thereby improving its barrier properties [49].

The ALD-deposited Al_2O_3 barrier layer has demonstrated its efficacy in providing superior encapsulation for OLEDs and OPVs across various environments, thereby significantly prolonging their operational lifespan. A notable example involves depositing a 200 nm-thick Al_2O_3 layer via ALD to encapsulate a pentacene/C_{60}-based organic solar cell, which exhibited stable operation for 6000 hours under standard conditions with minimal performance degradation, as depicted in Figure 2.12 [45, 49]. Encapsulating devices using ALD at 100 °C did not introduce any adverse effects on device performance, except for the previously reported thermal annealing effect [50]. For instance, an Al_2O_3-encapsulated top-emitting OLED deposited by ALD at 100 °C demonstrated exceptional performance for over 1000 hours under conditions of 85 °C and 85% RH following encapsulation [47]. Although these films exhibit outstanding encapsulation performance, the ALD deposition method has a relatively slow deposition rate compared to PECVD, which poses challenges for its application in low-cost large-scale manufacturing [50]. To overcome this limitation, a mixed structure that combines the high-efficiency deposition of PECVD and high-quality encapsulation of ALD was proposed. In this approach, they initially deposited SiO_x or SiN_x thin films on the device surface using PECVD and subsequently encapsulated it with an

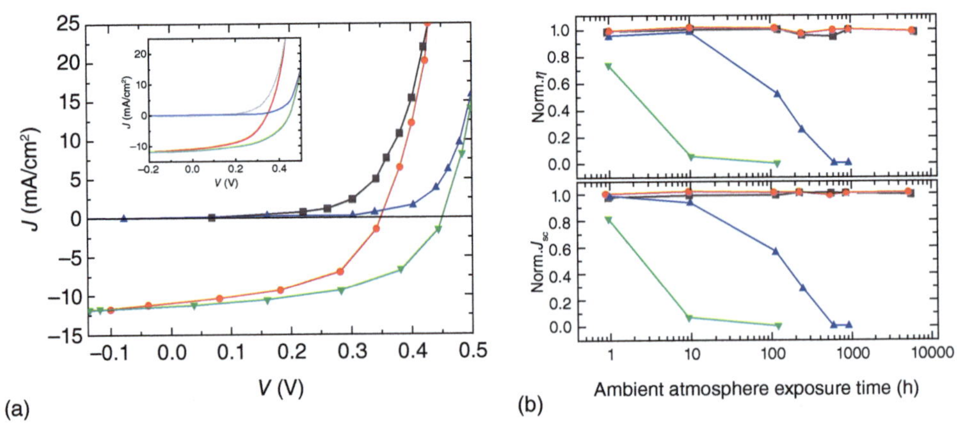

Figure 2.12 (a) The short-circuit current (J_{SC}) of the pentacene/C_{60}-based solar cells before (inset) and after ALD of a Al_2O_3 layer; (b) performance comparison illustration of the relative changes in solar cells encapsulated with UV epoxy (diamonds), Al_2O_3 (squares), and complex of Al_2O_3 and UV epoxy (circles) and the unencapsulated reference sample (triangles).

ALD-deposited Al_2O_3 thin film of 10–50 nm thickness. The organic solar cell based on pentacene/C_{60} encapsulated using this hybrid method exhibited exceptional performance, maintaining it for over 6000 hours. The effective WVTR of the device, as measured through a Ca corrosion test, was approximately 10^{-5} g/m^2/day.

In addition to the widely used PECVD and ALD deposition techniques, researchers have also attempted other methods for single-layer encapsulation of organic devices. For example, Jung et al. used vacuum evaporation to directly deposit a 200 nm-thick layer of tetratetracontane (TTC) [n-$C_{44}H_{90}$] (96%) on OTFTs, which can maintain stability for up to 40 days under standard conditions [51, 52]. Cho et al. used ion-beam-assisted deposition (IBAD) to prepare a 100 nm-thick SnO_2 layer for encapsulating OTFTs. The stability of the device was approximately 30 days. OFETs encapsulated with P3HT and PCBM had lifetimes exceeding 500 and 2000 hours, respectively. TiO_x prepared by spin-coating can be used as a protective layer for OFETs packages and for passivating device surface [53]. Figure 2.13a–c demonstrates the passivation effect of the thin film by comparing the transport performance of the device with and without coating TiO_x. However, for organic devices encapsulated with TTC, SnO_2, or TiO_x to achieve practical application suitability, their lifespans need to be extended to more than 10,000 hours. To gain a comprehensive understanding of the specific material requirements for desirable barrier properties, a thorough and systematic investigation of the relationship between barrier properties and the lifespan of organic device encapsulation is essential.

2.3.4.2 Multilayer Film Packaging

The multilayer encapsulated film plays a crucial role in reducing defects present in single-layer films and minimizing the permeability of impurities, such as water vapor, by employing alternating material structures. This enhances the isolation properties greatly (Figure 2.14) [54–59]. Generally, the isolation properties of multilayer encapsulated films is about three to four orders of magnitude higher than monolayer films, making them widely employed for encapsulating organic electronic devices. These films typically contain alternating organic and

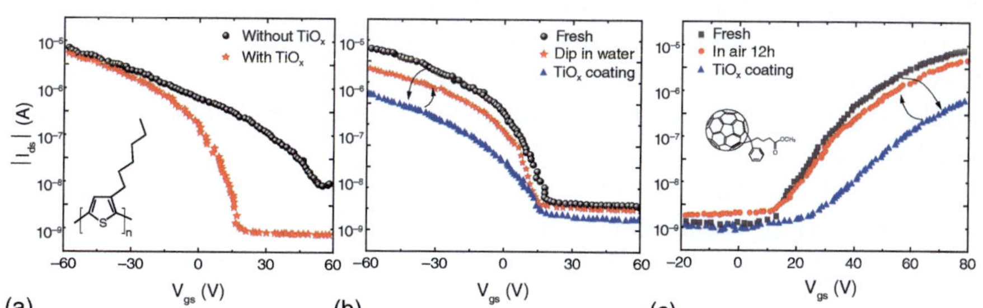

Figure 2.13 (a) Transport performance of P3HT FETs with and without spin-coatingTiO_x in air. (b) Influence of TiO_x passivation on degraded P3HT FETs by immersion in deionized water for 15 s. (c) Influence of TiO_x Passivation on degraded PCBM FETs after 12 h exposure to air.

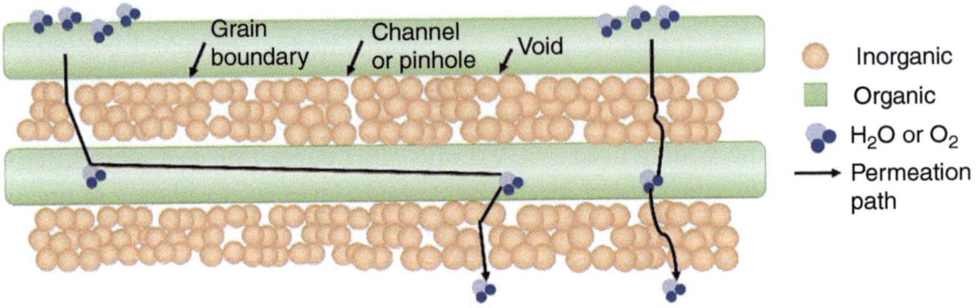

Figure 2.14 The convoluted permeation path created by the organic/inorganic multilayer structure.

inorganic layers, including diverse combinations of inorganic materials. Representative examples of inorganic material encapsulation films are SiN_x and SiO_x multilayer films deposited by PECVD, as well as Al_2O_3/SiO_x films deposited by ALD [48, 49].

In multilayer thin films comprising organic and inorganic layers, the organic part primarily acts as a defect interruption layer, aimed at reducing defects within the film. However, the properties and characteristics of the organic layer in multilayer thin films are yet to be fully established. Most organic/inorganic layer encapsulation films can achieve remarkably low effective WVTRs ($<10^{-4}$ g/m^2/day). The relationship between the thickness of organic layers and the overall barrier properties was investigated by numerical simulations [60]. Initially, it was believed that water vapor cannot permeate the inorganic layer and the penetration solely takes place at the internal defects, as shown in Figure 2.8a. According to the size ratio of polymer thickness to defect size, the WVTR figure can be separated into two different regions, as illustrated in Figure 2.15a, b. Theoretical simulations revealed that in Region 1, WVTR relies on the polymer thickness, whereas for Region 2, WVTR was not affected by the polymer thickness anymore [61]. The latter case indicates the absence of the added winding path effects, thereby making the design of multilayer structures feasible in Region 1 [62]. However, a balance must be struck between the polymer thickness and its capacity to decouple defects. Simulation results indicate that a thinner organic layer contributes to improved barrier performance of the device, but it may not effectively separate defects within the inorganic layer. Therefore, when designing multilayer encapsulation films, the aforementioned factors should be taken into consideration.

Commercial companies, for example, VITEX has successfully developed multilayer barrier coatings (BarixTM) with low effective WVTR. The calcium corrosion test conducted under standard conditions demonstrates an effective WVTR of $\sim 2 \times 10^{-6}$ g/m^2/day for this film. According to Chen et al., at 25 °C and 45% RH, the combined SiO_x/SiN_x/poly (para-xylylene) barrier structure on a polycarbonate substrate exhibits an effective WVTR of 2.5×10^{-7} g/m^2/day [59]. Currently, such values represent the lowest reported effective WVTR values, which are not less than two orders of magnitude lower than those of mono-layer packaging ones. PECVD

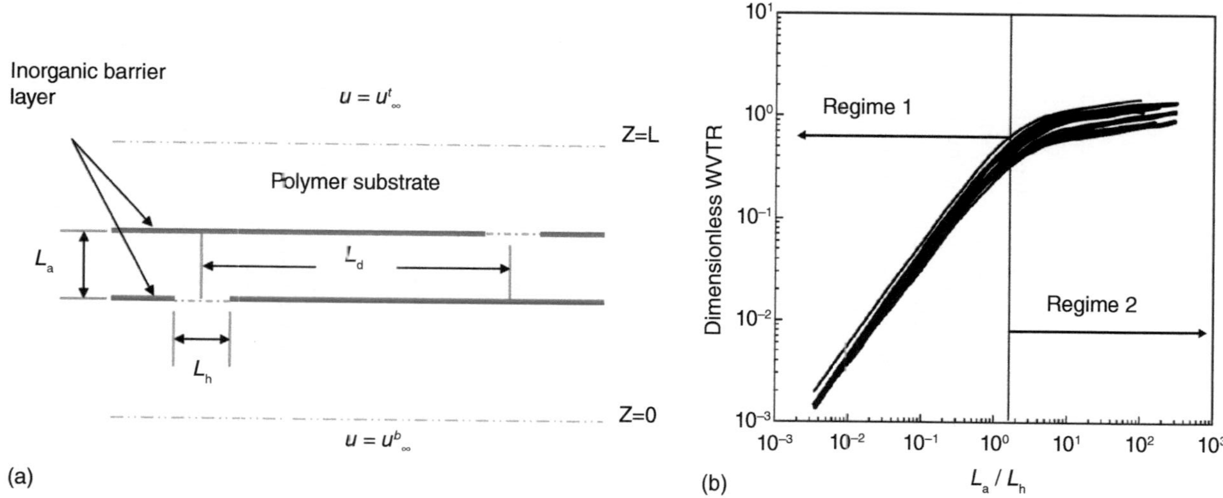

Figure 2.15 (a) The geometric shape of a layered structure made from organic adhesives. L_h and L_a represent the defect size and the polymer thickness, respectively. (b) Simulation outcome of the polymer thickness to defect size ratio dependence of dimensionless WVTR.

based gradient potential barrier layers has been developed by General Electric [63]. This structure comprises alternating single gradient layers (organic & inorganic) coated by PECVD, with an overall thickness lower than 1 μm. The organic layer employs a blend of organosilicon and Ar gas, resulting in an organic material (SiO_xC_y) with a modulus of lower than 10 GPa. An inorganic coating employs a combination of silane (2% in helium), NH_3, and O_2 to generate SiO_xN_y with a modulus ranging from 1 to 5×10^2 GPa. The reported effective WVTR range for various gradient films falls between 5 to 50×10^{-6} g/m²/day. Although multilayer films exhibit excellent isolation properties, specific items regarding the effect of their construction on packaging performance are lacking [63, 64]. Consequently, establishing connections between structure, permeability, and device lifetime requirements to develop advanced packaging films remains challenging.

It is crucial to emphasize that when considering the isolation performance of multilayer structures, the reported results of multilayer encapsulation may not comprehensively depict the equilibrium-state permeation rate, as the time lag of multilayer barriers has not been extensively studied [20]. Recent analyses and calculations have indicated that the time lag in multilayer barrier systems can persist for several years. Graff et al. highlighted in their report that the WVTR results are not primarily governed by equilibrium-state diffusion, but rather by the time lag effect arising from an exceptionally long effective moisture penetration pathway, particularly in cases involving multiple barrier layers, as depicted in Figure 2.16. Employing an analytical solution, they calculated the time lag of a five-layer structure with a defect spacing of 100 μm to be 1752 hours (approximately 73 days), as shown in Figure 2.16b [20]. Nevertheless, this computation was grounded on the presumption that permeation is predominantly regulated by micro-sized defects and an ample spacing between such defects (i.e. defect size of approximately 2 μm and defect spacing greater than 100 μm), as depicted in Figure 2.16a, and it assumes that the inorganic coating is impermeable. The extension of delay time primarily stems from the necessity for moisture, permeating through defects within the initial inorganic layer, to traverse the polymer layer ahead of reaching the subsequent defect within the next inorganic layer. On the other hand, according to Tropsha et al., the mitigation of most micro-scale defects requires meticulous substrate cleanliness and the deposition of sufficient thickness (above the critical thickness) [18]. The majority of water permeation occurs in "imperfect" structures featuring irregular surface arrays (such as particle-like structures), resulting from thermodynamic nonequilibrium deposition processes [18, 22, 61, 65]. Nevertheless, the permeation mechanism remains a subject of debate, and there are differing views on the primary sources of permeation through single-layer inorganic materials and multilayer structures. Therefore, at this stage, it is advisable to combine permeability measurements with the lifetime evaluation of encapsulated devices to comprehensively investigate the barrier properties of multilayer encapsulation [66].

Similar to single-layer films, multilayer films also extend the lifetime of encased organic devices. One study [58] reported that multilayer encapsulated phosphorescent OLEDs (PHOLEDs) prepared on glass substrate and coated plastic

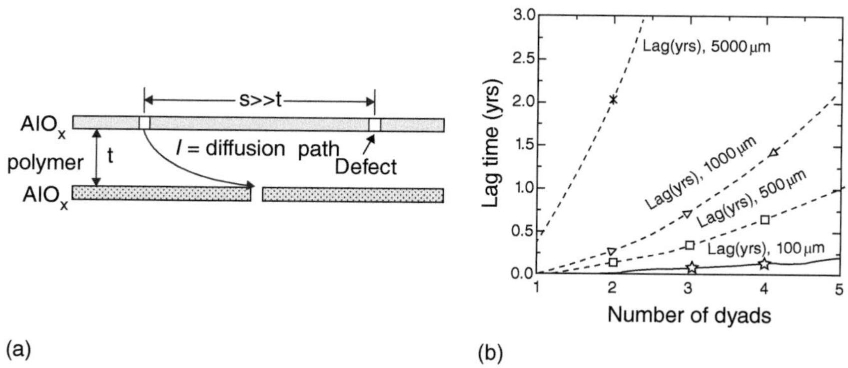

Figure 2.16 (a) A schematic diagram of the effective diffusion distance, l, in the organic layer (about half of the defect interval in the alumina layer). (b) The derived time lag as a function of the organic/alumina dyads number and defect spacing.

substrate shown in Figure 2.17a had lifetimes of 3700 and 2500 hours, respectively. The encapsulation layer consists of 4–5 pairs of alternating pairs of Al_2O_3 and poly(methyl methacrylate) (Barix^{TM} encapsulation), with a total thickness of 5–7 μm [58]. Another study [64] coated SiO_x and organosilicon alternately on PEN substrates using PECVD. SiO_x deposition was carried out employing volatile organosilicon compounds in the presence of oxidizing plasma conditions, whereas the preparation of organosilicon involved the utilization of similar compounds under non-oxidizing plasma conditions. Encapsulation of organic solar cells utilizing MDMO-PPV resulted in full-power operation for up to 3000 hours. As shown in Figure 2.17b, devices with the same packaging structure can be extended to about 6000 hours of lifetime by using more stable P3HT. These results indicate that the development of ultrahigh barrier packaging layers and the use of more stable materials are crucial for extending the lifetime of organic devices [66, 67].

Significant attention has been directed towards the development of packaging methods for OLED devices, driven by their strict requirements and commercial potential. However, it is important to note a notable technical distinction between packaging OLEDs and OPVs, namely the diverse environmental conditions they encounter. OPVs are directly exposed to sunlight, necessitating a barrier layer that can sustain its performance even under UV exposure, unless a UV barrier agent is utilized. Under AM 1.5G conditions, the degradation of barrier films in OPV devices, particularly those with polymer interlayers, can occur due to the combined effects of UV irradiation and substantial heat generated by radiation. Nevertheless, comprehensive reports addressing the enduring barrier performance of encapsulated devices or encapsulations under direct solar radiation or AM 1.5G conditions are currently lacking. One study conducted a test on a P3HT:PCMB based solar cell packaged in a barrier film with a WVTR of 0.03 g/m/day. The examination entailed exposing the encapsulated organic solar cells to AM 1.5G illumination under conditions of 38 °C and 100% relative humidity, as well as at 65 °C with humidity ranging from 30% to 60% [68]. Although the research revealed

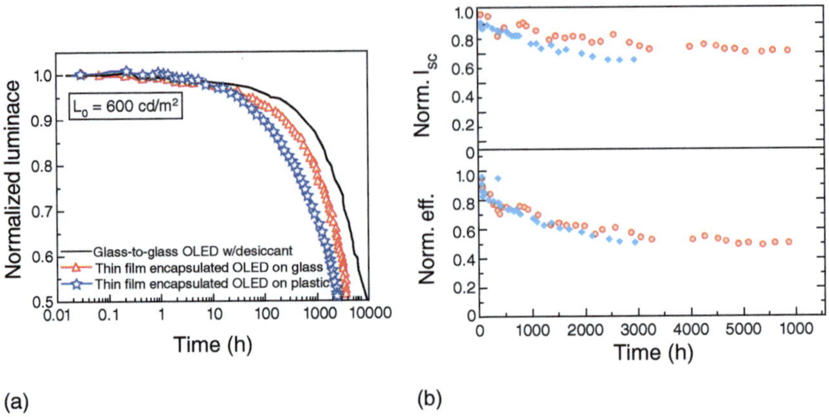

Figure 2.17 (a) The standardized luminance of PHOLEDs packaged in glass cover and desiccant, thin films packaged on glass substrates, and thin films packaged on plastic substrates coated with barrier layers. (b) The standardized I_{SC} and total efficiency of packaged solar cells utilizing MDMO-PPV:PCBM mixtures (diamond) and P3HT:PCBM mixture (circle) relative to the storage time in ambient air.

a performance loss of approximately 15% in the encased organic solar cells after 1200 hours under the aforementioned conditions, independent research on the deterioration of isolation performance in the presence of AM 1.5G conditions remains scarce. Consequently, to ensure the dependable safeguarding of organic devices operating in solar radiation environments, it is imperative to scrutinize the decline in barrier performance under identical conditions.

2.3.5 Film Encapsulation Mechanics

In addition to the stringent requirements for low permeability, the encapsulation film is required to sustain its isolation properties consistently, both during and after experiencing deformation. While inorganic layers can barrier H_2O vapor and O_2 well, they tend to restrict the flexibility, and consequently, reducing the device lifespan. Encapsulation with brittle inorganic layers exhibits a limited strain capacity, typically ranging from 1.2% to 2.0% before fracture occurs [67, 70]. As water vapor and oxygen permeation rates increase through cracks in the encapsulation, the device under protection would fail quickly. Film failure can manifest as the generation of parallel fissure as well as delamination. They may arise from preexisting flaws in lower-grade films. These cracks serve as route for the accelerated transmission of O_2 and H_2O vapor, leading to equipment damage [35]. Hence, for devices that require ultra-low permeability, depositing an inorganic layer is necessary to achieve such levels. Furthermore, understanding how thin films respond to deformation and developing packaging structures with enhanced mechanical stability are vital steps toward enabling the commercialization of flexible organic electronic products.

While inorganic layers play a crucial role in achieving exceptional barrier performance, their lack of flexibility introduces additional risks of barrier failure when subjected to bending. Consequently, to render organic electronic devices flexible, both the devices and packaging layers must possess the ability to endure repeated bending stresses and strains. Furthermore, these thin films must retain their isolation properties even under tension and pressure while also exhibiting sufficient adhesion to the device. Currently, significant research efforts have been devoted to the exploration of the brittle failure observed in transparent electrodes made of indium tin oxide (ITO) [78]. However, there has been comparatively limited investigation into mechanical properties of encasement films. Although conductivity measurement can identify cracks or failures in ITO electrodes, detecting cracks in transparent thin films presents a formidable challenge [69]. Even minutely damaged areas that remain undetectable through visual inspection can swiftly accelerate device failure within the brittle encapsulation layers. Optical microscopy proves effective in detecting such cracks. Nevertheless, there remains a scarcity of reported findings on the generation of crack in transparent thin films subjected to structural deformation. Studies employing AFM or SEM have documented images of cracks induced by uniaxial or biaxial stress, but drawing reliable conclusions from such investigations necessitates a significant amount of statistical analysis [70–72].

In order to minimize the number of samples required for analysis, Sobrinho et al. proposed the use of oxygen plasma treatment as a means to detect defects and visualize cracks induced by bending [26, 73]. This approach involves precision etching and controlled undercutting of the organics underlying the defect. By exposing the sample to AO for a specific duration, cracks become apparent even by a light microscope. However, it remains challenging to conduct on-site observation of cracks to ascertain the initiation strain during deformation. Another method to render cracks induced by mechanical deformation visible involves the utilization of a calcium (Ca) layer. In a manner akin to the oxygen plasma treatment methodology, this technique relies on observable changes in optical transparency. Initially, Ca exhibits opacity and conductivity; however, upon oxidation, it undergoes a transformation into optical transparency with non-conductive properties. By depositing a Ca layer onto a transparent polymer substrate and subsequently sealing it with a barrier film, cracks arising from mechanical deformation can be easily identified through changes in optical transparency. Consequently, by modifying the curvature radius and adjusting the applied strain, the formation of cracks can be observed.

In typical evaluations of encapsulation films, bending deformation tests are commonly performed using a two-point bending configuration. While these experiments are straightforward to conduct, the curvature radius is non-uniform, leading to an uneven distribution of strain across the entire thin film. Therefore, caution must be exercised when establishing a correlation between film wrinkles and nonuniform curvature-induced strain [74].

In their research, Grego and colleagues explored the mechanical failure of a SiO_xN_y film of 100 nm thick, deposited on a PEN substrate of 130 μm thick using a X–Y–θ bending architecture [75]. This advanced system enables the attainment of the bending radius depicted in Figure 2.18a. The results depicted in Figure 2.18 b indicate that cracks begin to emerge at a curvature bending radius of 7 mm, with the crack density augmenting as the radius decreases. The test also highlighted

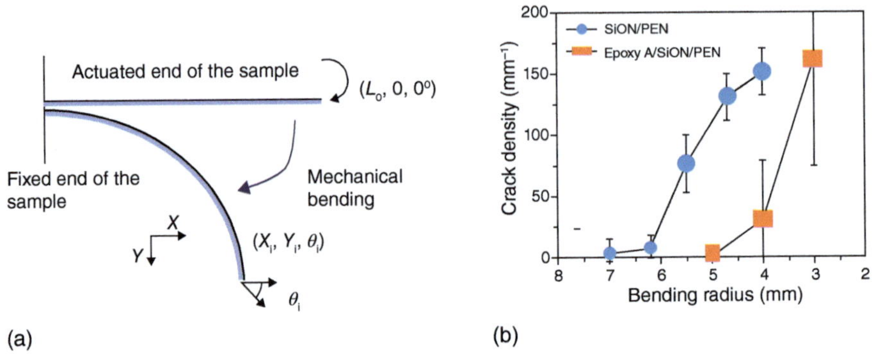

Figure 2.18 (a) A schematic representation of the bending test conducted in the $X-Y-\theta$ system; (b) Crack density as a function of the bending radius.

that the mechanical strength of the film can be enhanced by applying a topcoat polymer, which passivates the mechanical defects present in the inorganic coating. Specifically, the initial crack radius of the SiO_xN_y film, coated with a surface epoxy layer, decreased from 7 to 5 mm [73, 74]. Moreover, the initial fracture strain of the inorganic film and the occurrence of film failure resulting from repeated bending stresses should be taken into account in practical applications of flexible organic electronic devices. During this experiment, no cracks were observed when the SiO_xN_y film, with a surface epoxy layer, was subjected to bending with a curvature radius of 6 mm. Nevertheless, following a sequence of bending cycles (ranging from 10 to 100 and then to 1000), low-density cracks emerged, and the crack density seemed to have plateaued, indicating a state of saturation. Hence, when designing mechanically stable packaging, the potential breakdown of isolation layers due to single or multiple bends should be carefully considered.

Gaining insight into the mechanical behavior of thin-film encapsulation is crucial for determining the strain distribution that limits the maximum strain capacity of the barrier film, consequently improving its flexibility. As discussed earlier, the mechanical stability of encapsulation films primarily stems from the finite strain exhibited by the inorganic layer. The strain experienced by the inorganic layer is influenced by the curvature radius during bending and its position in relation to the device's neutral axis, which can be computed by the following equation:

$$\varepsilon_z = \frac{z}{R} \tag{2.13}$$

Equation (2.13) incorporates the variable z, which represents the distance from the interior or exterior surface to the neutral axis, while R denotes the curvature radius, as illustrated in Figure 2.19. By placing the inorganic layer nearby the neutral axis, the variable 'z' can be minimized, resulting in a reduction in strain [76]. To provide a more comprehensive description of strain in heterogeneous structures, Suo et al. introduced an elaborated expression, denoted by Eq. (2.14):

$$\varepsilon_{\text{film}} = \left(\frac{d_1 + d_2}{2R}\right) \frac{(\chi\eta^2 + 2\eta + 1)}{(1 + \chi\eta)(1 + \eta)} \tag{2.14}$$

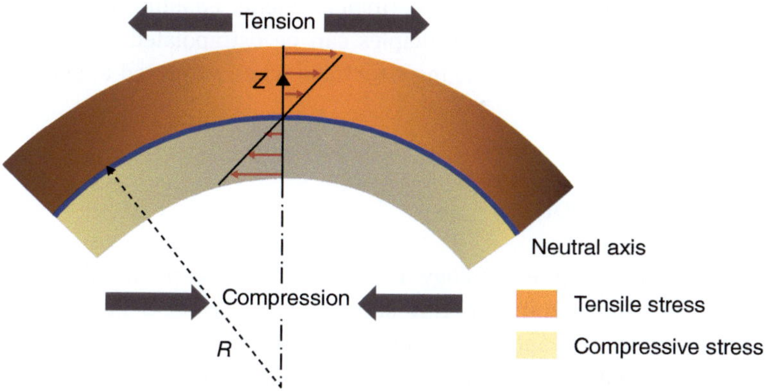

Figure 2.19 The strain distribution of a film with a bending radius of R. Under the neutral axis, compression stress is applied; whereas, above the neutral axis, tension stress is formed.

where, $\chi = E_1/E_2$, and $\eta = d_1/E_2$. E_1 is the modulus of the film, E_2 is the modulus of the substrate, d_1 is the film thickness, and d_2 is the substrate thickness. In this particular system, the position of the neutral axis can be adjusted according to the material's modulus. The researchers propose that by manipulating the neutral axis to shift closer to the brittle film and incorporating materials with varying moduli, the magnitude of strain can be mitigated by a factor of five for a given curvature radius, film thickness, and substrate thickness [77]. Furthermore, Park et al. conducted experiments that showcased a reduction in strain by introducing flexible materials between the substrate and the isolation layer [78].

Moreover, Sekitan and coworkers achieved successful functionality of a layered organic field-effect transistor (OFET) depicted in Figure 2.20a even under highly restrictive bending radii (less than 1mm), as illustrated in Figure 2.20b [77]. The device was fabricated on a PI film with a thickness of 13 μm and subsequently

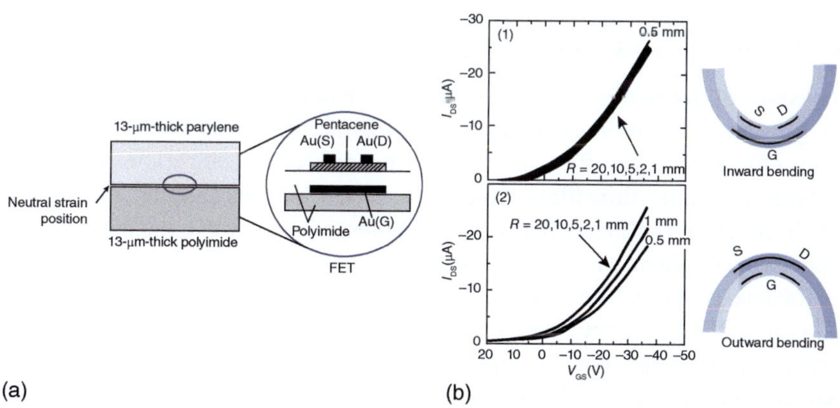

Figure 2.20 (a) A cross-section of an organic transistor sandwiched between a plastic film and a poly(p-xylylene) passivation layer. (b) The transport characteristics of the transistor examined at different bending radii, R, relative to inward and outward bending strains.

coated with a 13 μm-thick layer of polychlorinated xylene, enabling embedding within a neutral environment. These principles can be extrapolated to enhance the mechanical characteristics of packaging structures when subjected to bending deformations.

2.4 Packaging Technology

Flexible electronic packaging technology refers to the process of fabricating electronic devices on thin metal or plastic substrates that possess flexibility or stretchability. This innovative approach aims to overcome the limitations of traditional rigid silicon-based chip materials, which are thick and brittle. As a result, a new paradigm in process technology has emerged, wherein microstructures are integrated onto flexible substrates. Flexible electronic packaging technology offers numerous benefits, including flexibility, lightweight, shock resistance, cost-effectiveness, and ease of operation. Several prevalent packaging techniques are detailed below [79]:

2.4.1 Local Multilayer Packaging

The non-coplanar thin film-substrate structure is widely employed in flexible electronic packaging technology. However, it faces challenges in maintaining its integrity under significant loads after packaging, resulting in reduced ductility. To overcome this limitation, a novel approach known as localized multilayer packaging technology has been introduced. This innovative technique focuses on selectively softening the electronic packaging region above the thin film substrate while reinforcing the region below it. This localized approach effectively enhances the overall flexibility of the packaging structure, addressing the issue of load resistance and preserving ductility [80].

2.4.2 Multilayer Barrier Film Packaging

The objective of developing multilayer barrier membranes is to bridge the performance gap that exists in most single-layer inorganic thin films, thereby achieving a highly efficient barrier film that meets the desired specifications. Presently, single-layer inorganic thin films typically have thicknesses ranging from 0.1 to 0.2 μm. In contrast, multilayer barrier membranes, which involve the deposition of multiple layers of organic/inorganic materials, exhibit an organic layer thickness of approximately 1–20 μm. The increased thickness of the organic layer offers several advantages, including a smoother surface for the barrier film, effective coverage of particles and foreign objects, elimination of micropores, and reduced stress, thereby enhancing the overall flexibility of the barrier film [81]. This approach has successfully demonstrated the required performance on small-scale samples. However, several challenges need to be addressed to enable efficient, high-volume, and large-scale R2R production. One of the main challenges stems from the high sensitivity of multilayer barrier membranes to process conditions. First, the production environment must maintain a high level of cleanliness, and specific measures must be implemented to control electrostatic elimination and reduce surface contaminants. Second, minimizing contact between the roll/winding machine and the surface of the barrier membrane is essential during R2R operation. Third, it is crucial to maintain a slow and consistent speed while ensuring a narrow reel width to achieve uniform and precise film formation. Additionally, proper adhesive usage is necessary for effective film adhesion.

2.4.3 Online Thin-Film Encapsulation

The online thin film encapsulation method can be considered as a progressive advancement of multilayer barrier film technology [82]. With this method, the encapsulation film is directly formed on the device during online production, shifting the responsibility and risk primarily to equipment manufacturers. Moreover, this technology eliminates the need for separate substrates and additional adhesives. By reducing the thickness of the barrier film, enhanced product flexibility is attained. Poor-performing barrier films not only compromise the film itself but also jeopardize the integrity of the entire device. In approximately 2014, this technology was commercially implemented in an OLED display featuring a curved, hard plastic substrate. Compared to the original approach, the number of barrier film layers generated through this method was significantly reduced. The preparation process involved using techniques like PECVD and inkjet printing to deposit inorganic and organic layers, respectively, replacing PVD and evaporation deposition, resulting in substantial optimization [83]. Furthermore, material enhancements have been made to maximize the film's performance without exceeding the temperature limits that the organic multilayer structure can endure. Consequently, online film packaging has laid the groundwork for achieving bending radii of 2mm or less, facilitating the creation of thin devices. Despite the successes achieved through online film packaging technology, research and development in this field persist. Currently, efforts are focused on the development of the all-PECVD process and spatial ALD technology, which aim to prepare large-sized and highly effective ultrathin inorganic layers.

Presently, the application of online thin film encapsulation primarily focuses on the top-level packaging of devices, while bottom-level packaging necessitates relatively thick continuous inorganic layers and additional layers like transition connection layers. Nevertheless, the inorganic layers commonly employed as the bottom barrier layer lack the flexibility required to accommodate tighter bending radii. Consequently, further advancements are required in the development of organic material structures that exhibit higher temperature resistance at a lower cost. These advancements will contribute to enhancing the longevity of devices.

2.4.4 Atomic Layer Deposition (ALD) Encapsulation

ALD is a gas-phase technology that employs a series of controlled chemical reactions to form thin films. This process involves exposing the substrate surface to a sequence of gas-phase reactions, wherein each reaction deposits a sub-monolayer of material in the form of single atoms. However, it is important to note that ALD is not a continuous process, as it consists of two half-cycle reactions, necessitating a carefully designed deposition recipe for the reaction process. ALD exhibits several distinctive characteristics: Uniformity: By operating under a self-limiting growth mechanism, each layer is deposited atom by atom, resulting in a highly uniform and densely packed thin film. Controllability: The thickness of the thin film can be precisely controlled by adjusting the number of cycles, providing a high level of control over the film's dimensions. Stability: The precursor materials employed in ALD possess high reactivity, enabling deposition at low temperatures without undergoing decomposition at reaction temperatures, thereby ensuring the stability of the process [84, 85].

ALD encompasses various binary reactions that utilize the aforementioned method to enable controlled and quantitative deposition by employing two reactants. This approach ensures the formation of smooth, continuous, and defect-free thin films [86]. As a result, the utilization of ALD encapsulation technology enables effective prevention of water and oxygen corrosion on the device, as well as reduction in crack formation by mitigating excessive stress. Thus, ALD technology plays an indispensable role in the packaging of flexible display devices, facilitating the structural design of ultrathin films and the integration of inorganic and organic materials [87].

Despite the advantages of ALD technology, the thinness of the barrier layer it produces poses challenges for passivating surface particles and pollutants, which can lead to a decline in overall film performance. Consequently, the growth process of the film layer is considerably slow. To overcome these limitations, extensive research is focused on the development of R2R spatial ALD processes. Achieving successful implementation of this technology in R2R systems necessitates innovative and pioneering equipment designs that prevent damage to the ultrathin coating during the winding process of the paper reel. Furthermore, it is crucial to extend the applicability of this technology to broader and larger R2R applications without compromising its performance, thereby addressing potential WVTR concerns [88, 89].

2.4.5 Inkjet Packaging

Inkjet printing packaging, as an economical, non-contact, and non-vacuum technique for thin film deposition, involves the controlled spraying of a solution onto the sample surface at a frequency of hundreds of times per second, with a volume of tens of microliters per second. Subsequently, the solvent is evaporated, leaving behind a deposited film. This technology enables the creation of packaging films in various shapes and patterns using conductive ink [90]. Compared to traditional deposition processes, inkjet printing packaging offers advantages such as reduced material and energy consumption, as well as simplified processing steps. It facilitates the integration of interconnects and antennas within or on the package while also enabling vertical integration in a three-dimensional manner. Furthermore, inkjet printing provides a wide range of material options, including index-tunable thermoplastic formulations, UV-curable adhesives, optical adhesives, and other specialized adhesives [91].

2.4.6 Flexible Glass Packaging

Glass is renowned for its high-temperature processability and excellent barrier properties. However, traditional glass is rigid and lacks flexibility. Efforts from numerous researchers led to the introduction of flexible glass to the market approximately a decade ago. Nevertheless, the flexibility of these glasses is contingent upon their thickness, and achieving ultrathin glass is feasible through techniques like etching. Despite this progress, the commercial utilization of flexible glass for barrier packaging remains uncertain. Presently, only a small fraction of flexible glass is employed as a substrate for curved OLED lighting. The primary obstacle lies in the fragility of flexible glass, which hampers its flexibility compared to other materials. Additionally, the supply of flexible glass presents challenges as relevant companies have not yet invested in expanding production scale, leaving supply and cost considerations unanswered. Nonetheless, in the long run, as a high-temperature resistant and high-performance solution for flexible barrier and substrate integration, the market demand for flexible glass remains robust. However, despite the strong market demand, the commercial prospects of flexible glass are significantly constrained by technological factors such as flexibility and operation, as well as business aspects like final price and production capacity [92–94].

2.5 Packaging Stability

In the past decade, significant strides have been made in the advancement of organic semiconductor materials and their corresponding processing techniques, resulting in unprecedented enhancements in the performance of various devices including OLEDs, OPVs, and organic thin-film transistors (OTFTs) (Figure 2.21). These organic electronic devices have garnered considerable attention due to their remarkable tunability, multifunctionality, and widespread suitability for extensive manufacturing processes on a large scale, including techniques like inkjet

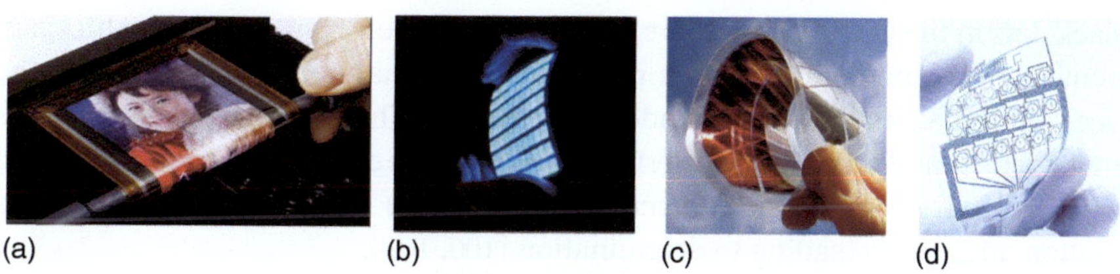

(a) (b) (c) (d)

Figure 2.21 (a) Full-color flexible OLEDs mode developed by Sony; (b) OLEDs for solid-state lighting developed by General Electric and Konica Minolta; (c) OPVs based on pentacene/C_{60}; (d) OTFT arrays fabricated on plastic film by printing technology. Source: Figures are from the Internet. Unidentified.

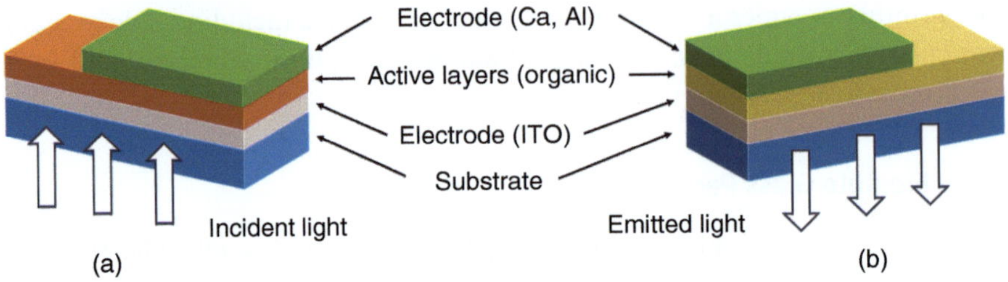

Figure 2.22 A schematic illustration depicting the fundamental structure of an organic diode: (a) Organic Photovoltaics and (b) Organic Light Emitting Diode.

or gravure printing, as well as their potential utilization in flexible electronics [95, 96]. According to existing literature, the distinctive properties exhibited by organic semiconductor materials have the potential to propel the development of a novel category of cost-effective electronic devices that offer superior flexibility and enhanced performance, surpassing what can be achieved with conventional crystalline semiconductor materials.

Despite notable advancements in the field of organic devices, the issue of stability continues to pose a significant challenge, hindering their widespread commercialization. The device architectures of OPV and OLED are illustrated in Figure 2.22, comprising a cathode of low work function, an active layer, an anode of high work function, and a substrate; the cathode or anode is usually transparent. Furthermore, to safeguard these devices against environmental factors that can impact their stability, they are often enclosed within a packaging layer or sealed. This protective measure aims to shield them from detrimental exposure to the surrounding environment.

The stability concerns associated with the device stem, in part, from the unfavorable conditions in which the active material layer and electrode of low work function are situated [97]. In practical applications, cathodes utilizing metals of low work function like Ca and Li are commonly employed to enhance the efficiency of electron injection or collection within the device. However, these low-work-function metals exhibit high reactivity towards oxygen and water vapor, leading to rapid oxidation [98]. Consequently, an insulating oxide barrier layer is formed, compromising the efficiency of carrier infusion and collection. The exposure to moisture and oxygen in ambient conditions further leads to the generation of black dots in the OLEDs [99], thereby diminishing their light output and lifespan. Moreover, environmental oxidation may induce internal delamination within the device, serving as an additional undesirable factor. When water vapor permeates through defects and reaches the interface formed by the cathode and active material layers, it can trigger chemical reactions resulting in volume expansion or gas formation, ultimately leading to delamination [100, 101].

In addition to the electrode degradation induced by environmental factors, the exposure of the active material layer to the ambient environment can significantly compromise the property of organic electronic devices. As illustrated in Figure 2.23a, OFETs utilizing materials such as C_{60} exhibit direct reactivity with water vapor

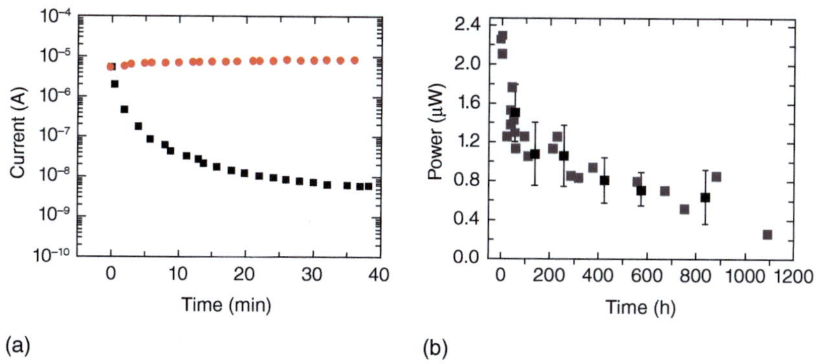

Figure 2.23 (a) The conductivity reduction of a C_{60} thin film FET in ambient air with no aluminum oxide coverage (square) and aluminum oxide coverage (circle) under a source-drain voltage of 30 V. (b) Time dependence of light power for sealed OLEDs.

and oxygen in the ambient air, resulting in rapid deterioration of their electrical performance, in contrast to transistors employing aluminum oxide [102–104]. Previous OLED applications, as depicted in Figure 2.23b, demonstrated a significant decline in light output power within a 10-hour environmental exposure for unsealed devices, while sealed devices exhibited an extension in lifespan by at least one order of magnitude under identical conditions [102]. Consequently, the advancement of encapsulation technology assumes paramount importance in enhancing the longevity of organic electronic devices.

2.6 Encapsulated Products

Commonly encountered electronic encapsulated products encompass a wide range of devices, such as OLEDs, PCB multilayer boards, field-effect transistors, electrochromic devices, photodetectors, fundamental components, planar supercapacitors, gas sensors, humidity sensors, handwriting light-emitting flexible screens, temperature sensors, thermochromic devices, organic solar cells, and more.

Taking OLED as an illustrative example, OLED technology, also referred to as the third generation of lighting and display technology, is currently experiencing substantial investments from global manufacturers, leading to rapid market demand growth and the maturation of production technologies for OLED flat panel displays. The fundamental material crucial to flexible OLED devices is the ultrathin organic electroluminescent layer, which exhibits high sensitivity to water, oxygen, and heat, resulting in inherent stability challenges. Oxygen has the capability to quench triplet excitons, thereby significantly diminishing the luminous quantum efficiency, as well as oxidizing the organic materials within the hole transport layer and the emitting layer, leading to the breakage of unsaturated double bonds and a subsequent reduction in luminous efficiency and electronic hole transport capacity. Water vapor can engage in hydrolysis reactions with the organic semiconductors

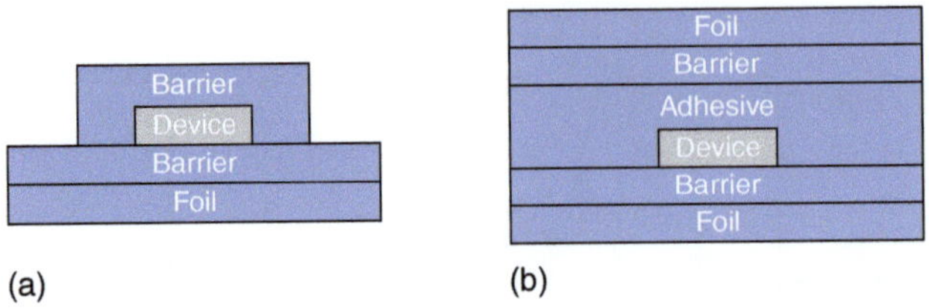

Figure 2.24 Structures of flexible OLEDs with inline packaging (a) and imprint packaging (b).

present in the emitting layer, thereby impacting the overall device performance. Moreover, the metals utilized for the electrodes are highly reactive and prone to oxidation, reacting with permeating water vapor and inducing hydrolysis. Furthermore, the water and oxygen barrier performance of the substrate materials employed in flexible devices are typically inferior to those of rigid materials, necessitating more stringent requirements for the encapsulation process. Consequently, stringent encapsulation of flexible OLED devices is an essential prerequisite for extending their lifespan and enhancing their stability and represents a critical step towards achieving mass production of flexible OLED devices.

OLED devices impose stringent packaging requirements to ensure their optimal performance. Practical OLED devices typically demand a WVTR of less than 10^{-6} g/m^2/day and an OVTR of less than 10^{-5} cm^3/m^2/day [103, 105]. As depicted in Figure 2.24, there are currently two predominant methods employed for packaging flexible OLED devices. One method, previously mentioned as inline packaging (Figure 2.24a), entails the direct deposition of alternating inorganic and organic barrier layers onto the surface of the metal electrode of the flexible OLED device. This method represents the prevailing packaging approach, as it offers the advantage of safeguarding the metal electrode against impurities such as water, oxygen, and dust, obviating the need to break the vacuum of the OLED device, and ensuring that the device meets the longevity requirements. However, the inline packaging process is overall more intricate and challenging to implement for R2R production. Furthermore, the inline processing temperature (≥90 °C) during the barrier layer preparation can potentially impair the OLED device, thereby affecting its lifespan. Presently, both domestic and international companies predominantly employ inline packaging technology. The alternative method is imprint packaging, which involves the application of a high-barrier film onto the surface of the metal electrode of the OLED device through adhesive/film lamination, as shown in Figure 2.24b. The imprint packaging process is comparatively simpler, enabling the separation of the OLED device preparation process from that of the high barrier layer, thus facilitating individual performance optimization. However, during the packaging process, the OLED device may come into contact with water, oxygen, and dust, while the packaging material may also cause damage to the metal electrode. Since the adhesive film employed does not provide adequate barrier properties, there exists a risk of lateral leakage, whereby water and oxygen can penetrate the device's interior through the adhesive layer, leading to potential damage. Additionally, incompletely reacted adhesive may also react with the OLED device, adversely affecting its lifespan.

2.7 Chapter Summary

In summary, flexible electronic packaging technology embodies strong interdisciplinary fusion characteristics, necessitating the integration of cutting-edge technologies from diverse disciplines such as biology, electronics, and materials science to achieve technological breakthroughs and advancements. For instance, the selection of different materials for motion-flexible sensors is based on the distinctive movement characteristics of various body parts. The sensor materials employed in a motion vest would possess different strength requirements compared to those used in the sole of a shoe. Over recent years, there have been notable advancements in the degree and scope of flexible requirements, encompassing stretching, bending, and folding. Device flexibility can be achieved through two primary means: material flexibility and structural flexibility. In the near term, flexible electronics are not poised to compete directly with silicon-based electronics in the realm of high-speed, high-performance devices. Consequently, the objective is not to replace traditional microelectronics technology but rather to foster the creation of disruptive new electronic devices and products.

References

1 Gates, B.D. (2009). Flexible electronics. *Science* 323 (5921): 1566–1567.

2 Yv, C.P., Liu, P.A., Li, L.Y. et al. (2019). Applications of flexible electronic materials and devices. *Chinese Journal on Internet of Things* 3 (3): 100876.

3 Yin, Z.P., Huang, Y.A., Rong, C. et al. (2020). Key technologies and applications in flexible electronics manufacturing. *China Basic Science* 22 (5): 21–34.

4 Dimitrakopoulos, C.D. and Malenfant, P.R.L. (2002). Organic thin film transistors for large area electronics. *Advanced Materials* 14 (2): 99–117.

5 Knipp, D. and Street, R.A. (2003). Pentacene thin film transistors on inorganic dielectrics: morphology, structural properties, and electronic transport. *Journal of Applied Physics* 93 (1): 347–355.

6 Frey, S.T., Haque, A.B., Tutika, R. et al. (2022). Octopus-inspired adhesive skins for intelligent and rapidly switchable underwater adhesion. *Science Advances* 8 (28): 1905.

7 Yin, Z.P. and Huang, Y.A. (2016). *Manufacturing of Flexible Electronics: Materials, Devices, and Processes.* Beijing Science and Technology Press.

8 Cannella, V., Izu, M., and Jones, S. (2005). Flexible stainless-steel substrates. *SID Information Display* 21 (6): 24–27.

9 Song, J., Jiang, H., Huang, Y. et al. (2009). Mechanics of stretchable inorganic electronic materials. *Journal of Vacuum Science & Technology* 27 (5): 1107–1126.

10 Kim, D.H., Xiao, J., Song, J. et al. (2010). Stretchable, curvilinear electronics based on inorganic Materials. *ChemInform* 22 (19): 2108–2124.

11 Wang, S.D., Huang, Y.G., and Rogers, J.A. (2015). Mechanical designs for inorganic stretchable circuits in soft electronics. *IEEE Transactions on Components, Packaging and Manufacturing Technology* 5 (9): 1201–1218.

12 Su, Y., Ping, X., Yu, K.J. et al. (2017). Stretchable electronics: in plane deformation mechanics for highly stretchable electronics. *Advanced Materials* 29 (8): 1604989.

13 Kim, D.H., Ghaffari, R., Lu, N. et al. (2012). Electronic sensor and actuator webs for large-area complex geometry cardiac mapping and therapy. *Proceedings of the National Academy of Sciences* 109 (49): 19910–19915.

14 Zhang, Y.H., Wang, S.D., Li, X.T. et al. (2014). Experimental and theoretical studies of serpentine microstructures bonded to prestrained elastomers for stretchable electronics. *Advanced Electronic Materials* 24 (14): 2028–2037.

15 Hanika, M., Langowski, H.C., Moosheimer, U. et al. (2003). Inorganic layers on polymeric films – influence of defects and morphology on barrier properties. *Chemical Engineering & Technology* 26 (5): 605–614.

16 Roberts, A.P., Henry, B.M., Sutton, A.P. et al. (2002). Gas permeation in silicon-oxide/polymer (SiO_x/PET) barrier films: role of the oxide lattice, nano-defects and macro-defects. *Journal of Membrane Science* 208 (1–2): 75–88.

17 Tropsha, Y.G. and Harvey, N.G. (1997). Activated rate theory treatment of oxygen and water transport through silicon oxide poly(ethylene terephthalate) composite barrier structures. *Journal of Physical Chemistry B* 101 (13): 2259–2266.

18 Erlat, A.G., Spontak, R.J., Clarke, R.P. et al. (1999). SiO_x gas barrier coatings on polymer substrates: morphology and gas transport considerations. *Journal of Physical Chemistry B* 103 (29): 6047–6055.

19 Graff, G.L., Williford, R.E., Burrows, P.E. et al. (2004). Mechanisms of vapor permeation through multilayer barrier films: lag time versus equilibrium permeation. *Journal of Applied Physics* 96 (4): 1840–1849.

20 Crank, J. (1975). *The Mathematics of Diffusion*. Clarendon University Press.

21 Henry, B.M., Dinelli, F., Zhao, K.Y. et al. (1999). A microstructural study of transparent metal oxide gas barrier films. *Thin Solid Films* 355: 500–505.

22 Doremus, R.H. (1973). *Glass Science*. New York: Wiley.

23 Kim, N., Potscavage, W.J., Domercq, B. et al. (2009). A hybrid encapsulation method for organic electronics. *Applied Physics Letters* 94: 163308.

24 Erlat, A.G., Wang, B.C., Spontak, R.J. et al. (2000). Morphology and gas barrier properties of thin SiO_x coatings on polycarbonate: correlations with plasma-enhanced chemical vapor deposition conditions. *Journal of Materials Research* 15 (3): 704–717.

25 Sobrinho, A.S.D., Czeremuszkin, G., Latreche, M. et al. (2000). Defect-permeation correlation for ultrathin transparent barrier coatings on polymers. *Journal of Vacuum Science & Technology A* 18 (1): 149–157.

26 Sobrinho, A.S.D., Latreche, M., Czeremuszkin, G. et al. (1998). Transparent barrier coatings on polyethylene terephthalate by single- and dual-frequency plasma-enhanced chemical vapor deposition. *Journal of Vacuum Science & Technology A* 16 (6): 3190–3198.

27 Hanika, M., Langowski, H.C., Moosheimer, U. et al. (2003) 46th Annual Tech. Conf., San Francisco. 26 (Year) 592.

28 Perkins, W.G. and Begeal, D.R. (1971). Diffusion and Permeation of He, Ne, Ar, Kr, and D2 through silicon oxide thin films. *The Journal of Chemical Physics* 54: 1683.

29 Carcia, P.F., McLean, R.S., and Reilly, M.H. (2006). Ca test of Al_2O_3 gas diffusion barriers grown by atomic layer deposition on polymers. *Applied Physics Letters* 89: 3.

30 Stevens, M., Tuomela, S., Mayer, D. (2005). 48th Annual Technical Conference, 505, 189.

31 Crank, J. (1975). *The Mathematics of Diffusion*. Clarendon Press.

32 Paetzold, R., Winnacker, A., Hensele, D. et al. (2003). Permeation rate measurements by electrical analysis of calcium corrosion. *Review of Scientific Instruments* 74 (12): 5147–5150.

33 Nisato, G., Bouten, P.C.P., Slikkerveer, P.J., et al. (2001). Evaluating high performance diffusion barriers: the calcium test. *The 19th International Display Workshops in conjunction with Asia Display, 61*, 1435.

34 Paetzold, R., Winnacker, A., Hensele, D. et al. (2003). Permeation rate measurements by electrical analysis of calcium corrosion. *Review of Scientific Instruments* 74 (12): 5147–5150.

35 Lewis, J.S. and Weaver, M.S. (2004). Thin-film permeation-barrier technology for flexible organic light-emitting devices. *IEEE Journal of Selected Topics in Quantum Electronics* 10 (1): 45–57.

36 Dennler, G., Lungenschmied, C., Neugebauer, H. et al. (2005). Flexible, conjugated polymer-fullerene-based bulk-heterojunction solar cells: basics, encapsulation, and integration. *Journal of Materials Research* 20 (12): 3224–3233.

37 Weaver, M.S., Michalski, L.A., Rajan, K. et al. (2001). Organic light-emitting devices with extended operating lifetimes on plastic substrates. *Applied Physics Letters* 81 (16): 2929–2931.

38 Moro, L., Rutherford, N.M., and Visser, R.J. (2006). Barix multilayer barrier technology for organic solar cells. *The International Society for Optical Engineering* 6334: 63340M1–63340M9.

39 Park, S.H.K., Oh, J., Hwang, C.S. et al. (2005). Ultra thin film encapsulation of organic light emitting diode on a plastic substrate. *ETRI Journal* 27 (5): 545–550.

40 Henry, B.M., Dinelli, F., Zhao, K.Y. et al. (1999). A microstructural study of transparent metal oxide gas barrier films. *Thin Solid Films* 355: 500–505.

41 Leterrier, Y. (2003). Durability of nanosized oxygen-barrier coatings on polymers – internal stresses. *Progress in Materials Science* 48 (1): 1–55.

42 Sobrinho, A.S.D., Latreche, M., Czeremuszkin, G. et al. (1998). Transparent barrier coatings on polyethylene terephthalate by single- and dual-frequency plasma-enhanced chemical vapor deposition. *Journal of Vacuum Science & Technology A* 16 (6): 3190–3198.

43 Huang, W.D., Wang, X.H., Sheng, M. et al. (2003). Low temperature PECVD SiN_x films applied in OLED packaging. *Materials Science & Engineering, B: Advanced Functional Solid-State Materials* 98 (3): 248–254.

44 Mandlik, P., Gartside, J., Han, L. et al. (2008). A single-layer permeation barrier for organic light-emitting displays. *Applied Physics Letters* 92 (10): 103309.

45 Potscavage, W.J., Yoo, S., Domercq, B. et al. (2007). Encapsulation of pentacene/C60 organic solar cells with Al_2O_3 deposited by atomic layer deposition. *Applied Physics Letters* 90: 253511.

46 Park, S.H.K., Oh, J., Hwang, C.S. et al. (2005). Ultrathin film encapsulation of an OLED by ALD. *Electrochemical and Solid-State Letters* 8 (2): H21–H23.

47 Ghosh, A.P., Gerenser, L.J., Jarman, C.M. et al. (2005). Thin film encapsulation of organic light emitting devices. *Applied Physics Letters* 86: 223503.

48 Carcia, P.F., MacLean, R.S., Reilly, M.H. et al. (2006). Catalyst of Al_2O_3 gas diffusion barriers grown by atomic layer deposition on polymer. *Applied Physics Letters* 89: 031915.

49 Dameron, A.A., Davidson, S.D., Burton, B.B. et al. (2008). Gas diffusion barriers on polymer using multi layers fabricated by Al_2O_3 and rapid SiO_2 atomic layer deposition. *Journal of Physical Chemistry C* 112: 4573.

50 Yoo, S., Potscavage, W.J., Domercq, B. et al. (2007). Analysis of improved photovoltaic properties of pentacene/C_{60} organic solar cells: effects of exciton blocking layer thickness and thermal annealing. *Solid-State Electronics* 51: 1367.

51 Fenrych, J., Reynhardt, E.C., and Basson, I. (1997). Structures and molecular dynamics of binary mixtures of n-alkanes: C46:C38 and C46:C44. *Chemical Physics Letters* 12: 49.

52 Jung, H., Lim, T., Choi, Y. et al. (2008). Lifetime enhancement of organic thin-film transistors protected with organic layer. *Applied Physics Letters* 92: 16.

53 Cho, S., Lee, K., and Heeger, A.J. (2009). Extended lifetime of organic field-effect transistors encapsulated with titanium sub-oxide as an 'active' passivation/barrier layer. *Advanced Materials* 21 (19): 1941–1944.

54 Chiang, C.C., Wuu, D.S., Lin, H.B. et al. (2006). Deposition and permeation properties of SiN_x/parylene multilayers on polymeric substrates. *Surface and Coatings Technology* 200 (20–21): 5843–5848.

55 Chen, T.N., Wuu, D.S., Wu, C.C. et al. (2007). Improvements of permeation barrier coatings using encapsulated parylene interlayers for flexible electronic applications. *Plasma Processes and Polymers* 4 (2): 180–185.

56 Lewis, J.S. and Weaver, M.S. (2004). Thin film permeation-barrier technology for flexible organic light emitting devices. *IEEE Journal of Selected Topics in Quantum Electronics* 10: 45–57.

57 Weaver, M.S., Chwang, A.B., Rothman, M.A., et al. (2002). Recent progress in flexible displays. *The International Society for Optical Engineering, 4712.*

58 Chwang, A.B., Rothman, M.A., Mao, S.Y. et al. (2003). Thin film encapsulated flexible organic electroluminescent displays. *Applied Physics Letters* 83 (3): 413–415.

59 Chen, T.N., Wuu, D.S., Wu, C.C. et al. (2006). High-performance transparent barrier films of SiO_x/SiN_x stacks on flexible polymer substrates. *Journal of the Electrochemical Society* 153 (10): F244–F248.

60 Greener, J., Ng, K.C., Vaeth, K.M. et al. (2007). Moisture permeability through multilayered barrier films as applied to flexible OLED display. *Journal of Applied Polymer Science* 106 (5): 3534–3542.

61 Henry, B.M., Dinelli, F., Zhao, K.Y. et al. (1999). A microstructural study of transparent metal oxide gas barrier films. *Thin Solid Films* 355: 500–505.

62 Affinito, J.D., Gross, M.E., Coronado, C.A. et al. (1996). A new method for fabricating transparent barrier layers. *Thin Solid Films* 291: 63.

63 Kim, T.W., Yan, M., Erlat, G. et al. (2005). Transparent hybrid inorganic/organic barrier coatings for plastic organic light-emitting diode substrates. *Journal of Vacuum Science & Technology A* 23 (4): 971–977.

64 Lungenschmied, C., Dennler, G., Neugebauer, H. et al. (2007). Flexible, long-lived, large-area, organic solar cells. *Solar Energy Materials & Solar Cells* 91 (5): 379–384.

65 Erlat, A.G., Wang, B.C., Spontak, R.J. et al. (2000). Morphology and gas barrier properties of thin SiO_x coatings on polycarbonate: correlations with plasma-enhanced chemical vapor deposition conditions. *Journal of Materials Research* 15 (3): 704–717.

66 Yan, M., Kim, T.W., Erlat, A.G. et al. (2005). A transparent, high barrier, and high heat substrate for organic electronics. *Proceedings of the IEEE* 93 (8): 1468–1477.

67 Yanaka, M., Kato, Y., Tsukahara, Y. et al. (1999). Effects of temperature on the multiple cracking progress of sub-micron thick glass films deposited on a polymer substrate. *Thin Solid Films* 355: 337–342.

68 Lungenschmied, C., Chwang, A.B., Rothman, M.A., Mao, S.Y. et al. (2003). Thin film encapsulated flexible organic electroluminescent displays. *Applied Physics Letters* 83 (3): 413–415.

69 Grego, S., Lewis, J., Vick, E. et al. (2005). Development and evaluation of bend-testing techniques for flexible-display applications. *Journal of the Society for Information Display* 13 (7): 575–581.

70 Rochat, G., Leterrier, Y., Fayet, P. et al. (2003). Mechanical analysis of ultrathin oxide coatings on polymer substrates in situ in a scanning electron microscope. *Thin Solid Films* 437 (1–2): 204–210.

71 Yanaka, M., Henry, B.M., Roberts, A.P. et al. (2001). How cracks in SiO_x-coated polyester films affect gas permeation. *Thin Solid Films* 397 (1–2): 176–185.

72 Andersons, J., Leterrier, Y., and Fescenko, I. (2003). Analysis of the initial fragmentation stage of oxide coatings on polymer substrates under biaxial tension. *Thin Solid Films* 434 (1–2): 203–215.

73 Rego, S., Lewis, J., Vick, E. et al. (2005). Development and evaluation of bend-testing techniques for flexible-display applications. *Journal of the Society for Information Display* 13 (7): 575–581.

74 Leterrier, Y., Bouten, P., and Jiang, X. (2001). *Layer Mechanics: Experimental Methods and Models*, 1. Information Society Technologies.

75 Ghosh, A.P., Gerenser, L.J., Jarman, C.M. et al. (2005). Thin-film encapsulation of organic light-emitting devices. *Applied Physics Letters* 86: 223503.

76 Suo, Z., Ma, E.Y., Gleskova, H. et al. (1999). Mechanics of rollable and foldable film-on-foil electronics. *Applied Physics Letters* 74 (8): 1177–1179.

77 Someya, T., Kato, Y., Sekitani, T. et al. (2005). Conformable, flexible, large-area networks of pressure and thermal sensors with organic transistor active matrixes. *Proceedings of the National Academy of Sciences of the United States of America* 102 (35): 12321–12325.

78 Park, S.K., Han, J.I., Moon, D.G. et al. (2003). Mechanical stability of externally deformed indium-tin-oxide films on polymer substrates. *Japanese Journal of Applied Physics* 42 (2A): 623–629.

79 Yuan, J. (2017). Research progress and outlook of flexible electronic packaging technology. *Journal of Shandong Industrial Technology* 15: 107–276.

80 Li, Z.W. (2012). Multi-laver encapsulation and viscoelastic effect analvsis for flexible electronics Chinese master's theses full-text database.

81 Woo, H.J., Seo, D.S., and Kim, J.Y. (2006). Improvement of permeation of solvent-free multi-layer encapsulation of thin films on polyethylene terephthalate (PET). *Journal of the Korean Institute of Electrical and Electronic Material Engineers* 19 (8): 754–757.

82 Zeng, K.J., Liu, J.J., Liu, Y.W. et al. (2016). Recent research progress in thin-film encapsulation for FOLED. *Journal of Hunan University of Technology* 30 (6): 44–49.

83 Zhou, C.J., Sun, Y.J., Tong, Q.G. et al. (2015). Preparation technology of high barrier films. *China Packaging Industry* 7: 47–48.

84 Yasmeen, S., Ryu, S.W., Lee, S.H. et al. (2022). *Atomic Layer Deposition Beyond Thin Film Deposition Technology*. Advanced Materials Technologies.

85 Kim, H., Lee, H.B.R., and Maeng, W.J. (2009). Applications of atomic layer deposition to nanofabrication and emerging nanodevices. *Thin Solid Films* 517 (8): 2563–2580.

86 Huang, Y., Song, K.H., and Feng, H. (2016). Atomic layer deposition and its development of the application on the powder materials. *New Chemical Materials* 44 (5): 16–18. 21.

87 Rahman, F. and Runyon, J.C. (2021). Atomic layer processes for material growth and etching—a review. *IEEE Transactions on Semiconductor Manufacturing* 34 (4): 500–512.

88 Zhou, Z.W., Li, M., Xv, M. et al. (2016). Low-temperature multi-layer Al_2O_3/TiO_2 composite encapsulation thin film by atomic layer deposition. *Chinese Journal of Liquid Crystals and Displays* 31 (6): 532–539.

89 Jong G, Hyun G, Sung S. Defect-sealing of Al_2O_3/ZrO_2 multilayer for barrier coating by plasma-enhanced atomic layer deposition process. *Thin Solid Films*, 2015, 577:143-148.

90 Liu, Y., Zhang, B.X., and Huang, C.J. (2017). Analysis of thin film encapsulation technology of flexible OLED. *Optoelectronic Technology* 37 (2): 139–115.

91 Accelerated breakthrough process of OLED inkjet printing technology. http://www.chinafpd.net/news/27934.html, 1 June 2016.

92 Ming, C. (2020). The study on degradable controllable monitoring design of implantable device package structure. Chinese master's theses full-text database.

93 Woo, J.H., Park, S.Y., Koo, D. et al. (2022). Highly elastic and corrosion-resistive metallic glass thin films for flexible encapsulation. *ACS Applied Materials & Interfaces* 14 (4): 5578–5585.

94 Jiang, J., Cheng, Y., Sun, X.C. et al. (2022). Flexible full-surface conformal encapsulation or each fiber in graphene glass fiber fabric against thermal oxidation. *ACS Applied Materials & Interfaces* 14 (17): 19889–19896.

95 Shen, J., Wang, D., Langlois, E. et al. (2000). Degradation mechanisms in organic light emitting diodes. *Synthetic Metals* 111: 233–236.

96 Popovic, Z.D., Aziz, H., Hu, N.X. et al. (2000). Long-term degradation mechanism of tris(8-hydroxyquinoline) aluminum-based organic light-emitting devices. *Synthetic Metals* 111: 229–232.

97 Clemens, W., Fix, I., Ficker, J. et al. (2004). From polymer transistors toward printed electronics. *Journal of Materials Research* 19 (7): 1963–1973.

98 Brabec, C.J., Cravino, A., Meissner, D. et al. (2001). Origin of the open circuit voltage of plastic solar cells. *Advanced Functional Materials* 11 (5): 374–380.

99 Do, L.M., Han, E.M., Niidome, Y. et al. (1994). Observation of degradation processes of Al electrodes in organic electroluminescence devices by electroluminescence microscopy, atomic force microscopy, scanning electron microscopy, and Auger electron spectroscopy. *Journal of Applied Physics* 76 (9): 5118–5121.

100 Aziz, H., Popovic, Z., Xie, S. et al. (1998). Humidity-induced crystallization of tris (8-hydroxyquinoline) aluminum layers in organic light-emitting devices. *Applied Physics Letters* 72 (7): 756–758.

101 Schaer, M., Nuesch, F., Berner, D. et al. (2001). Water vapor and oxygen degradation mechanisms in organic light emitting diodes. *Advanced Functional Materials* 11 (3): 158–158.

102 Burrows, P.E., Bulovic, V., Forrest, S.R. et al. (1994). Reliability and degradation of organic light emitting devices. *Applied Physics Letters* 65 (23): 2922–2924.

103 Sun, F.B., Duan, Y., Yang, Y.Q. et al. (2014). Fabrication of tunable [Al_2O_3:alucone] thin-film encapsulations for top-emitting organic light-emitting diodes with high performance optical and barrier properties. *Organic Electronics* 15 (10): 2546–2552.

104 Horiuchi, K., Nakada, K., Uchino, S. et al. (2002). Passivation effects of alumina insulating layer on C60 thin-film field-effect transistors. *Applied Physics Letters* 81 (10): 1911–1912.

105 Han, P., Lai, T.C., Wan, M. et al. (2019). Outstanding memory characteristics with atomic layer deposited $Ta_2O_5/Al_2O_3/TiO_2/Al_2O_3/Ta_2O_5$ nanocomposite structures as the charge trapping layer. *Applied Surface Science* 467–468: 423–424.

3

Flexible Substrates

Yanhui Chen, Xian Zhang, and Zhiqiang Wu

Northwestern Polytechnical University, School of Chemistry and Chemical Engineering, Dongxiang Road, 710129 Xi'an, China

3.1 Concept and Connotation of Flexible Substrates

Flexible electronics is a rapidly emerging interdisciplinary technology that involves the deposition of organic, inorganic, or organic–inorganic composite (hybrid) materials on flexible substrates such as ductile plastic or thin metal substrates. The end result of this process is the creation of various electronic, optoelectronic, and photonic components, which can be used to assemble integrated systems such as circuits (though it is not limited to circuits). Now in its infancy, it is an emerging discipline [1]. According to the different processes, they are also called plastic electronics, printed electronics, organic electronics, polymer electronics, soft electronics, etc. Flexible electronics realize the flexibility of electronic devices through flexibilization technology, which improves and overcomes the shortcomings of traditional electronic devices that are rigid and fragile [2], and its appearance overturns the original information carrier form, comprehensively enhances the ability of information perception, storage, and display, realizes the miniaturization, flexibility, and ubiquity of electronic systems, and can realize the high integration of human–computer-object. So far, flexible electronics have demonstrated broad potential applications in the fields of biomedicine [3], flexible displays [4], energy storage and conversion devices [5], and human–computer interaction [6], as shown in Figure 3.1. The basic structure of flexible electronics mainly includes electronic components, flexible substrates, cross-linked conductors, and adhesive layers. The flexible substrate is a thin film that can be rolled and used as a coated substrate [7], which usually controls the overall performance of flexible electronic devices, and it plays a very important role in the integrated system of flexible electronic devices.

Flexible substrate is the most prominent place where flexible electronics differ from traditional electronics; it not only has the characteristics of traditional rigid substrates: insulation, inexpensiveness, and high strength, but also has its own unique characteristics: flexibility, thin film, easy deformation, etc. Different flexible electronic devices use different flexible substrates to adapt to different materials

Flexible Electronic Packaging and Encapsulation Technology, First Edition.
Edited by Hong Meng and Wei Huang.

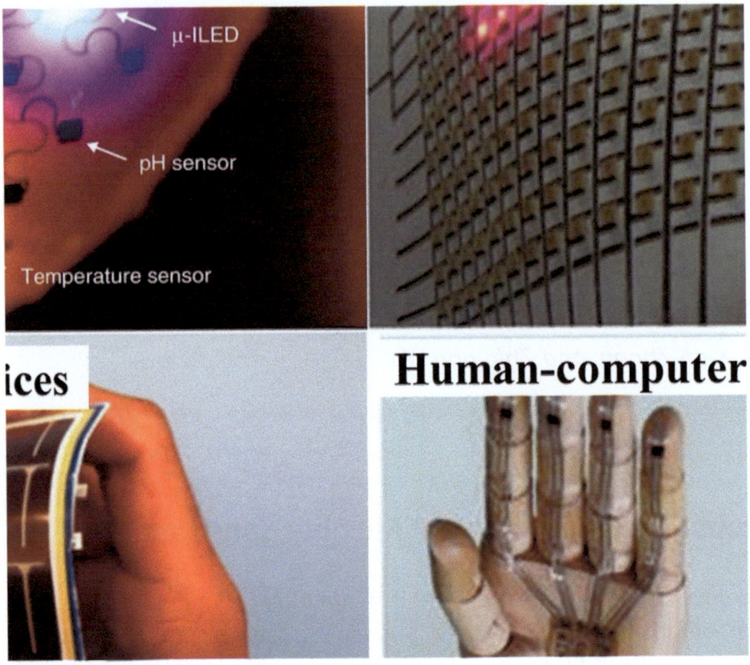

Figure 3.1 Potential applications of flexible electronics. Source: Xu et al. [3]; © 2014, Reproduced with permission from Springer Nature; Russo et al. [4]; © 2011, Reproduced with permission from John Wiley & Sons, Inc.; Ostfeld et al. [5]; © 2016, Reproduced with permission from Springer Nature / CC BY 4.0.; Chortos et al. [6]; © 2015, Reproduced with permission from Springer Nature.

and different application scenarios to meet the requirements of flexible electronic devices: light, thin, transparent, flexible and good stretch, insulation and corrosion resistance, and other properties. Common flexible substrate materials are polydimethylsiloxane (PDMS) [8], polyvinyl alcohol (PVA) [9], polycarbonate (PC) [10, 11], polyethylene terephthalate (PET) [12, 13], polyimide (PI) [14–16], polyethylene naphthalate (PEN) [17–19], copolyesters (PCTN) [20], thermoplastic polyurethane elastomer rubber (TPU) [21, 22], polymer hydrogels [23–25], polymer foams [26], paper sheets [27–29], textile materials [30–33], etc., as shown in Figure 3.2. All of the above materials have the advantages of easy availability, chemical stability, transparency, and thermal stability [34], among which PI shows more excellent heat resistance, dimensional stability, and mechanical properties due to the rigid structure of the imide ring contained in the main chain, and has unique advantages in terms of heat resistance of flexible substrates compared with polymeric materials such as polyester and silicone rubber, further expanding the application of polymer materials for flexible substrates.

3.2 Development History of Flexible Substrates

Flexible substrates have a wide range of promising applications in healthcare, environmental monitoring, display and human–computer interaction, energy

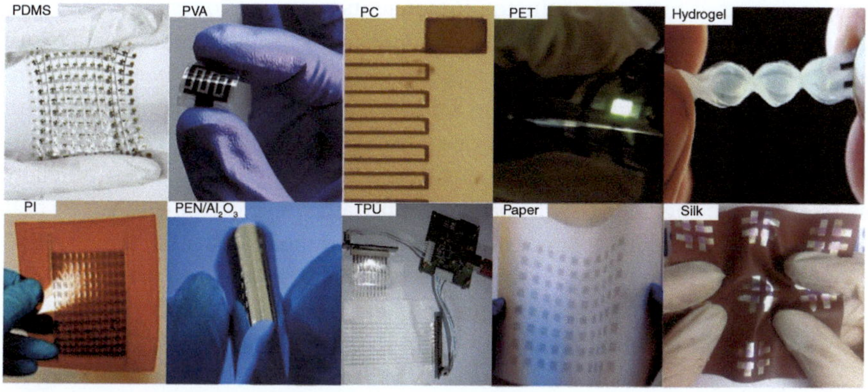

Figure 3.2 Common flexible substrate forms and material types. Source: Li et al. [8]; © 2019, Reproduced with permission from springer Nature / CC BY 4.0.; Bellani et al. [9]; © 2019, Reproduced with permission from John Wiley & Sons, Inc.; Kusaka et al. [10]; © 2021, Reproduced with permission from Elsevier; Zhou et al. [13]; © 2018, Reproduced with permission from John Wiley & Sons, Inc.; Xu et al. [23]; © 2018, Reproduced with permission from John Wiley & Sons, Inc.; Zhang et al. [14]; © 2019, Reproduced with permission from Royal Society of Chemistry; Duan et al. [18]; © 2019, Reproduced with permission from John Wiley & Sons, Inc.; Yang et al. [21]; © 2017, Reproduced with permission from John Wiley & Sons, Inc.; Zhang et al. [27]; © 2015, Reproduced with permission from Royal Society of Chemistry; Kim et al. [31]; © 2013, Reproduced with permission from Elsevier.

conversion, management and storage, communication, and wireless networks. Strong market demand for applications has driven the in-depth and rapid development of flexible substrate research. In the 1970s–1990s, the flexibility of electronics was inspired by science fiction. In 1984, the Terminator movie series depicted humanoid robots and even a self-healing robot [35]. These virtual representations of flexible electronic skin occurred against the backdrop of the reality of vibrant microelectronics research, which brought flexible electronics to the general public and the science fiction community. Researchers attempted to link science fiction with scientific reality. In the 1990s, scientists began to use flexible materials to fabricate large-area, low-cost, and printable sensor sheets. By etching thin Si wafers and integrating them on flexible PI films, thereby creating silicon (Si) microelectromechanical (MEM) islands. Jiang prepared the first haptic shear force sensing for flexible sensor sheets [35]. The twentieth century has seen significant advances in the development and advancement of flexible electronic substrates, particularly the substantial increase in mechanical compatibility and sensitivity of flexible substrates. Su developed stretchable electrodes based on flexible substrate materials [36], and Rogers modified a typical brittle material, Si. By using ultrathin (100 nm) flexible substrate films connected by stretchable interconnects to make flexible, high-performance electronics [37]. Hammock has developed a flexible conductive composite substrate with repeatable self-healing and mechanical force sensing capabilities for high-sensitivity capacitive pressure sensors [38]. Other teams have developed scalable optoelectronic products based on flexible substrate materials, including light-emitting diodes (LEDs) [39] and organic photovoltaics (OPVs) [40].

Flexible substrates are a signature component of flexible electronics, and their development is mainly in the area of electronics packaging. The term "electronic packaging" is used for semiconductor components such as triodes and ICs. Their appearance has changed the history of electronic engineering. In order to make full use of the functions of semiconductor electronic components, they need to be reinforced, sealed, and enlarged in order to achieve reliable electrical connection with external circuits, as well as to get effective mechanical support and protection during signal transmission, and since then the term "packaging" has been officially used in electronic engineering. The package substrate is an electronic substrate that can provide electrical connection, protection, support, heat dissipation, and assembly for chips and electronic components to realize multi-pin, reduce the size of the package, improve electrical performance and heat dissipation, ultrahigh density or multi-chip modularity, and high reliability, so the package substrate can be understood as a printed circuit board (PCB) or thin thick film circuit substrate with higher performance or special performance substrates [41].

Electronic substrate materials for packaging can be divided into several categories, such as inorganic substrate materials, organic rigid substrate materials, and composite substrate materials, according to their electrical insulation or mechanical support materials.

(1) **Inorganic substrate materials**: The traditional inorganic substrate materials are Al_2O_3, SiC, BeO, AlN, and other substrates, which are widely used in high-power devices such as hybrid integrated circuits (HIC) and multi-chip assemblies (MCA) due to their excellent characteristics in thermal conductivity, flexural strength, and thermal expansion coefficient. Inorganic substrate materials can be divided into general ceramic substrate materials (HTCC), such as alumina, mullite, aluminum nitride, silicon carbide, and beryllium oxide substrates; low-temperature co-fired ceramic (LTCC) multilayer substrate materials; and other types of inorganic substrate materials, such as glass substrate materials for LCD, glass substrate materials for PDP, and glass-based copper-clad laminates. Inorganic substrate materials, despite the small proportion of electronic packaging, have seen a further decline in recent years (replaced by organic substrate materials), but in many areas, such as supercomputers with high-density multilayer substrate, aerospace computers with MCM substrate, crystal oscillator carrier, SAW carrier, and power devices, they still have an irreplaceable role.

(2) **Organic rigid substrate materials**: that is, PCB commonly used in various types of copper-clad board. Copper laminate also known as copper-clad laminate, can be divided into rigid copper-clad laminate and flexible copper-clad laminate. Rigid copper-clad foil board includes paper-reinforced phenolic resin copper-clad foil board, paper-reinforced epoxy resin copper-clad foil board, paper-reinforced polyester resin copper-clad foil board, glass fiber-reinforced epoxy resin copper-clad foil board, and glass fiber-reinforced polytetrafluoroethylene resin copper-clad foil board. Flexible copper-clad foil board includes polyester resin copper-clad foil board and PI resin copper-clad foil board.

(3) **Composite substrate materials**: with the development of high-density packaging technology, a single material or a single structure of the substrate

can no longer meet the requirements of use. Composite substrate materials can achieve multifunction and high performance. For example, in addition to the substrate carrying components, wiring connections, and thermal conductivity functions, there are also additional resistance, capacitance, inductance, electromagnetic shielding, optical structure body, and other functions; in terms of high performance, to achieve lower resistance, impedance matching, improved transmission performance, improved mechanical properties, more convenient for high-density packaging, etc. In order to achieve the above objectives, it is necessary to take the necessary measures in terms of structure, performance, and process, for example: in terms of structure, multilayer, thin film, micro-porous micro-wiring, etc.; in terms of materials, the use of different materials in terms of strength, thermal conductivity, thermal expansion coefficient, dielectric constant, insulation properties, etc. complement each other. Accordingly, a variety of different processes are required. Composite substrate materials have a wide variety of functions, a wide variety of materials and structures, and a variety of processes, and are still developing rapidly, and there is currently no unified naming standard. Generally, according to the purpose of the composite, it will be divided into functional composite, structural composite, and material composite. Functional composite: such as the inclusion of multilayer printing substrate materials, the inclusion of raw laminated substrate materials, etc. Structural composite: such as resin–ceramic composite substrate materials, resin–porous ceramic composite substrate materials, resin–silicon composite substrate materials; material composite: such as metal-based composite substrate materials, and metal core composite substrate materials.

With the development of mechanics, materials science, and technology, the ductility of electronic materials has put forward new requirements and challenges. Flexible electronic devices are a typical representative of ductile electronic materials. In recent years, with foldable cell phones, computers, and other rapid developments, the traditional copper-clad laminate and composite substrate materials, although containing organic polymer materials, the final preparation of the substrate material is hard, so the above inorganic substrate materials, copper-clad laminate, and composite substrate materials are difficult to achieve flexible electronic devices in the ductility of the hard requirements, thus limiting their application in flexible electronic devices. Flexible substrate materials are the key materials for flexible electronic devices, which are lined with anodes to achieve flexibility and are also demanding as import and export channels for light [42, 43]. Many commercial flexible polymer materials with excellent properties are available for flexible substrate materials, such as PDMS, PET, PI, polyethylene (PE), and polyurethane (PU) [34]. However, most polymers such as PET suffer from good optical properties and poor heat resistance. Nowadays, the developed colorless polyimide (CPI) has excellent optical and thermal properties, but the monomers used are very expensive and the performance indexes are yet to be improved, so there are still difficulties in large-scale promotion. Therefore, the development of flexible, transparent, and heat-resistant polymer substrate materials is very important for the development and promotion of flexible electronic devices, and there is still a lot of room for improvement for polymer flexible substrate materials.

3.3 Flexible Substrate Materials

Flexible substrates are an essential component of flexible electronic devices, and in order to fabricate flexible devices, polymer films are often used as flexible substrates [44]. The commonly used flexible substrate materials are natural polymers and synthetic polymers. Natural polymers include textiles [30, 31], silk [32, 33], and paper [27–29], which are abundantly stored, inexpensive, and easy to use but have poor heat and chemical resistance. Synthetic polymers include PET, PI, PEN, PC, and PDMS. Other flexible substrate materials include metal foils [45, 46] and shape memory polymers [47–49]. Table 3.1 shows the common flexible substrate materials and their basic physical properties.

3.3.1 Polydimethylsiloxane

Depending on its relative molecular mass, PDMS can vary in appearance from a colorless and transparent volatile liquid to a high-viscosity liquid or silicone. It boasts several advantages such as being odorless, highly transparent, resistant to both heat and cold, low in viscosity changes with temperature, water-resistant, having a low surface tension, and showing good biocompatibility and flexibility. PDMS can be molded into various shapes by utilizing the template method, making it ideal for use as a substrate in flexible electronic devices [50]. Therefore, a large number of flexible electronic devices are based on PDMS and have good conductivity and sensitivity by coating different conductive modification layers (e.g. carbon nanotubes, graphene, metal nanomaterials, conductive polymers, etc.). PDMS is usually obtained by a stepwise polymerization reaction, where dimethylsilane is hydrolyzed, dehydrated, or dehydrogenated, as shown in Figure 3.3. The molecular structure of the polymer backbone of –Si–O–Si– gives it great flexibility, and the flexible substrate films prepared from it have great promise for flexible electronic devices, such as flexible wearable electronic devices [51] and stretchable planar inverted cone liquid metal antennas [52]. PDMS is also widely used in a variety of biomedical applications, such as soft tissue implants, microfluidic devices, and heart valves [53].

3.3.2 Polyvinyl Alcohol

PVA is a white powder, flake, or flocculent solid with a glass transition temperature of 60–85 °C [54]. PVA cannot be made via the polymerization of its monomer-vinyl alcohol, as the free vinyl alcohol is very unstable and easily isomerizes to acetaldehyde or ethylene oxide. Instead, it is generally produced by hydrolyzing vinyl acetate after the polymerization process. The preparation of PVA is divided into three steps. Firstly, polyvinyl acetate is produced by the polymerization of vinyl acetate. Then, the alcoholization of polyvinyl acetate is carried out to produce PVA while recovering methyl acetate (acetic acid and methanol). The specific reaction equation is shown in Figure 3.4. Due to the presence of numerous hydroxyl groups in its molecular chain, PVA is polar and can form hydrogen bonds with water, resulting in good water solubility. Additionally, it is soluble in hot

Table 3.1 Different flexible substrate materials and their basic physical properties.

Substrate material	PI	PET	PC	PEN	PDMS	PU	Cellulosic fiber	Cotton textile
T_g (°C)	155–366	70–110	145	120–155	−125	80	–	–
T_m (°C)	250–452	115–258	115–160	269	–	180	–	–
CTE (ppm/°C)	8–20	15–33	75	20	310	153	–	–
Density (g/cm)	1.35–1.43	1.39	1.20–1.22	1.36	1.03	1.18	1.4	1.55
Working temperature (°C)	400	−50 to 150	−40 to 130	–	−45 to 200	130	RT	RT
Water absorption (%)	1.3–3	0.4–0.6	0.16–0.35	0.3–0.4	>0.1	0.2	–	8.5
Solvent resistance	+	+	–	+	–	+	–	–
Light transmittance performance	–	+	+	+	+	+	–	–
Dimensional stability	–	+	–	+	+	+	–	–
Elongation at break	80	90	1–2	85	50–500	300–800	5–15	7–10
Surface roughness	30	30	74	15	–	–	50–100	–

+: excellent performance, −: Moderate performance, –: Poor performance.

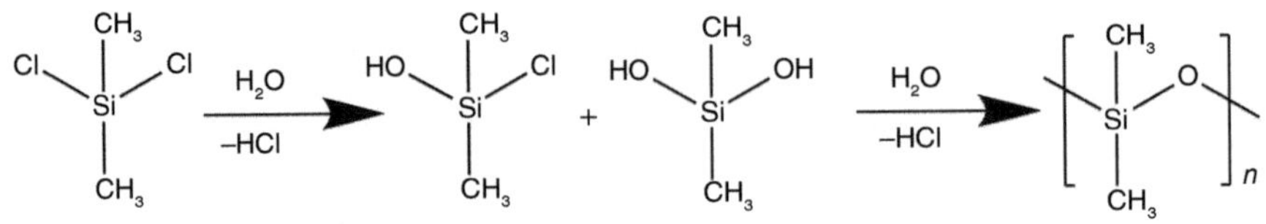

Figure 3.3 PDMS synthesis route.

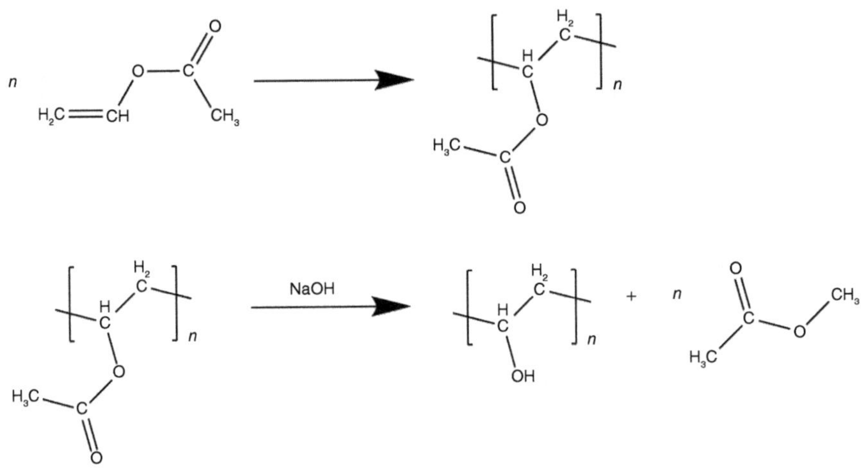

Figure 3.4 PVA synthesis route.

hydroxyl-containing solvents such as glycerol and phenol but remains insoluble in general organic solvents such as methanol, benzene, acetone, and gasoline. PVA flexible substrates can be used in stretchable energy storage devices [55], and for stretchable energy storage devices, it is critical to be able to withstand high stretchability. PVA/H_3PO_4 is used in a stretchable energy storage device after 100 stretch cycles with a maximum strain of 200%.

3.3.3 Polycarbonate

PC is a polymer containing carbonate groups in the molecular chain [56], mainly composed of phenyl, ether, carbonyl, ester, and isopropyl groups. PC is characterized by colorless transparency, heat resistance, impact resistance, high refractive index, and easy processing. It generally does not require additives to have UL94V-2 level flame retardant properties and good mechanical properties at ordinary use temperatures. There are two methods of synthesis: Phosgene method (interfacial polycondensation) and ester exchange method (melt polycondensation). The specific process is shown in Figure 3.5. Phosgene method mainly through three reaction processes: the first reaction process is the Phosgene reaction, as the name implies; that is, under the stirring of the phosgene into the inert solvent (dichloroethane, dichloromethane) and bisphenol in a NaOH aqueous solution, the reaction at the interface of the two phases, the production of end groups for the acyl chloride of the simple polymer. The second reaction course is the interfacial polycondensation reaction, which involves adding alkali, adding catalyst (tertiary amine, quaternary ammonium salt), molecular weight adjuster (phenol, *p-tert*-butylphenol), and controlling the molecular weight within 30–100 000, to facilitate the subsequent molding process. The third reaction course involves post-processing activities that include neutralizing the reaction solution using cresol to reduce residual alkali levels, PH = 3–5. The upper aqueous phase is then separated, and the salts in the reaction solution are washed with water. The resin is precipitated from the solvent using precipitating agents (methanol, acetone, xylene), dried, and extruded into pellets that form PC resin for flexible substrates. During the ester exchange method, only oligomers are formed, and phenol is generated as a by-product. To increase the molecular weight significantly, polycondensation must occur under nitrogen protection. The polycondensation process produces diphenyl carbonate as a second by-product, which needs to be acid-washed and water-washed to neutrality before being dried at 100–110 °C to obtain PC resins for flexible substrates. Although manufacturing conductive pathways by inkjet printing on PC flexible substrates has been reported, other devices with complex functions have not been extensively studied [10].

3.3.4 Polyester

Polyester is a general term for polymers obtained by condensation of polyols and polyacids, mainly referring to PET), whose main chain consists of benzyl groups, ester groups, and methylene groups and is a crystallizable thermoplastic resin [57] with a milky white or light-yellow color. The films made from PET possess exceptional toughness and boast outstanding heat resistance – their heat deflection temperature and long-term use temperature are the highest among thermoplastic general engineering plastics. PET films also exhibit excellent mechanical properties, including a bending strength of up to 200 MPa, a modulus of elasticity of up to 4 GPa, high surface hardness, and resistance to creep and fatigue. These films remain stable in the majority of organic solvents and inorganic acids, although they are not resistant to alkalis. PET films also feature excellent weather resistance, with a glass transition temperature of 69 °C, a softening range of 230–240 °C, and a melting point of 255–260 °C. Additionally, they offer low gas and water vapor permeability and excel in blocking properties against gases, water, oil, and odors. PET films also feature high levels of transparency, contain UV blockers, and feature outstanding glossiness. Moreover, they are nontoxic, tasteless, and safe to use for food packaging applications.

PET can be prepared by the synthesis of the ester exchange method and the direct esterification method, the synthesis process is shown in Figure 3.6. The ester exchange method is to add dimethyl terephthalate (DMT) and ethylene glycol in a certain ratio into the ester exchange reactor to produce hydroxyethyl terephthalate in the presence of a catalyst, and then carry out the polycondensation reaction to produce PET resin. The key to the preparation of PET using the ester exchange process is the selection of a suitable catalyst. Ester exchange catalysts are mainly inorganic and organic acid catalysts, organozinc catalysts, organotin catalysts, and

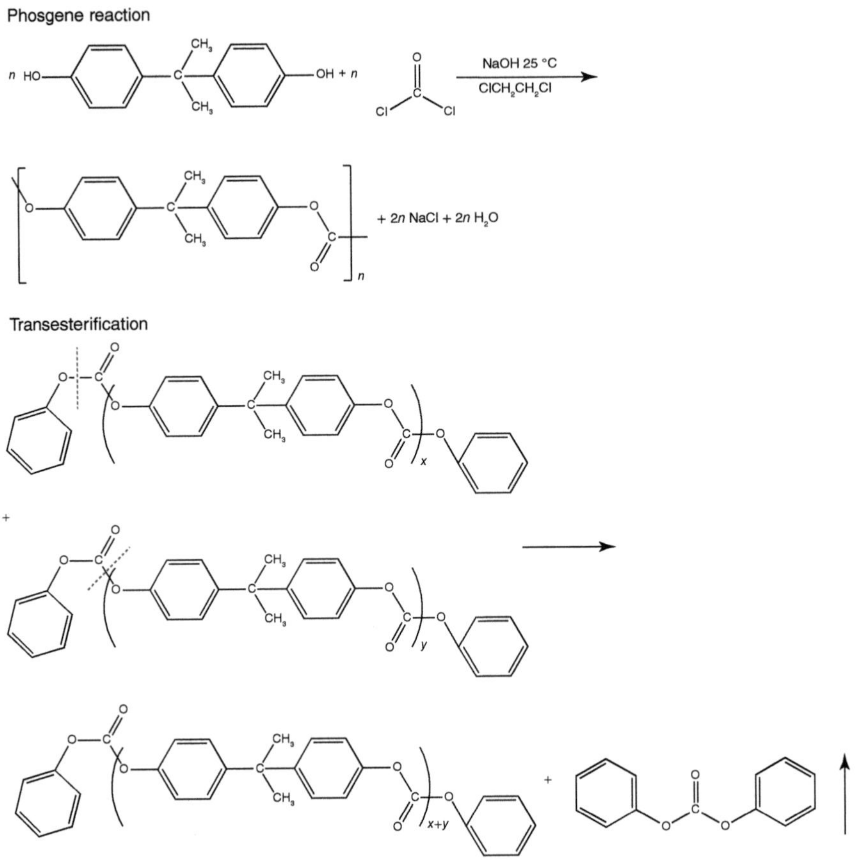

Figure 3.5 PC synthesis route.

phthalate catalysts. The ester exchange method has mild reaction conditions, a small amount of high boiling material generation, high product yield, and product purity. The direct esterification method first formulates terephthalic acid (TPA) into a slurry in ethylene glycol and then reacts at 220–250 °C under pressurized conditions. The direct esterification method requires more ethylene glycol, and the amount of diethylene glycol generated at high temperatures is also higher, which tends to lower the melting point of the polymerization product. For this reason, it is possible to add weakly basic substances and control the particle size of TPA to the reaction system to reduce the amount of ethylene glycol and inhibit the generation of diethylene glycol. With these measures, the molar ratio of TPA to ethylene glycol can be brought to a level close to 1 : 1. At the same time, the water of the product is removed from the reaction system, so that the esterification reaction moves in the positive direction and the yield is improved. The direct esterification method is simple to operate, but it is prone to the generation of high boiling products, low product recovery and purity, and adverse effects on subsequent products. PET flexible substrates can be prepared by inkjet printing for the preparation of flexible sensor devices for biological applications [12].

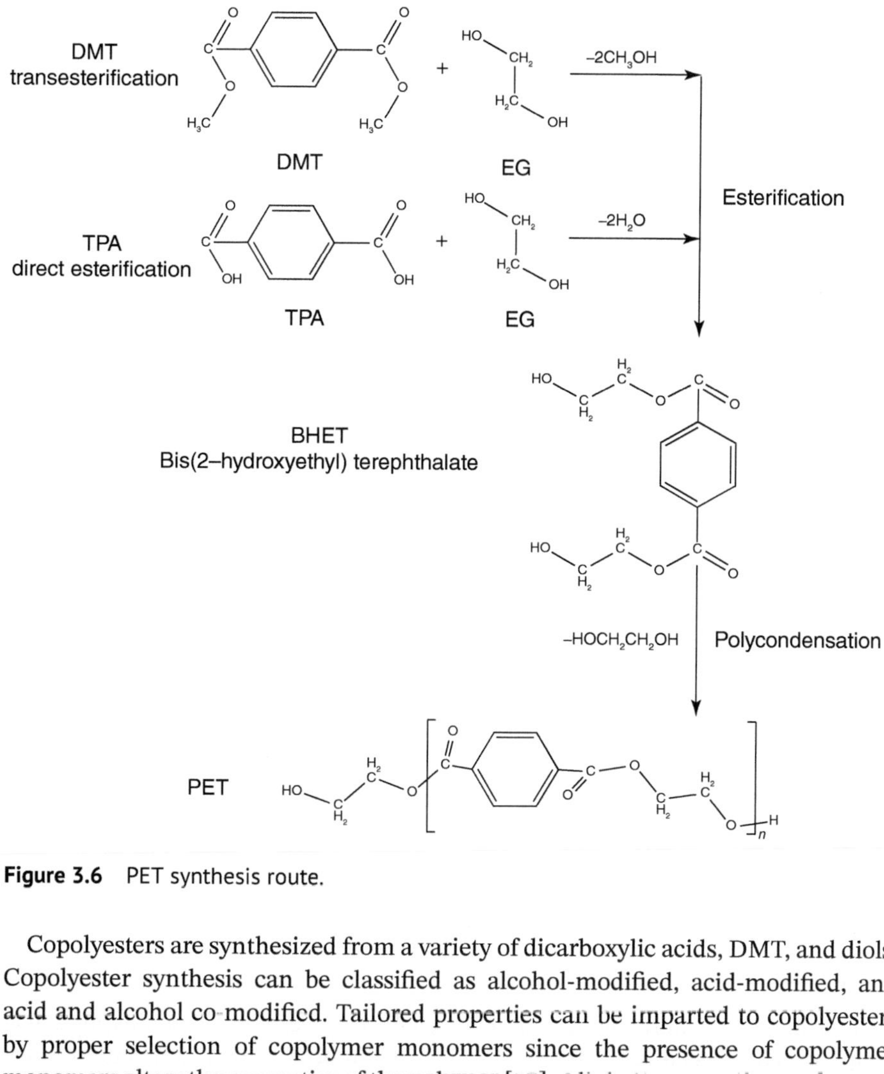

Figure 3.6 PET synthesis route.

Copolyesters are synthesized from a variety of dicarboxylic acids, DMT, and diols. Copolyester synthesis can be classified as alcohol-modified, acid-modified, and acid and alcohol co-modified. Tailored properties can be imparted to copolyesters by proper selection of copolymer monomers since the presence of copolymer monomers alters the properties of the polymer [58]. Aliphatic aromatic copolyesters are prepared by introducing appropriate aromatic chain segments in the main or side chains of aliphatic polymers to effectively improve their mechanical properties under the control of suitable biodegradability. Copolyesters in general exhibit superior viscosity compared to poly(methyl methacrylate) (PMMA). They also boast high transparency, excellent toughness, chemical resistance, and stress-whitening resistance. These materials feature exceptional processing and molding capabilities, enabling designers to craft them into virtually any shape they desire. Copolyesters are compatible with traditional extrusion, injection molding, blow molding, blister molding, and other molding techniques and are extensively utilized in markets such as sheeting, high-performance shrink films, bottles, and profiles. Additionally, copolyesters are highly processable and can be easily finished through conventional machining techniques. By copolymerizing different diol and diacid

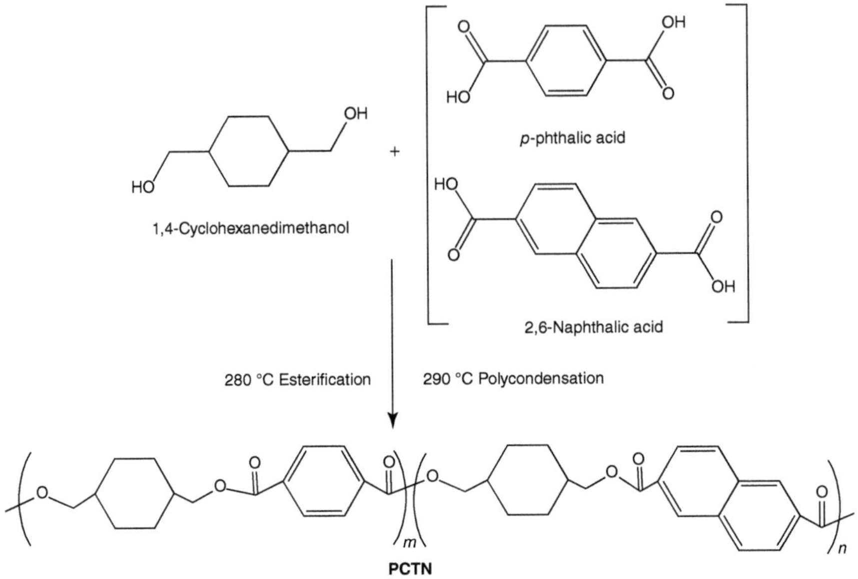

Figure 3.7 PCTN copolyester synthesis route.

monomers, it is possible to tailor these materials for a vast array of applications. Fiaz Hussain [20] synthesized a novel semi-crystalline copolyester (PCTN) using TPA, 2,6-naphthalene dicarboxylic acid (NDA), and 1,4-cyclohexanedimethanol (CHDM) with the reaction equation shown in Figure 3.7. The corresponding copolyester films were prepared by melt extrusion and uniaxial stretching, and the properties of the prepared PCTN films are comparable to those of conventional polymer films and can be used as substrate materials for flexible electronic devices.

3.3.5 Polyimide

PI is a polymer whose molecular main chain or side chain contains an imide ring [59]. The special rigid structure and strong intermolecular interactions make the glass transition temperature (T_g) of PI generally above 200 °C and can be maintained in an air atmosphere from 200 to 250 °C for a long time, with excellent thermal stability, excellent mechanical properties and dimensional stability, excellent irradiation resistance, and corrosion resistance. These superior characteristics over other polymers expand the range of applications of PIs. PI synthesis can be performed in three ways [60], as shown in Figure 3.8. The most common synthesis process for PIs is a two-step polycondensation process (Figure 3.8a,b). The first step is a rapid exothermic polymerization of dianhydride and diamine in DMAc or NMP at a relatively low temperature. The nucleophilic attack of the carbonyl carbon of the amino to the anhydride group leads to the formation of PAA. The viscous PAA solution is cast on a thin layer on a glass or stainless steel substrate and heated from room temperature to a high temperature at least above 300 °C in an inert

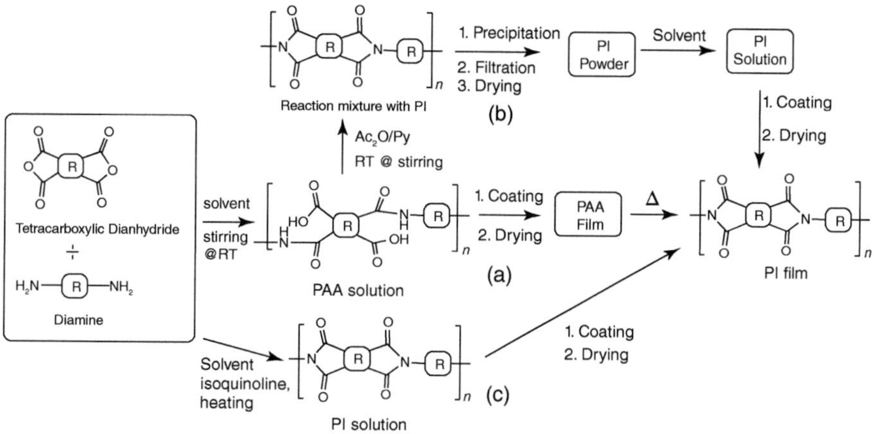

Figure 3.8 Synthesis and preparation of PIs by different routes: (a) thermal imidization, (b) chemical imidization, (c) one-step method.

atmosphere under programmed curing conditions; this PI obtained by thermal imidization methods is usually insoluble in organic solvents [61] and has poor solubility. Imidization can also be achieved chemically via acetic anhydride/pyridine, specifically, chemical imidization by slowly adding an excess of dehydrating reagent (acetic anhydride/pyridine) to the PAA solution and stirring for 12–24 hours at room temperature. The imidated homogeneous reaction mixture is precipitated by slowly pouring a large amount of undesirable solvent (e.g. methanol, water, or aqueous methanol) into the stirring. The PIs are separated into fibrous powders after filtration, washing, and drying. Chemically imidized powder samples are typically highly soluble in common organic solvents and form stable PI solutions at room temperature with a high solids content sufficient for subsequent solution casting processes. Chemically imidized PI films typically show a lower yellowness index and higher transmittance in the UV–vis wavelength range than films prepared by conventional two-step thermal imidization, and they are optically more transparent. This may be due to the fact that the unstable end amino groups of the PI chains may develop color after high-temperature thermal imidization and may be capped by Ac_2O during the chemical imidization process. Another advantage of the chemical imidization method is that PI can be coated and dried to form a film at a lower temperature (≤250 °C) than in the thermal imidization process (≥300–350 °C) [60]. Flexible PI substrates are generally used in flexible display applications, such as flexible display support materials for flexible organic light-emitting diodes (FOLEDs) [62].

3.3.6 Polyurethane

Polyurethane, abbreviated as PU, is a polymer a containing carbamate group (-NH–CO–O–) in the molecular chain. It is typically made through a gradual polymerization reaction between polyisocyanate and compounds or polymers containing polyhydroxyl groups. The resulting polyurethane macromolecule can contain ether, ester, urea, diurea, ureido formate, and other groups in addition to carbamate. Polyurethane materials possess excellent tensile strength, tear strength, impact resistance, abrasion resistance, weather resistance, hydrolysis resistance, and oil resistance. They are also chemically stable, nonpolluting, nontoxic, and tasteless [63]. The chemical reaction of synthetic polyurethane is complex, and the specific reaction equation is shown in Figure 3.9. It includes primary reactions and secondary reactions. The primary reactions include pre-polymerization and

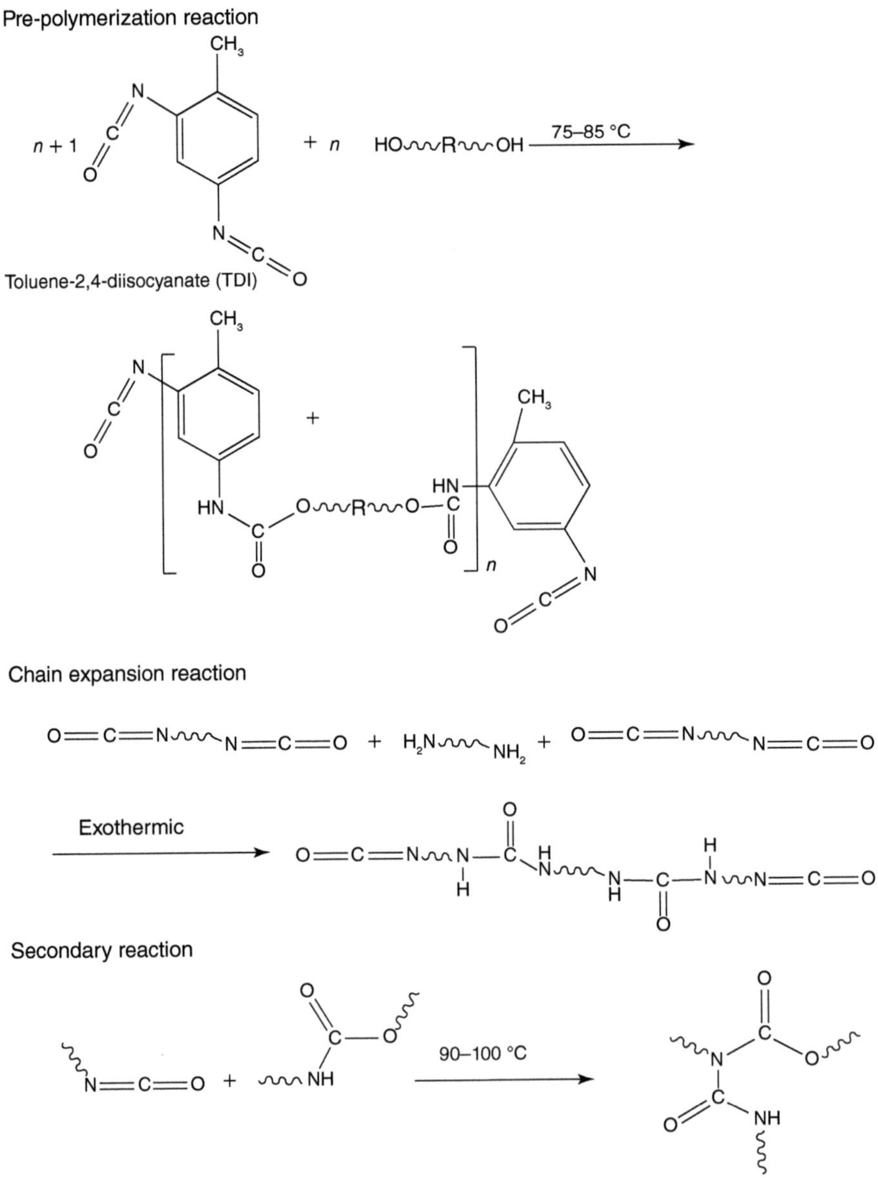

Figure 3.9 PU synthesis route.

chain expansion reactions for the synthesis of prepolymers. The pre-polymerization reaction refers to the reaction in which the terminal hydroxyl compound and excess diisocyanate are reacted by stepwise addition of polymerization to produce oligomers containing isocyanate end groups. The chain expansion reaction refers to the reaction of the prepolymer with compounds containing active hydrogen (commonly water, amines, hydrazines, and benzidine) to produce substituted urea groups, which increase the relative molecular mass. The secondary reaction refers to the reaction between the excess –N—C—O end group present in the system and the carbamate group –NH–CO–O– on the main chain to produce ureido formate group and crosslink; through the secondary reaction, the molecular structure of the polymer changes from linear to bulk structure to obtain PU resin. PU flexible substrates can be used as electronic skin substrate materials [21], stretchable electrode materials [22], and flexible sensing materials [64].

3.3.7 Parylene

Poly(parylene), commonly known as (Parylene) parylene, first developed and produced by Uinon Carbide Company (UCC) in the United States in the mid-1960s, is a linear polymeric material made from the polymerization of parylene monomers and can reach molecular weights of more than 500 000. In 1969, UCC industrialized three types of chlorine-based parylene (N, C, D), whose molecular structures are shown in Table 3.2. Because the benzene ring is in the main chain, it has a good heat resistance and can be used at 100–130 °C for 10 years; excellent mechanical and physical properties; low temperature resistance (−270 °C for N-type and −160 °C for C-type); good radiation resistance under vacuum; no significant change in tensile strength and electrical properties after irradiation with γ-rays at a total dose of 100 Mrad at a rate of 1.6 Mrad/h. Excellent solvent resistance, generally insoluble in all organic solvents at 150 °C; acid and alkali resistant; good gas and water resistance; and dielectric properties (frequency from 60 to 10^6 Hz; the dielectric constant is always maintained at 2.65). PPXC film still shows good dimensional stability in the range of 40 °C and 35–95% relative humidity, so it has more application space and can be used as the substrate for bioelectronics [65]. The fabrication of organic electrochemical transistor arrays doped with poly(3,4-ethylenedioxythiophene) polystyrene sulfonate (PEDOT-PSS) on poly(parylene) substrates using a photolithographic patterning method can be used to measure *in vivo* cortical activity in human epilepsy patients [66].

Table 3.2 Molecular structure formulae of PPXN, PPXC, and PPXD.

Name	PPXN	PPXC	PPXD
Molecular structure formula	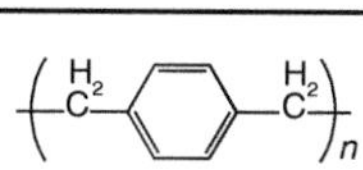	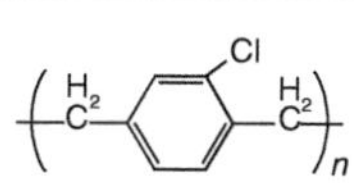	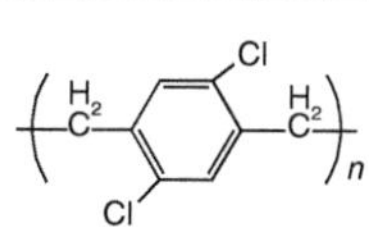

The preparation of poly-*p*-xylene is divided into two parts: one is the preparation of *p*-xylenebenzene cyclic dimer from *p*-xylene monomer; the other is the preparation of poly-*p*-xylene by cyclic dimer. The *p*-xylene cyclic dimer or its substituted derivatives are the monomers used to produce different types of poly(*p*-xylene), and there are two usual methods used to prepare cyclic dimer: one is obtained by the Hofmann elimination reaction of (*p*-methylbenzyl)trimethylammonium hydroxide [67]; the second is the Gorham synthesis, which is based on *p*-xylene, which is pyrolyzed with water vapor at atmospheric pressure (950–1000 °C), then the pyrolysis gas is cooled to 150–400 °C, absorbed with xylene or other solvents, and the dimer is obtained after concentration and cooling, and white crystals are obtained after recrystallization in xylene [68, 69], which is characterized by simplicity and high yield. The cyclic dimer of *p*-xylene ($C_{16}H_{16}$) is sublimated to a gaseous cyclic dimer in the solid state at 150 °C under vacuum and cleaved to an active intermediate *p*-dimethylene benzene near 650 °C, which agglomerates on the solid surface at a vacuum less than 1.33×10^2 Pa and a temperature below 300 K, while auto-polymerization occurs and is accompanied by crystallization [70]. The specific process is shown in Figure 3.10.

3.3.8 Liquid Crystal Polymer

Liquid Crystal Polymer is abbreviated as LCP, and the basic structure can be expressed as (–linked group–liquid crystal group–linked group–spacer group–), where the liquid crystal group includes more than two benzene rings, which are generally rigid chain groups; the linking group can be ester, ether, or carbonate, etc.; and the spacer group is generally a flexible chain group, such as methylene

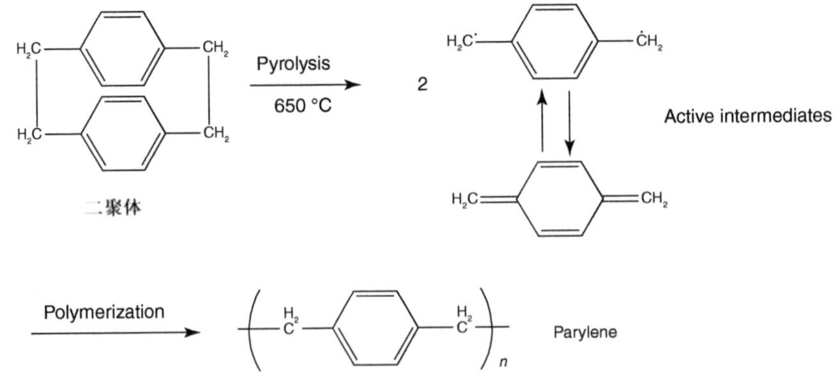

Figure 3.10 Synthesis route of poly-*p*-xylene.

Table 3.3 Classification and molecular structure formula of liquid crystal polymers.

Type	Characteristic	Molecular structure formula
I	*p*-Hydroxybenzoic acid + biphenyl rigid structure copolymerization	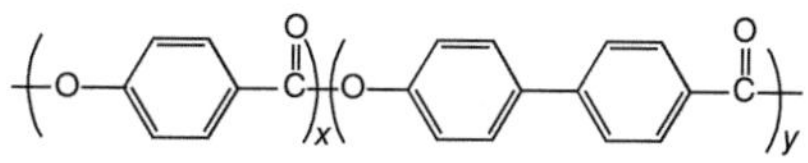
II	*p*-Hydroxybenzoic acid + naphthalene ring copolymerization	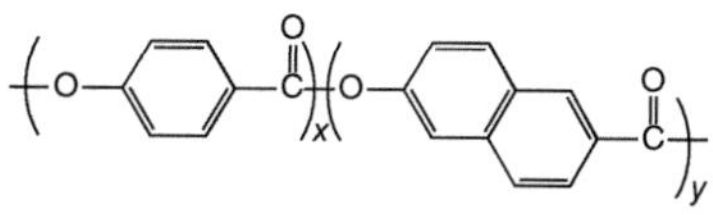
III	*p*-Hydroxybenzoic acid + copolymerization of aliphatic ring structure	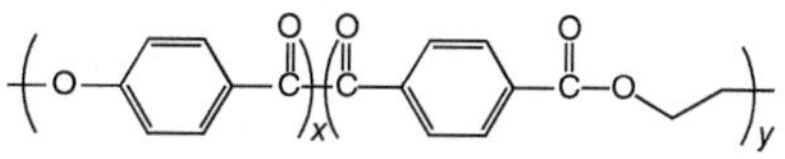

and oxyethylene bond. It can be divided into three categories, and the structural formula is shown in Table 3.3. After being melted by heat or dissolved by solvent, it forms a liquid crystal state with the properties of both solid and liquid parts, and the molecular arrangement is between liquid and crystal with a certain order (one-dimensional or two-dimensional), which is a new polymer material developed in the 1980s. According to the origin of the substance, it can be divided into natural liquid crystal polymers and synthetic liquid crystal polymers, and according to the conditions of liquid crystal formation, it can be divided into thermogenic liquid crystal polymers and solvogenic liquid crystal polymers [71, 72]. LCP has high strength, a high modulus ($E \approx 10\,\text{GPa}$), outstanding high temperature resistance ($T_{\text{deg}} \approx 380\,°\text{C}$), hot and cold alternating properties, excellent flame retardancy, corrosion resistance, wear resistance, and electrical properties. It is known as a super engineering material because of its excellent overall properties, such as flame retardancy, corrosion resistance, wear resistance, electrical properties, moisture absorption (<0.04 wt%), molding and processing properties, small linear coefficient of thermal expansion (CTE) and friction coefficient, high dimensional stability, radiation resistance, and microwave resistance [73]. They can be used as encapsulation materials in microelectronic devices and as substrate materials in flexible bioelectronic devices, for example, in neural interfaces and retinal prostheses [74].

3.3.9 Hydrogel

Hydrogel is a class of extremely hydrophilic three-dimensional network-structured gels that swell rapidly in water and, in this swollen state, can hold a large volume of water without dissolving. They are usually prepared by chemical or physical crosslinking of water-soluble or hydrophilic macromolecules, which can be classified into two categories: natural and synthetic, according to their origin. Natural hydrophilic polymers include polysaccharides (starch, cellulose, alginate, hyaluronic acid, chitosan, etc.) and peptides (collagen, poly-L-lysine, poly-L-glutamic acid, etc.). Synthetic hydrophilic polymers include alcohols, acrylic acids, and their derivatives (polyacrylic acid, polymethacrylic acid, polyacrylamide, poly-*N*-polyacrylamide, etc.) [75]. Their aggregate state structures are neither completely solid nor completely liquid; solids exhibit that they can maintain a certain shape and volume under certain conditions, and liquids exhibit that solutes can diffuse or permeate from the hydrogel. Because of its high water content, which is similar to the molecules of natural soft tissues, it has a wide range of biomedical applications, including electronic skin, cardiovascular tissue engineering, and implantable or wearable biosensors for human health monitoring. However, it is also due to the high content of water hydrogel films and the corresponding low percentage of macromolecules that they have a low elastic modulus and poor mechanical properties, which limit their application in flexible electronic devices [76]. Approaches to develop strong hydrogels include supramolecular polymer networks [77], nanoclay hydrogel composites [78], and other combinations of interpenetrating networks [79]. Although these strategies enable the development of hydrogel materials with promising bioelectronic applications, their use as structural materials in the practical application of flexible bioelectronic devices is not widely adopted, and there is still much room for development.

3.4 Molding Technology of Flexible Substrate

Flexible substrates are essential for the base material of flexible electronic devices and can be obtained through molding processes. These processes include the coating method, melt extrusion cast method, melt extrusion blown film method, solution cast method, and bidirectional stretching method. Among them, the coating method is popular due to its relatively low equipment cost and high stock utilization. The melt extrusion method can produce flexible substrates quickly with balanced cross and longitudinal properties. The melt extrusion blown film method and bidirectional stretching method produce substrates with uniform thickness and small weight tolerance. The solution method is suitable for preparing thin, flexible substrates with a thickness of 5–8 μm. Selecting the appropriate preparation process is crucial in practice for meeting the performance requirements of different application scenarios.

3.4.1 Coating Technology

The coating process has the advantages of simple processing principles, relatively low equipment costs, and high slurry utilization, and can therefore be used for the preparation of flexible substrate films. There are various coating methods, and the best coating method can be selected according to the characteristics and application requirements of each product in each industry.

3.4.1.1 Dip Coating Method

The dip coating method is an early and widely used coating method, due to its simple equipment structure. Easy to promote the application. As early as the end of the

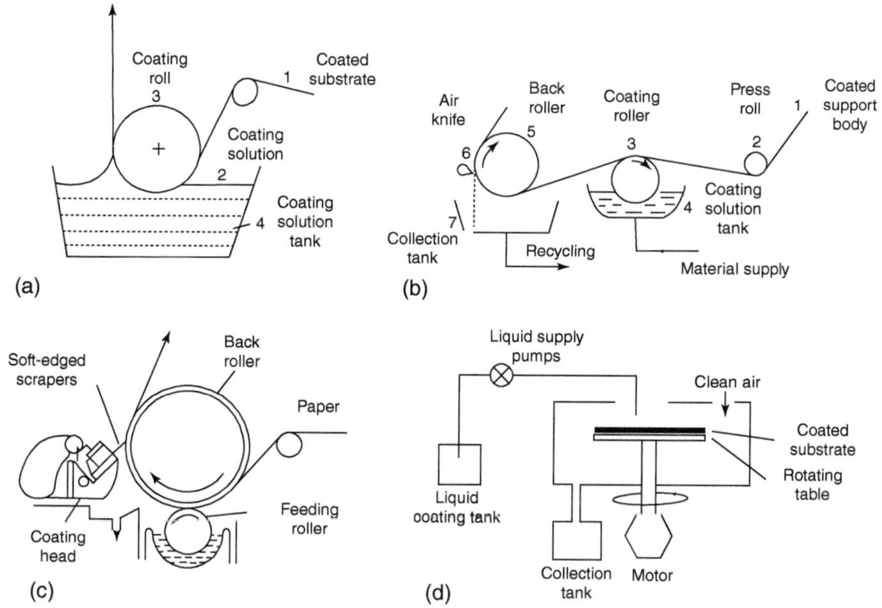

Figure 3.11 The diagram of the working principle of the coating method. Source: Adapted from https://www.researchgate.net/figure/Principleset-up-of-blade-coating-with-an-applicator-roll_fig5_271492399.

nineteenth century, dip coating and hanging rod drying were used in the continuous production of photographic paper. The principle of the dip coating method is shown in Figure 3.11a. The coated substrate 1 goes around the coating roll 3 into the coating solution tank 4 and comes into contact with the coating solution 2, and the coating adheres to its surface as it is pulled upward to form a coating. PDMS films are generally prepared by the dip coating method. The polysiloxane (solution) is coated and attached to the carrier film by the flow casting method or dip coating method through the coupling reaction between the large PDMS end-active groups and the curing agent (platinum), which is in contact with air; at a certain temperature, the polysiloxane layer formed on the carrier film is cross-linked and cured into a film; after the cross-linking, the resulting polysiloxane film can be separated from the said carrier film to obtain the PDMS siloxane film.

3.4.1.2 Air Knife Coating Method

The principle of the air knife coating method is shown in Figure 3.11b. The coated support body 1 is carried by the coating roller 3 after the press roll 2, and the material in the coating tank 4 is carried by the coating roller 3, and the excess coating is blown down in the collection tank 7 by the airflow from the air knife 6 at the back roller 5. After the air–liquid separation in the collection tank, the returned coating can be sent back to the coating tank for recycling. The coating quality of air knife coating is directly related to the uniformity and stability of the airflow distribution, i.e. it is closely related to the structural design and processing accuracy of the air knife. The coating volume of air knife coating is related to the air pressure, jet angle, and distance between the air knife and the original paper, which are also the process parameters that can be properly regulated in actual operation.

3.4.1.3 Scraper Coating Method

Scraper coating is used to remove the excess coating liquid with a specially set scraper to achieve the required coating volume. Compared to air knife contactless coating. It is a direct contact coating method with high flatness of the coating, and its coating surface is not affected by the original surface roughness of the coated support body. The principle of the squeegee coating method is shown in Figure 3.11c. The car speed of flexible scraper coating can reach 30–800 mm/min, and the solid content of the coating can be as high as 50–60%. The coating volume can be adjusted in the range of 6–30 g/m^2, and the process can be adapted to a wide range.

3.4.1.4 Rotary Coating Method

The rotary coating is to spread the liquid material added to the surface of the support body with the help of centrifugal force generated by high-speed rotation to form a coating of the required thickness. The rotational speed of the support body can be regulated from a few hundred rpm to tens of thousands of rpm. The principle of the rotational coating method is shown in Figure 3.11d. When applying coating solutions with lower viscosity and solid content, the coating thickness can range from a few tens of nanometers to a few microns, with a thickness uniformity of up to a hundred meters to a few microns and a thickness uniformity of up to one-hundredth of a percent. The rotary coating is commonly used to coat various polymer solutions, and the thickness of the coating depends on the viscosity (solid content) of the coating solution, the speed of rotation, and the evaporation rate of the coating solution solvent.

3.4.2 Melt Extrusion Molding

The melt film forming method, also known as the melt extrusion method, involves melting the resin and then casting or calendaring it into film through extrusion, followed by a transformation process to obtain the final product. The equipment principle for this method is illustrated in Figure 3.12. This process is ideal for producing films with uniform thickness, high performance, and high production efficiency, making it suitable for batch production of thick films. Moreover, the production process is environmentally friendly, as there is no need for the use of solvents. However, the process cannot be used for the production of thin films due to its inherent characteristics.

3.4.3 Melt Extrusion Blow Molding

The melt extrusion blown film process involves first extruding the molten thermoplastic polymer through a tubular die. This results in an amorphous polymer tube that is released from the die. Under internal air pressure, the tube forms an air pocket or blown film, which is then deflated to create a flat film. This process is illustrated in

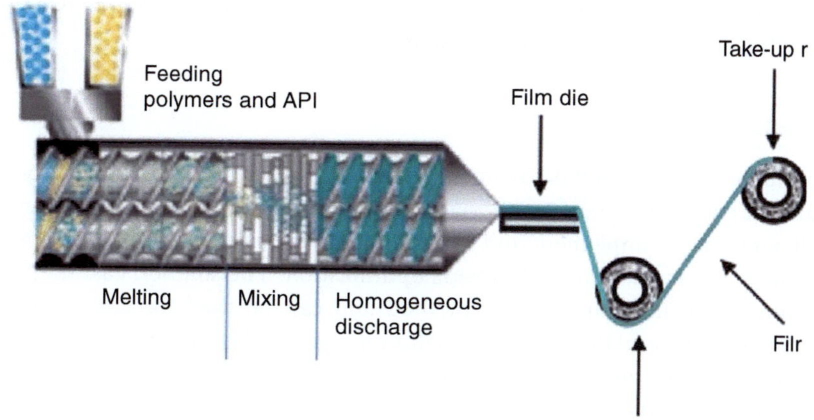

Figure 3.12 The schematic diagram of melt extrusion casting equipment. Source: ResearchGate GmbH / www.researchgate.net/figure/Schematic-of-hot-melt-extrusion-of-films_fig6_265174657 / last accessed 02 August 2023.

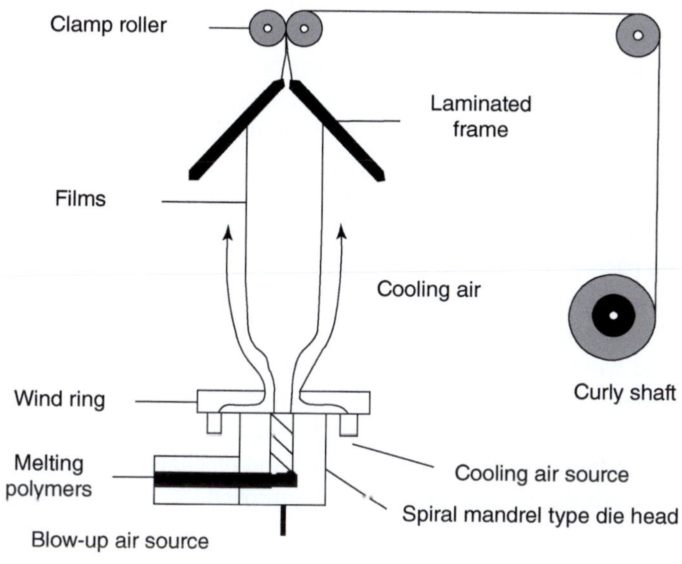

Figure 3.13 Working principle of the melt extrusion blown film method. Source: Adapted from Dongguan Baoding Precision Instrument Co.

Figure 3.13 and is commonly used for preparing flexible base films of PC, LCP, and PVA. Additionally, the extrusion blown film method (dry method) is used for manufacturing PVA films and is readily available from traditional plastic film production. The specific process includes mixing raw materials such as PVA, water, plasticizers, and lubricants (plus pigments) with forced circulation. The mixture is then extruded and granulated, melt-extruded and blown, modified, and finally rolled to obtain the final PVA film. For PC, the PC masterbatch is melted, and the PC flexible substrate is obtained using blow molding processing methods, typically involving film blowing with a slightly lower molecular weight polymer.

3.4.4 Solution Tape Casting

The solution casting method is to directly cast the polymer solution in a flat mold and make the solvent evaporate at a certain temperature to form a film. This method is simple and easy to implement and is mainly used for basic laboratory research, pre-commercial formulation, and process optimization. The solution cast method is an extension of the solution cast method and can be used for large-volume continuous production; therefore, the solution cast method is mostly used for commercialized products. Compared with the melt extrusion method, the process is longer, more complex, and requires solvent recovery, but it has the advantage of better performance and thinner film thickness.

PDMS, PVA, and LCP flexible substrate film preparation will use this method; manufacturing PVA film solution cast process: PVA raw materials and a certain amount of water, after two steps of cast coating, drying (heat treatment), and peeling and winding, can be obtained as PVA film finished products. The process is very simple and easy to operate and is an efficient production method. However, product quality is not easily controlled, film yield is low, energy consumption is high, and equipment investment is large. PI film preparation also adopts the solution calendering method, which is mainly related to the synthesis method of PI resin, because the PI resin can be synthesized in a non-protonic polar solvent, and the synthesized PI solution or intermediate polyamide acid (PAA) solution can be directly calendered into film. Specifically, the PAA solution is evenly flowed onto a continuously running metal belt and passes through a drying chamber with the movement of the belt to evaporate part of the solvent, and the partially dried PAA film can be peeled off from the belt, then imidized at high temperature by the action of a heated roller, and then quenched and wound to obtain a continuous length of the film. Such a process is called calendering. In the process of calendering, the solvent has to be evaporated, and the price of the solvent is high, which requires an additional solvent recovery system to reduce the cost. The uniformity and mechanical properties of the films produced by this method are significantly improved.

3.4.5 Bidirectional Drawing Molding

The bidirectional stretching method is to stretch the thick film or cast sheet longitudinally and transversely when it is below the melting point and above the glass transition temperature of the film material, and then to perform appropriate cooling or heat-setting treatment or special processing (such as corona and coating) under tension. PET, PU, and PI can all use bidirectional stretching process to prepare films of more uniform thickness. Specifically, PET is made into thick flakes by extrusion, and then the film material is made by biaxial stretching. Polyurethane resin prepared by synthesis is mixed with other additives, such as wear-resistant additives, antioxidants, and lubricants, and extruded by single-screw extruder. It is then melted, cooled, and biaxially stretched to obtain polyurethane film. PET, PU, and PI can all use the biaxial stretching process to prepare films of more uniform thickness. Specifically, PET is made into thick flakes by extrusion, and then the film material is made by biaxial stretching. The synthetically prepared polyurethane resin is mixed with other additives, such as abrasion-resistant additives, antioxidants, and lubricants, and then extruded by melt extrusion with a single-screw extruder, cooled, and biaxially stretched. The biaxial stretching method involves stretching the thick film or cast sheet longitudinally and transversely below the melting point of the film material and above the glass transition temperature, followed by appropriate cooling or heat-setting treatment or special processing (such as corona and coating) under tension. PET, PU, and PI can all be used in a bidirectional stretching process to prepare films of more uniform thickness. Specifically, PET is made into a thick sheet by extrusion and then a film material made by biaxial stretching. The polyurethane resin prepared by synthesis is mixed with other additives, such as wear-resistant additives, antioxidants, and lubricants, and extruded by melting in a single-screw extruder, cooled, and biaxially stretched to obtain polyurethane film. As for PI, the flow-drawing method is generally used. The film produced by the flow-drawing method (biaxially oriented method) has significantly improved performance, but the process is complicated with harsh production conditions, large investment, and high product price, which can obtain high-quality film products with high dimensional stability and low moisture absorption. The biaxially oriented cast and stretch method includes resin synthesis, cast, biaxially oriented, winding, and other processes. The flow chart of the biaxial stretching process is shown in Figure 3.14 to obtain polyurethane films. As for PI, the calendering-stretching method is generally used. The film produced by the calendering-stretching method (biaxially oriented method) has significantly improved performance, but the process is complicated with harsh production conditions, large investment, and high product price, which can obtain high-quality film products with high dimensional stability and low moisture absorption. The biaxially oriented cast and stretch method includes resin synthesis, casting, biaxially oriented, winding, and other processes. The flow diagram of the biaxial stretching process is shown in Figure 3.14, and the flow-drawing method is generally used. The film produced by the flow-drawing method (biaxially oriented method) has significantly improved performance, but the process is complicated with harsh production conditions, large investments, and high product prices, and high-quality film products such as high dimensional stability and low moisture absorption can be obtained.

3.4.6 Chemical Vapor Deposition

Chemical vapor deposition (CVD) is a technique that uses heating, plasma excitation, or optical radiation to introduce two or more gaseous raw materials into

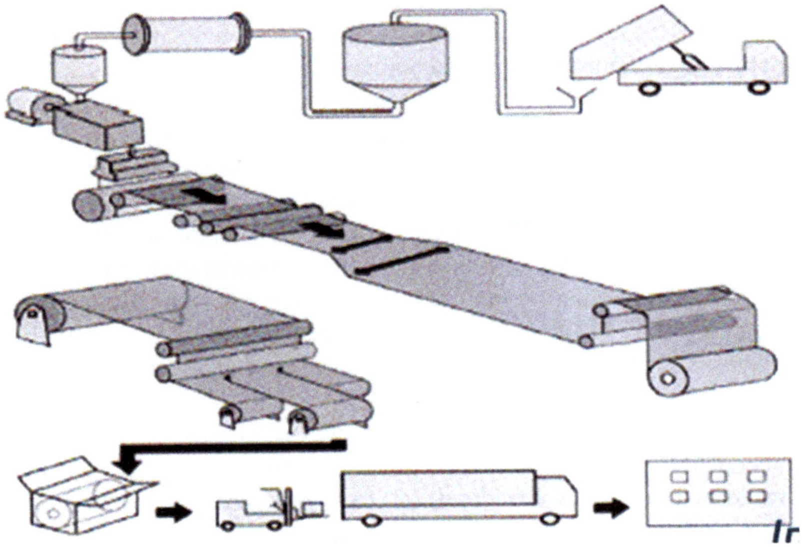

Figure 3.14 Process flow diagram for preparation of biaxially oriented stretch PI film. Source: Dongguan Mingda Adhesive Products Co., Ltd / http://www.mingdatape.com.cn / last accessed 02 August 2023.

a reactor and chemically react to form a new material that is deposited on the substrate surface. The CVD process is divided into three important stages: diffusion of the reaction gas to the substrate surface, adsorption of the reaction gas on the substrate surface, chemical reaction on the substrate surface to form a solid deposit, and the resulting gas phase by-products off the substrate surface. The most common CVD reactions are thermal decomposition reaction, redox reaction, chemical synthesis reaction, and chemical transport reaction. According to the reaction occurrence conditions, they can be classified as atmospheric pressure chemical vapor deposition (APCVD), low-pressure chemical vapor deposition (LPCVD), ultrahigh vacuum chemical vapor deposition (UHVCVD), laser chemical vapor deposition (LCVD), and plasma enhanced chemical vapor deposition (PECVD) [80]. In general, the CVD reaction temperature is too high, generally to be performed at 850–1100 °C, and many substrate materials cannot tolerate the high temperature of CVD. The deposition temperature can be reduced by using plasma or laser-assisted techniques. A common method to prepare poly(parylene) films (parylene) is vacuum vapor deposition. The vapor deposition method does not introduce solvents or catalysts in the intermediate process, and the films are free of impurities and internal stresses, pinholes, and film formation in the pores and seams. A typical device for the preparation of poly(parylene) films consists of three main parts: the cyclic dimer sublimation crucible, the high-temperature cracking chamber, and the deposition polymerization chamber. A schematic diagram of the system structure is shown in Figure 3.15. The deposition polymerization chamber can be equipped with an angle-adjustable or rotatable substrate holder or with a plasma generation device in which a specific gas is introduced, etc. [81].

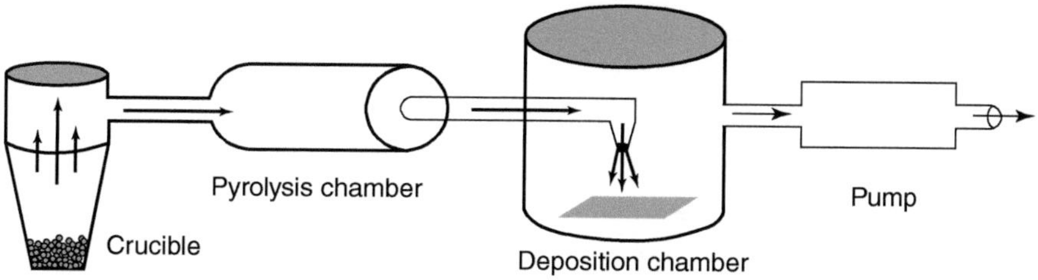

Figure 3.15 Schematic diagram of the system for the preparation of poly-*p*-xylene films by CVD. Source: http://www.mingdatape.com.cn/h-nd-43.html.

3.5 Performance Evaluation of Flexible Substrates

In flexible electronics, selecting substrates with superior performance is crucial to ensure proper functionality in specific application scenarios. Above all, a substrate's mechanical flexibility is its most crucial feature, yet maintaining functional and electronic performance parameters while retaining excellent mechanical flexibility is a great challenge. Substrates must also maintain strong adhesion and bonding properties during compression, tension, and torsion. Other important properties include thermal and chemical stability, surface smoothness, barrier properties, and optical clarity. For wearable electronic devices that come into contact with the human body, biocompatibility and bioabsorbability are essential considerations, and excellent biocompatibility ensures safe long-term use of the substrate material. In practical applications, substrate materials and preparation processes can be tailored to the specific performance requirements of the application scenario at hand.

3.5.1 Mechanical Flexibility

Mechanical flexibility refers to the "tolerance" of a material to various modes of mechanical deformation. For flexible electronics, the flexible substrate should deform at high strain levels while maintaining functional properties and electronic performance parameters. Flexible substrates exhibit inherent ductile plastic behavior to prevent mechanical fracture beyond their elastic limits. As ductility increases, the strain limit increases, resulting in an increase in the flexibility and elongation of the substrate. Therefore, the flexibility and ductility of flexible electronic device systems can be achieved by combining the mechanical requirements of low a Young's modulus, high strain, and high elastic limit of the substrate material. Among other requirements, the material stiffness is usually expressed in terms of Young's modulus, which is defined as the ratio of stress to strain in a linear elastic system with uniaxial deformation. Without deteriorating the mechanical and electrical properties, lower modulus values result in a wider strain range and thus exhibit desirable mechanical flexibility. In addition, as the thickness of polymeric plastics such as PET and PI substrates decreases, the modulus also decreases. The main flexible substrates are ultrathin glass, metal foil, and polymeric materials, of which polymeric materials have greater possibilities for batch production than the other

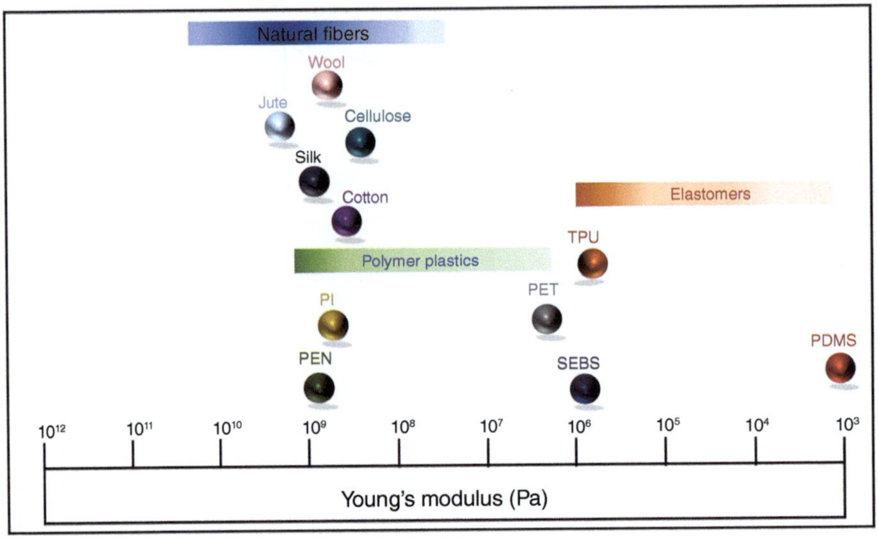

Figure 3.16 Young's modulus of flexible substrate materials (unit: Pa). Source: Reproduced with permission from Hassan et al. [34]; © 2021, John Wiley & Sons.

two due to the efficient and economical roll-to-roll (roll-to-roll) processing that can be applied to the preparation of polymeric substrates. The roll-to-roll process has significant winding tension, which makes low-modulus polymer substrates susceptible to internal deformation during processing and production, thus affecting film quality. This is a clearer requirement that the modulus of flexible substrate materials should not be considered solely in terms of application but also in terms of molding and processing.

When dealing with flexible systems, it is important for the substrate material to have a low modulus, allowing the electrode to be highly deformable above the plane while retaining significant tensile properties even after maximum doping. In substrate materials for flexible display devices, there are quantitative targets for flexibility, with a 1000-bend capability on a 1-in. diameter pipe being the norm. In the case of flexible electronic devices designed for different application scenarios, flexibility measurement varies. Young's modulus serves as an intuitive physical quantity that can serve as an initial indicator of the substrate material's flexibility. In Figure 3.16, the range of Young's modulus for commonly used substrate materials for flexible and wearable systems is shown [82]. Moreover, when processing substrates as electrodes for flexible and stretchable applications, the end stress of the electrode directly correlates with the Young's modulus of the substrate that encapsulates the conductive material. This highlights the significance of selecting and optimizing the substrate material, material structure, and processing method when aiming to enhance flexibility in relation to the elastic modulus.

3.5.2 Ductility

For flexible electronic materials, having a highly ductile flexible substrate can be advantageous in ensuring optimal electrode performance even when subjected to significant strains that cause edge failures. Ductility, as a property of electrodes, refers to their ability to maintain electrical conductivity despite mechanical deformation. In quantifying ductility, the focus is on the critical strain, which is the point at which conductive mode transitions to nonconductive mode. In discussions relating to flexible systems, this pertains to an electrode's capability to withstand the strain caused by the object's movement or deformation. Currently, electrodes can achieve a malleability of up to 1300%, thanks to different elastic components, such as polymeric elastomers like PDMS and other co-blocked polymers [83, 84]. Other structures such as polymeric hydrogels, sponges, and fiber structures exhibit excellent tensile properties.

3.5.3 Adhesive Property

The adhesion force is commonly assessed by measuring the force or energy exerted on the constituent materials during their separation. Due to differences in the physicochemical properties of flexible substrate materials and metals, adhesion can be limited, particularly between flexible substrate materials and membrane electrodes [85]. In flexible electronic devices, achieving strong adhesion between the substrate material and the deposited film is essential to withstand high external loading. Poor bonding can lead to device malfunction caused by energy input, which triggers material separation. Surface energy is typically the determining factor in material adhesion, which depends on the polarity and dispersion of the substrate material. Since the average surface energy of most adhesives is 34–36 dyn/cm^2, to improve adhesion, the surface energy of the substrate material should be 7–10 dyn/cm^2 higher than the surface energy of the adhesive material. In addition, during a strong bond between layers, bond damage (i.e. damage that occurs at the interface) occurs before cohesive damage (the overall material may be damaged individually) occurs.

To ensure strong adhesion between substrate and conductive materials in flexible implantable devices, two methods are commonly employed: chemical bonding and physical interlocking. Although chemical bonds can be unreliable, physical interlocking is effective, particularly when a transition layer is used to meet adhesion requirements and device stability. One successful approach is to create a composite interlocking layer through the use of nanoparticles (NPs). However, this technique can present adhesion challenges when attempting to disperse metal NPs on the gel/polymer. Hence, achieving excellent adhesion properties is crucial to ensure the reliable performance of flexible electronic devices.

3.5.4 Barrier Property

For flexible electronics, its components are in a complex environment of gases, liquids, and molecules during use. The infiltration of small molecules such as water vapor and oxygen within the flexible device can reduce the performance and sensitivity of the device due to environmental humidity, which can lead to a decrease in device life. In the case of wearable devices, it can also cause skin damage. For OLEDs with a lifetime >10 000 hours, the WVTR and OTR should

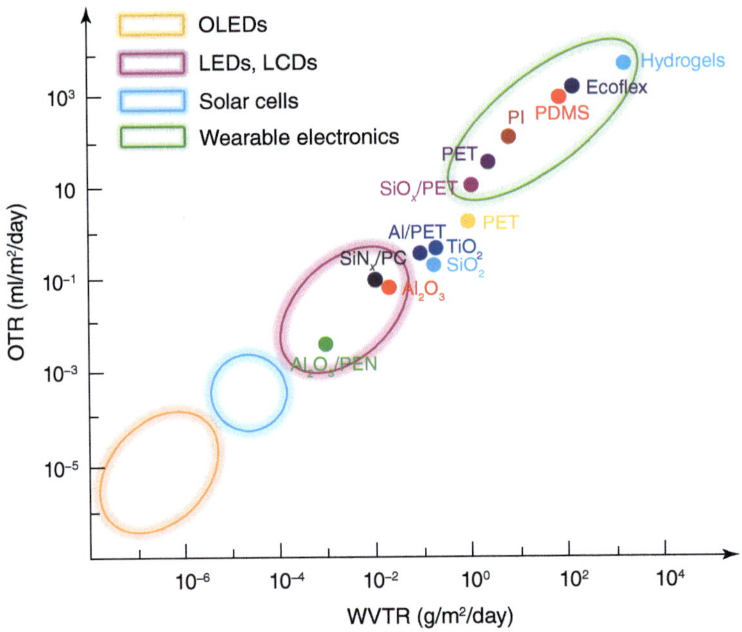

Figure 3.17 WVTR and OTR performance indicators required for flexible electronics. Source: Adapted from Zhang et al. [86].

not exceed 10^{-6} g/m^2/day and 10^{-5} ml/m^2/day, respectively. Figure 3.17 shows the water and oxygen barrier performance requirements for flexible electronic devices and the water and oxygen barrier performance of common substrate materials. Many hydrophobic polymer substrates, including PET, PDMS, and silicone rubber. Among them, PDMS is widely used, but its sealability in water vapor decreases with time. To solve this problem, hydrophobic elastomers such as *p*-(poly(styrene-isoprene-styrene)) with excellent barrier properties have entered the vision of researchers [87]. The long chains and larger pore sizes of stretchable polymeric substrates give them higher permeability to oxygen and water compared to small-molecule substrates [86]. Therefore, the modification of flexible substrate materials with chemical methods or structural treatments has become increasingly desirable, as these treatments can lead to significantly improved barrier properties. In recent years, various methods have been developed both domestically and internationally to capitalize on the hydrophilicity of flexible substrate materials. Interestingly, when thinking about these methods from a reverse perspective, all of them actually contribute toward enhancing the waterproofing properties of flexible substrate materials. Furthermore, achieving a superhydrophobic, flexible substrate can lead to even better device performance and sensitivity. Coating a barrier layer onto a flexible substrate represents a highly effective means of obtaining a substrate with exceptional barrier properties. In particular, poly(parylene) coatings can be produced through CVD, providing a perfect fit for a range of complex shapes and delivering a uniform and complete barrier layer [88].

3.5.5 Electrical Property

Flexible electronic packaging substrate materials, which generally have a lower dielectric constant (ε_r), are more suitable for high-frequency signal transmission than ceramic substrate materials [41] and are therefore more adapted to the development trend of high-speed signal. Nevertheless, with the encapsulation of high-speed circuit technology and the increase in signal transmission frequency, higher requirements are placed on the reduction of the dielectric constant of organic packaging substrate materials. For flexible electrode materials, the flexible substrate is required to have a high permittivity, which is generally in the range of 10^2–10^7 S/cm for different application scenarios.

3.5.6 Chemical Stability

The production procedure of flexible electronic devices typically calls for the use of solvents or chemicals that may negatively impact the flexible substrate material. To counteract this issue, it is necessary for the substrate material to be capable of tolerating acidic, alkaline, and other low boiling point solvents. Additionally, the creation of a substrate material barrier serves to protect it from chemical damage. As a viable solution, a chemical hard coating applied to one side of the substrate material can provide ample protection and prevent the infiltration of harmful substances into the substrate material [89]. For instance, a PC flexible substrate that features an acrylic silicone coating exhibits notable enhancements in both solvent resistance and resistance to contamination. This modified PC substrate is capable of withstanding a range of solvents, such as acetone, toluene, carbon tetrachloride, methanol, ethanol, butanol, gasoline, kerosene, 10% ammonia, and 5% phenol [90]. The corrosion of PC substrates by organic solvents such as toluene and tetrahydrofuran can be significantly reduced if the surface of PC is modified with polyethyleneimine coating [91].

3.5.7 Dimensional Stability

In the world of flexible electronics, the high-temperature stability of the substrate material plays a critical role in both the manufacturing process and application environment. When it comes to flexible display devices, the temperature required to produce thin-film transistors (TFTs) on flexible substrates typically surpass 300 °C, while the operating temperature can exceed 150 °C. This indicates the importance of maintaining thermal dimensional stability at temperatures of up to 300 °C for the flexible substrate. Put simply, the reliability and quality of flexible devices heavily hinge on the dimensional stability of the flexible substrate material [41]. To optimize the thermal and dimensional stability, a tolerable temperature difference ($|\Delta\text{CTE} - \Delta T| \leq 0.1$–$0.3\%$) is used to measure the high-temperature dimensional stability of flexible substrates, where the variation of the CTE between the substrate material and the device layer and the temperature deviation during fabrication are denoted by ΔCTE and ΔT, respectively. For flexible electronic devices, the in-plane linear CTE of the electronic material deposited on top of the flexible substrate is generally less than 20 ppm/°C. In order to match the thermal expansion behavior of the substrate material with the subsequently deposited layers, the substrate material should also meet this requirement [92]. Furthermore, any temperature disparity between the device film and substrate material can cause the substrate material to bend in a convex manner. This could lead to multiple issues in the device fabrication process, such as cracking and delamination. Thus, to avoid thermal cracking due to thermal cycling, it is imperative to develop device films with the ability to endure thermal expansion mismatches.

3.5.8 Surface Smoothness and Thickness Uniformity

In order to prevent negative impacts on the performance of flexible electronic devices, the surface weave and morphology of the substrate material must be carefully controlled. The substrate surface should be ultrasmooth with a roughness of only 1 nm × rms and no local surface anomalies (such as spikes) larger than a few tens of nanometers. The quality of flexible substrate surfaces is especially crucial for more demanding flexible display applications, as subsequent fabrication of conductive or barrier layers can create defects or debris that protrude from the surface. Although minor roughness is acceptable over long ranges, unevenness and roughness should be avoided over short ranges. The surface quality of flexible substrates is affected by two factors: First, the inherent surface smoothness, which heavily depends on the internal cleanliness of the film substrate and is usually characterized by peaks tens to hundreds of nanometers in size detected by white light interferometry. Second, surface cleanliness, such as dust and surface scratches. These unavoidable defects can be tens of microns in lateral or longitudinal dimensions for substrate films that are handled and cut in a non-cleanroom environment. A certain percentage of dust particles can be removed by surface cleaning, but to eliminate these types of defects, it is necessary to apply a flat coating, which is preferably done in a cleanroom environment. These coatings have a dual function, providing both surface smoothness and a degree of surface hardness that helps prevent scratching during film treatment [93]. In general, metallic substrate materials exhibit both short-range and long-range surface roughness properties, while polymeric substrate materials may exhibit only long-range surface roughness properties [94]. The excellent surface smoothness and thickness uniformity ensure that the surface is free of artifacts and thus does not interfere with the fabrication of organic TFTs on top of flexible substrates.

3.5.9 Optical Clarity (Transmittance)

Optically transparent flexible substrate materials are crucial for low birefringence transmissive or bottom-emitting displays, particularly liquid crystal displays (LCDs). In flexible display applications, the necessary transmittance should exceed 85% in the wavelength range of 400–800 nm, with haze values less than 0.7%.

Table 3.4 Light transmittance of common flexible substrate materials.

Materials	Thickness (μm)	Transmittance (%)
Glass	100	>90 [99]
PEN	100	87.02 [92]
PI	100	30–60 [92]
PET	100	90.42 [92]
PDMS	1.5	82.0 [100]

For instance, polyester films are excellent for improving the optical transparency of LCDs and are applicable as brightness enhancement films. Furthermore, amorphous polymer films have lower birefringence values and are considered more fitting as optically transparent flexible substrate materials for LCDs [87]. Many metal-containing PDMS composites exhibit high transparency in the visible region, including PDMS/(Au NWs/Ag NWs) with 86% transparency [95], Ag NWs/PDMS with 88.3% transparency [96], and Ag NPs/PDMS with 86% transparency [97], in addition to PU/(NPs) showing 77% transparency [98]. Flexible substrate materials with different optical clarity can be selected for flexible electronic devices in different application scenarios, and the light transmission of common flexible substrate materials is shown in Table 3.4.

3.5.10 Biocompatibility

The biocompatibility of a material refers to its ability to elicit an appropriate host response in a given situation. In the case of wearable devices, the biocompatibility of the flexible substrate material is particularly critical since it comes into direct contact with the biological interface. Factors such as surface charge, chemical composition, and pH can induce cytotoxicity and reduce biocompatibility. Although numerous factors affect biocompatibility, PDMS, cellulose substrates, and silk-based textile substrates are examples of biocompatible materials. Natural materials are also suitable for flexible substrate materials to improve biocompatibility in electronic devices. Materials such as chitin, cellulose, and silk, which can yield biomimetic nanofiber polymers with a natural biological repetitive structure, also hold promise for use in the area of flexible substrate materials with biocompatible properties [101]. Bacterial cellulose (BC) has recently been reported as a biocompatible fiber for wearable health monitoring applications [102].

3.5.11 Bioabsorbability

The idea of bioresorbable substrate materials that dissolve and break down toxic residues in the environment was first introduced in 2009 as a means to mitigate chemical and physical harm to the environment. Currently, this concept is being applied to integrate bioresorbable substrate materials into the design of flexible wearable electronics [103]. Researchers have further enhanced the dependent compatibility of devices with the human body by using bioresorbable silk substrate materials that can be used for deposition or implantation inside or on the surface of the brain [104], and many more in-depth studies are needed in developing biocompatible and bioresorbable substrate material devices for the purpose of device integration with the human body.

3.6 Application of Flexible Substrates

There has been a noteworthy boost in research and development concerning the use of flexible substrates, highlighting their exceptional versatility and potential applications. These substrates have found utility in a range of fields, including but not limited to, flexible electrodes for electronics, energy devices such as supercapacitors, batteries, and solar cells, as well as sensors and other electronic devices, such as photovoltaics, transistors, and display devices.

3.6.1 Flexible Display Substrates

Compared to traditional displays, flexible displays offer numerous advantages, including being thin, flexible, foldable, portable, and unbreakable. The emergence of flexible displays is transforming the design of electronic products and having a far-reaching impact on the future of human–computer interaction. Organic light-emitting diode (OLED) displays have several advantages over LCDs, including being all-solid-state, offering high brightness, high contrast, high response, active light-emitting, and having no perspective limitations. As a result, OLED displays are gradually replacing LCD as an important form of flat panel display [105, 106]. Flexible OLED displays are characterized by thinness, impact resistance, easy embedding, and bendability, which are important development directions for OLED displays at present [62]. There are three major types of substrate materials used in flexible display devices: ultrathin glass, metal foil, and polymer substrates. While ultrathin glass offers excellent potential for flexible displays, the technical difficulties associated with creating a thin, flexible glass with good toughness and impact resistance is a significant hurdle to overcome given the brittle nature of nonmetallic inorganic materials. As for metal foil, it boasts high temperature resistance and can withstand semiconductor processing temperatures. However, its high surface roughness of Ra 0.6 μm poses challenges for device processing, as it can easily lead to processing failure. Nevertheless, the material itself does not suffer from issues with blocking water vapor and oxygen, making it an attractive option for high-temperature applications [107]. The development of ultrathin glass and metal foil as substrate materials for OLED devices has been hampered by their inherent weaknesses in fabrication. Alternatively, polymer substrates, although possessing shortcomings related to temperature resistance and barrier properties, are organic polymers that can have their properties optimized through molecular

Figure 3.18 Flexible display substrate materials. Source: Perelaer et al. [106]; © 2012, Reproduced with permission from John Wiley & Sons, Inc.; Zyung et al. [105]; © 2005, Institute of Electrical and Electronics Engineers Inc.; Hassan et al. [34]; © 2021, Reproduced with permission from John Wiley & Sons, Inc.

structure design and preparation technique adjustment. As such, they represent a viable option for flexible OLED displays. Looking ahead, new process technologies such as printing or roll-to-roll processes could help reduce manufacturing costs and lead to mass production (Figure 3.18).

3.6.2 Flexible Electrode Substrates

Optoelectronic display devices, like LCDs and OLEDs, usually require expensive indium tin oxide (ITO) films that are deposited on glass substrates to act as transparent conductive electrodes. Indium is a rare material that drives up the cost of producing ITO films, and they require high-temperature processes to be deposited on glass substrates for good quality. However, the main drawback of ITO films is their rigidness, which makes them unsuitable for use in flexible electronic devices [92]. To ensure signal stability during transmission in flexible electronic devices, it is crucial for flexible electrode materials to maintain conductivity at normal usage levels, as well as during device deformation and stretching. In recent years, substantial progress has been made in the research of flexible electrode substrate materials. For example, Sekitani [108] mixed single-walled carbon nanotubes as well as imidazole ionic liquid into flexible fluoropolymer rubber, and after a jet milling process, the three were uniformly mixed to obtain a conductive elastomer, and the conductivity was maintained at 102 S/cm at an elongation of 118%. However, it is clear that the elongation as well as the conductivity of this conductive elastomer is relatively low, which is not conducive to practical applications. To address this issue, Lee [109] blended the conductive polymer PEDOT:PSS with the block polymer PEO_{20}-PPO_{70}-PEO_{20} and successfully prepared a flexible composite substrate material with high electrical conductivity (~1700 S/cm) and

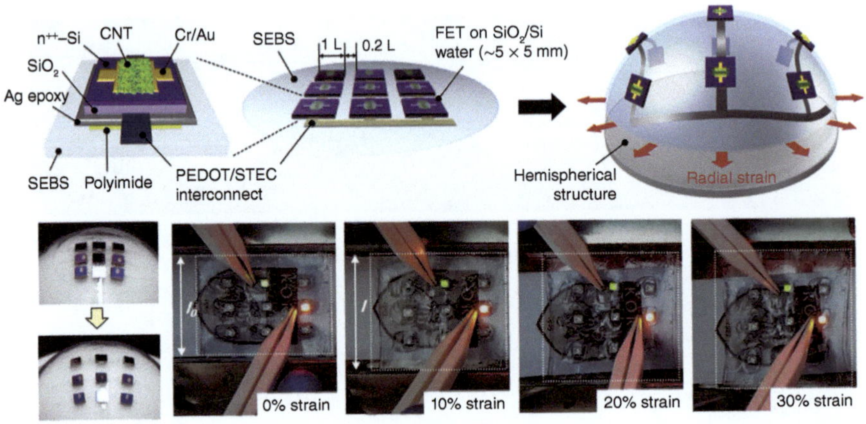

Figure 3.19 Flexible electrode substrate material. Source: Reprinted with permission from Lee et al. [109]; © 2018, American Chemical Society.

exhibited only 4% resistance over 1000 stretch-release cycles Bao's group [110] mixed PEDOT:PSS with ionic organic molecules to obtain a polymer film with a conductivity of more than 3100 S/cm and a strain at break of 800% by improving the chemical bonding within the conducting polymer as well as the molecular composition, and after 1000 cycles (100%). In this field, to meet the future needs of flexible electronics, the most difficult research task is to achieve stable resistance and maintain high conductivity under large strains to obtain flexible electrode substrates with promising applications. Designing from a molecular perspective and combining it with condensed matter structure studies will help to achieve this goal (Figure 3.19).

3.6.3 Flexible Sensing Substrates

Flexible sensing substrates are an important component in electronic skins, and it is clear that flexible sensing substrate materials need to have not only excellent mechanical properties but also specific sensing properties. To meet this need, flexible sensing substrate materials can be designed with micro and nanostructures [111–113] to enhance the response of the signal upon deformation. For example, Kang [64] deposited a 20 nm platinum film on a flexible urethane acrylate substrate and prepared a microcrack structure similar to that on a spider's foot by stretching, resulting in a sensing coefficient of more than 2000 for the entire flexible device at a tensile strain of 0–2%; Liu [114] mixed carbon nanotubes and other additives into a natural latex system, which, due to the film-forming properties of natural latex The resulting three-dimensional self-isolated conductive pathway inside the prepared flexible conductor, which is extremely sensitive to stretching, achieves an average sensing coefficient of 100 within 40% strain. Miniaturized smart sensor modules that can be applied in the medical field, fabricated by conventional flexible circuit films and standard assembly processes and equipment (Figure 3.20).

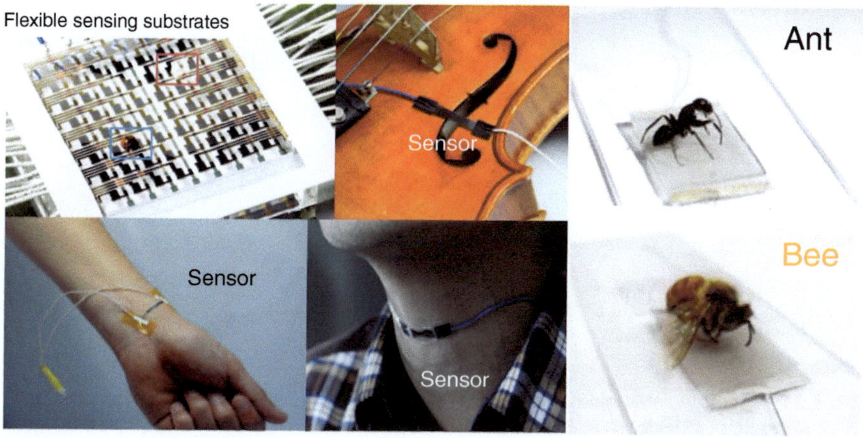

Figure 3.20 Flexible sensing substrate materials. Source: Reprinted with permission from Wang et al. [112]; © 2013, Reproduced with permission from John Wiley & Sons, Inc.

3.7 Development Trend of Flexible Substrates

In recent years, the flexible electronics market has experienced rapid growth, driven by its diverse range of applications, including foldable smartphones, automotive, medical, smart home, Internet of Everything, artificial intelligence, and big data. Flexible electronics not only has a larger market than microelectronics but also has the potential to enhance the value-added of traditional industries, revolutionizing industrial structure and human life. This has led flexible electronics to be known as the new infrastructure of "steel and cement." At present, polymer thin film materials are the primary source of flexible substrates, and functional polymer materials have expanded the application of flexible substrates. Polymer functional materials not only enable traditional electronics to achieve flexibility and ductility but also provide flexible substrates with different functionalities, making them an indispensable part of flexible electronic materials. Furthermore, they drive the advancement of flexible electronics in areas such as biomedical, flexible display, energy storage and conversion devices, displays, soft robotics, and human–computer interaction. However, with the ongoing development of flexible electronics technology, expectations for flexible electronics have increased. The future development trend of flexible substrates, as an integral part of flexible electronics, is closely related to the promotion and application of flexible electronics [115].

3.7.1 Intelligent and Functional Flexible Substrates

In the development of flexible, malleable, and functionalized electronic devices, polymer functional materials have emerged as a critical source of substrate materials. These materials play an essential role in advancing the intelligence and functionality of these devices. To meet the increasing demands for flexible electronic devices, the development of a new generation of polymer functional

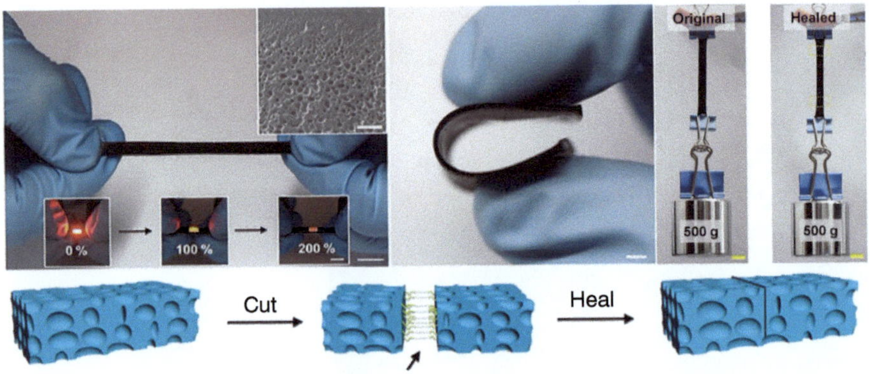

Figure 3.21 Flexible self-healing substrate material. Source: Reprinted with permission from Wang et al. [118]; © 2018, Reproduced with permission from John Wiley & Sons, Inc.

materials has become necessary. These materials offer significant potential for functionalizing flexible electronic devices and are characterized by their mechanical properties as well as additional features like electrical conductivity, electromagnetic properties, catalytic properties, and bioactivity. Meanwhile, flexible polymeric substrate materials, such as PI, polyester, and silicone rubber, are known for their mechanical and electrical properties, including ease of processing, flexibility, ductility, suitable modulus, and insulation [43, 116, 117]. Despite their many advantages, these materials still have certain limitations that must be addressed. For example, once processed, they are unable to change shape, and there is a risk of damage like small cracks in flexible electrode materials and sensing materials during processing or use. Such damage can cause degradation and even lead to a loss of performance. To cope with this scientific problem, research on self-healing flexible sensing materials inspired by the self-healing ability of human skin [118], as shown in Figure 3.21, has gradually become the focus of research in the field of electronic skin in recent years [119]. At present, the main role of flexible substrate materials is only as device substrates, and the realization of intelligent and functional substrate materials will greatly expand the application areas of flexible electronics.

3.7.2 Green Degradable Flexible Substrates

Flexible electronics still encounter numerous problems and challenges when it comes to practical applications. With the widespread use of electronic devices, the issue of electronic pollution is inevitably on the rise [120, 121]. The use of functionalized flexible substrate materials for the controlled degradation of flexible electronic devices can significantly improve the environmental sustainability of the technology. In turn, this can allow for the expansion of applications of flexible electronics in various areas, including bioelectronics and smart electronics. As a result, the development of flexible, degradable substrate materials based on functional polymer materials has emerged as a crucial research focus in the field

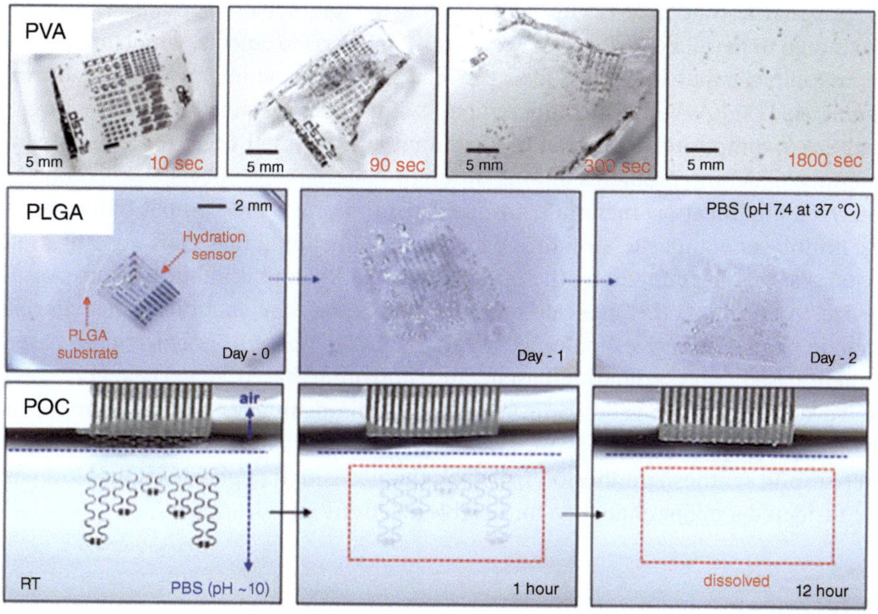

Figure 3.22 Dissolution behavior of water-soluble polymer substrates. Source: Reprinted with permission from Fu et al. [122]; © 2016, Reproduced with permission from American Chemical Society.

of flexible electronics in recent years. The functional polymers that are widely used in degradable substrate materials can be classified into three categories according to their material properties [122]: natural polymers, water-soluble polymers, and sub-stable polymers. Cellulose nanofiber (CNF) is a natural polymer material derived entirely from wood with high transparency [123] and flexibility, as well as desirable electrical properties [124]. Jung [125] investigated the microwave RF properties of high-performance electronic devices based on biodegradable CNF substrates and the fungal biodegradation process. The development of flexible electronic devices based on biofriendly substrates needs a long exploration (Figure 3.22).

3.7.3 Optimization of Interface Compatibility of Flexible Substrates

To address the rapid development of flexible electronics for applications such as biomedical, flexible display, and human–computer interaction, there is a need for flexible substrates with unique properties such as high transparency, dimensional stability, thermal stability, barrier properties, solvent resistance, low coefficient of thermal expansion (LCTE), as well as smooth surfaces. In some cases, a conductive layer might be necessary for specific structures. However, very few single-polymer substrate films can meet all of these complex requirements. Therefore, the development of polymer flexible film substrates with multilayer composite structures represents a practical solution to fulfill these essential demands [93]. Meanwhile, the mismatch that may occur between the flexible substrate material and the functional material deposited on it can result in interface defects, which are caused by modulus mismatch. Such defects can lead to device spalling and damage during actual use [126]. Whether it is the structure of the flexible substrate that features a multilayer composite, or the electronic circuit layer printed on it, there are always interface matching problems among different materials. To alleviate this problem, interface modification materials are needed to improve the compatibility within the multilayer composite structure between the functional materials and substrate materials, thereby enhancing the operational stability and durability of the overall device. At present, the most commonly used interfacial modification materials include some adhesive materials such as acrylic-based, polyurethane-based, protein-based, polysaccharide-based, and dopamine-based materials [127–129]. However, existing adhesion materials still have disadvantages such as poor performance and single function, as well as being expensive. Therefore, finding a way to prepare and synthesize adhesion materials that have high applicability and low cost is considered a major challenge for flexible electronic applications.

References

1 Huang, W. (2021). The development of flexible electronic technology to accelerate the pace of future industrial development. *Digital Economy* Z2: 15–18.

2 Wallenberger, F.T. and Bingham, P.A. (2010). *Fiberglass and Glass Technology: Energy-Friendly Compositions and Applications*. New York: Springer.

3 Xu, L., Gutbrod, S.R., Bonifas, A.P. et al. (2014). 3D multifunctional integumentary membranes for spatiotemporal cardiac measurements and stimulation across the entire epicardium. *Nature Communications* 5: 1–10.

4 Russo, A., Ahn, B.Y., and Adams, J.J. (2011). PenonPaper flexible electronics. *Advanced Materials* 23: 3426–3430.

5 Ostfeld, A.E., Gaikwad, A.M., Khan, Y., and Arias, A.C. (2016). High-performance flexible energy storage and harvesting system for wearable electronics. *Scientific Reports* 6: 10.

6 Menard, H., Mammerickx, J., and Macdonald, K.C. (2015). References and notes. *Science* 350: 313–316.

7 Zhang, Y.J. (2012). Study on properties of common flexible substrate films. *Henan Science and Technology* 6: 75.

8 Li, W., Meredov, A., and Shamim, A. (2019). Coat-and-print patterning of silver nanowires for flexible and transparent electronics. *Flexible Electronics* 3: 7.

9 Bellani, S., Petroni, E., Martín-García, B. et al. (2019). Scalable production of graphene inks via wet-jet milling exfoliation for screen-printed micro-supercapacitors. *Advanced Functional Materials* 14: 1807659.

10 Kusaka, Y., Kawamura, T., Nakagawa, M. et al. (2021). Fabrication of extremely conductive high-aspect silver traces buried in hot-embossed polycarbonate films via the direct gravure doctoring method. *Advanced Powder Technology* 32: 764–770.

11 Beedasy, V. and Smith, P.J. (2020). Printed electronics as prepared by inkjet printing. *Materials*, Multidisciplinary Digital Publishing Institute 13 (3): 1–23.

12 Biswas, T., Yu, J., and Nierstrasz, V. (2021). Effective pretreatment routes of polyethylene terephthalate fabric for digital inkjet printing of enzyme. *Advanced Materials* 8: 1–10.

13 Zhou, L., Yu, M., Chen, X. et al. (2018). Screen-printed poly(3,4-ethylene-dioxythiophene): poly(styrene sulfonate) grids as ITO-free anodes for flexible organic light-emitting diodes. *Advanced Functional Materials* 28 (11): 1705955.

14 Zhang, Y.-Z., Wang, Y., Cheng, T. et al. (2019). Printed supercapacitors: materials, printing and applications. *Chemical Society Reviews* 1–36.

15 Hyun, D.W.J., Secor, E.B., Hersam, M.C. et al. (2015). High-resolution patterning of graphene by screen printing with a silicon stencil for highly flexible printed electronics. *Advanced Materials* 27: 17.

16 Siddik, A.B., Billah, M.M., Lee, S. et al. (2021). Blue laser annealed sub-micron channel P-type low temperature poly-Si TFT without kink effect for high-resolution display. *IEEE Electron Device Letters* 42 (2): 172–175.

17 Knapp, C.E., Chemin, J.-B., Douglas, S.P. et al. (2018). Room-temperature plasma-assisted inkjet printing of highly conductive silver on paper. *Advanced Materials Technology* 3: 1–6.

18 Duan, S., Gao, X., Wang, Y. et al. (2019). Scalable fabrication of highly crystalline organic semiconductor thin film by channel-restricted screen printing toward the low-cost fabrication of high-performance transistor arrays. *Advanced Materials* 31: 1–8.

19 Singh, S.B., Singh, T.I., Kim, N.H., and Lee, J.H. (2013). Core-shell MnO_2@Au nanofibers network as high-performance flexible transparent supercapacitor electrode. *Journal of Materials Chemistry A* 7: 14.

20 Hussain, F., Jeong, J., Park, S. et al. (2020). Fabrication and characterization of a novel terpolyester film: an alternative substrate polymer for flexible electronic devices. *Polymer* 210: 123019.

21 Yang, W., Li, N.-W., Zhao, S. et al. (2017). A breathable and screen-printed pressure sensor based on nanofiber membranes for electronic skins. *Advanced Science and Technology Letters* 3: 7.

22 Suikkola, J., Björninen, T., Mosallaei, M. et al. (2016). Screen-printing fabrication and characterization of stretchable electronics. *Scientific Reports* 6: 8.

23 Xu, S., Vogt, D.M., Hsu, W.-H. et al. (2018). Biocompatible soft fluidic strain and force sensors for wearable devices. *Advanced Functional Materials* 7: 14.

24 Lv, Z., Luo, Y., Tang, Y. et al. (2017). Editable Supercapacitors with customizable Stretchability based on mechanically strengthened ultralong MnO_2 nanowire composite. *Advanced Materials* 30: 1–9.

25 Pu, X., Liu, M., Chen, X. et al. (2017). Ultra stretchable, transparent triboelectric nanogenerator as electronic skin for biomechanical energy harvesting and tactile sensing. *Science Advances* 3: 11.

26 Ding, Y., Xu, T., Onyilagha, O. et al. (2019). Recent advances in flexible and wearable pressure sensors based on piezoresistive 3D monolithic conductive sponges. *ACS Applied Materials & Interfaces*, American Chemical Society 11 (7): 6685–6704.

27 Zhang, Y.-Z., Wang, Y., Cheng, T. et al. (2015). Flexible supercapacitors based on paper substrates: a new paradigm for low-cost energy storage. *Chemical Society Reviews* 44, 19.

28 Tobjörk, D. and Österbacka, R. (2011). Paper electronics. *Advanced Materials* 23: 27.

29 Fujisaki, Y., Koga, H., Nakajima, Y. et al. (2014). Transparent nanopaper based flexible organic thin film transistor array. *Advanced Functional Materials* 24: 1657–1663.

30 Matsuhisa, N., Kaltenbrunner, M., Yokota, T. et al. (2015). Printable elastic conductors with a high conductivity for electronic textile applications. *Nature Communications* 6: 11.

31 Kim, W., Kwon, S., Lee, S.M. et al. (2013). Soft fabric-based flexible organic light-emitting diodes. *Organic Electronics* 7: 3007–3013.

32 Liu, Y., Qi, N., Song, T. et al. (2014). Highly flexible and lightweight organic solar cells on biocompatible silk fibroin. *ACS Applied Materials & Interfaces* 6: 1–22.

33 Zhu, B., Wang, H., Leow, W.R. et al. (2015). Silk fibroin for flexible electronic devices. *Advanced Materials* 28: 16.

34 Hassan, M., Abbas, G., Li, N. et al. (2022). Significance of flexible substrates for wearable and implantable devices: recent advances and perspectives. *Advanced Materials Technologies* 7 (3): 2100773.

35 Hammock, M.L., Chortos, A., Tee, B.C.-K. et al. (2013). 25th anniversary article: the evolution of electronic skin (E-skin): a brief history, design considerations, and recent progress. *Advanced Materials* 25 (42): 5997–6038.

36 Lacour, S.P., Jones, J., Wagner, S. et al. (2005). Stretchable interconnects for elastic electronic surfaces. *Proceedings of the IEEE* 93 (8): 1459–1467.

37 Kim, D.-H., Ahn, J.-H., Choi, W.M. et al. (2008). Stretchable and foldable silicon integrated circuits. *Science* 320 (5875): 507–511.

38 Tee, B.C.-K., Wang, C., Allen, R., and Bao, Z. (2012). An electrically and mechanically self-healing composite with pressure- and flexion-sensitive properties for electronic skin applications. *Nature Nanotechnology* 7 (12): 825–832.

39 Yu, Z., Niu, X., Liu, Z., and Pei, Q. (2011). Intrinsically stretchable polymer light-emitting devices using carbon nanotube-polymer composite electrodes. *Advanced Materials* 23 (34): 3989–3994.

40 Kaltenbrunner, M., White, M.S., Głowacki, E.D. et al. (2012). Ultrathin and lightweight organic solar cells with high flexibility. *Nature Communications* 3 (1): 770.

41 Shi, J.Y. (2018). Analysis on the design and development of packaging substrate. *Proceedings of the 19th China Copper Clad Panel Technology Seminar*, Kunshan, Jiangsu. China: China Electronic Materials Industry Association Copper Clad Plate Material Branch, China Electronic Circuit Industry Association Substrate Materials Branch. pp. 29–44. 2018.10.25–2018.10.27.

42 Tao, J., Wang, R., Yu, H. et al. (2020). Highly transparent, highly thermally stable Nanocellulose/polymer hybrid substrates for flexible OLED devices. *ACS Applied Materials & Interfaces* 12: 9701–9709.

43 Spechler, J.A., Koh, T.-W., Herb, J.T. et al. (2015). A transparent, smooth, thermally robust, conductive polyimide for flexible electronics. *Advanced Functional Materials* 25 (48): 7428–7434.

44 Forrest, S.R. (2004). The path to ubiquitous and low-cost organic electronic appliances on plastic. *Nature Materials* 428 (29): 911–918.

45 Xie, Z., Hung, L.-S., and Zhu, F. (2003). A flexible top-emitting organic light-emitting diode on steel foil. *Chemical Physics Letters* 381: 6.

46 Nagashima, K., Koga, H., Celano, U. et al. (2014). Cellulose nanofiber paper as an ultra flexible nonvolatile memory. *Scientific Reports* 4: 1–7.

47 Schauer, S., Liu, X., Worgull, M. et al. (2015). Shape-memory polymers as flexible resonator substrates for continuously tunable organic DFB lasers. *Optical Materials Express* 5 (3): 9.

48 Avendano-Bolivar, A., Ware, T., Arreaga-Salas, D. et al. (2013). Mechanical cycling stability of organic thin film transistors on shape memory polymers. *Advanced Materials* 25: 3095–3099.

49 Reeder, J., Kaltenbrunner, M., Ware, T. et al. (2014). Mechanically adaptive organic transistors for implantable electronics. *Advanced Materials* 26: 7.

50 Qi, D., Zhang, K., and Tian, G. (2021). Stretchable electronics based on PDMS substrates. *Advanced Materials* 33 (6): 2003155.

51 Tian, W.S., Cao, H.Y., Gao, J. et al. Research progress of wearable flexible sensor based on polydimethylsiloxane. *Analytical Chemistry* 50: 1–13.

52 Cheng, S., Wu, Z., Hallbjorner, P. et al. (2009). Foldable and stretchable liquid metal planar inverted cone antenna. *IEEE Transactions on Antennas and Propagation* 57 (12): 3765–3771.

53 Razavi, M., Primavera, R., Vykunta, A., and Thakor, A.S. (2021). Silicone-based bioscaffolds for cellular therapies. *Materials Science and Engineering: C* 119: 111615.

54 Zhang, M. and Ma, W.Z. (2022). Preparation and properties of polyvinyl alcohol hydrogel film for soilless cultivation. *Petrochemical Application* 41 (6): 102–107.

55 Guo, B., Ma, Z., Pan, L., and Shi, Y. (2019). Properties of conductive polymer hydrogels and their application in sensors. *Journal of Polymer Science Part B: Polymer Physics* 57 (23): 1606–1621.

56 Chu, J.Y., Wang, W.W., and Wang, H. (2022). Research progress of polycarbonate synthesis technology. *Liaoning Chemical Industry* 51 (8): 1114–1117.

57 Li, J.G. (2011). Synthesis, modification and application of polyethylene terephthalate. *Chinese and Foreign Entrepreneurs* 2: 150–151.

58 Liu, R.H., Bai, C.Y., and Qi, X. (2022). Progress in synthesis and application of copolyesters. *Synthetic Resins and Plastics* 39 (4): 92–94.

59 Ding, M.X. (2012). *Polyimide Chemistry, Structure and Properties of Relations and Materials*, 2e. Science Press.

60 Hasegawa, M. (2017). Development of solution-Processable, optically transparent polyimides with ultra-low linear coefficients of thermal expansion. *Polymers*, Multidisciplinary Digital Publishing Institute 9 (10): 520.

61 Ghosh, A., Sen, S.K., Banerjee, S., and Voit, B. (2012). Solubility improvements in aromatic polyimides by macromolecular engineering. *RSC Advances* 2 (14): 5900.

62 Yang, Z.H., Kang, C.Q., Guo, H.Q. et al. (2021). Synthesis and properties of polyimide substrate for flexible OLED display. *Journal of Polymers* 52 (10): 1308–1315.

63 Zeng, X.Y., Wen, X., and Zhang, W.L. (2021). Synthesis, preparation and properties of heat-resistant polyurethane. *Functional Materials* 52 (7): 7191–7196.

64 Kang, D., Pikhitsa, P.V., Choi, Y.W. et al. (2014). Ultrasensitive mechanical crack-based sensor inspired by the spider sensory system. *Nature* 516: 5.

65 Mandelli, J.S. and Koepp, J. (2021). Cell viability and cytotoxicity of inkjet-printed flexible organic electrodes on parylene C. *Biomedical Microdevices* 23 (1): 2.

66 Khodagholy, D., Gelinas, J.N., Thesen, T. et al. (2015). NeuroGrid: recording action potentials from the surface of the brain. *Nature Neuroscience*, Nature Publishing Group 18 (2): 310–315.

67 Zan, L.N. (2008). Progress in preparation and application of poly(*p*-xylene). *Chemical Propellants and Polymer Materials* 5: 22–25.

68 Lee, C. and Bassett, D.R. (1988). Process for the preparation of the parylene dimer. *Journal of Polymer Science Part A: Polymer Chemistry* 47 (6): 1–8.

69 Waters, J.F., Sutter, J.K., Meador, M.A.B. et al. (1991). Addition curing thermosets endcapped with 4-amino paracyclophane. *Journal of Polymer Science Part A: Polymer Chemistry* 29 (13): 1917–1924.

70 Zhang, H.Y. (1983). A review of poly(p-xylene). *New Chemical Materials* 4: 1–4.

71 Northolt, M.G. and Sikkema, D.J. (1992). Lyotropic main chain liquid crystal polymers. In: A. A. Collyer, *Liquid Crystal Polymers: From Structures to Applications*, 273–348. Dordrecht: Springer.

72 Noël, C. and Navard, P. (1991). Liquid crystal polymers. *Progress in Polymer Science* 16 (1): 55–110.

73 Zhong, D.X. (2004). Super engineering plastics-liquid crystal polymer (LCP). *Proceedings of the 8th Academic Conference on Connectors and Switches, Components Branch of the Chinese Electronics Society*, Xi'an, China. Xi'an, China: Connector and Switch Technology Department, Components Branch, Chinese Institute of Electronics, China Electronic Components Industry Association Electrical Plug-in Components Branch,. pp. 96–103. 2004.10.01

74 Balakrishnan, G., Song, J., Mou, C., and Bettinger, C.J. (2022). Recent progress in materials chemistry to advance flexible bioelectronics in medicine. *Advanced Materials* 34 (10): 2106787.

75 Gao, T.T., Chang, D., Liu, F.Q. et al. (2016). Research progress of organic composite hydrogels. *Functional Materials* 47 (S1): 72–77.

76 Sun, J.-Y., Zhao, X., Illeperuma, W.R. et al. (2012). Highly stretchable and tough hydrogels. *Nature*, Nature Publishing Group 489 (7414): 133–136.

77 Okumura, Y. and Ito, K. (2001). The polyrotaxane gel: a topological gel by figure-of-eight cross-links. *Advanced Materials* 13 (7): 485–487.

78 Gaharwar, A.K., Dammu, S.A., Canter, J.M. et al. (2011). Highly extensible, tough, and elastomeric nanocomposite hydrogels from poly(ethylene glycol) and hydroxyapatite nanoparticles. *Biomacromolecules* 12 (5): 1641–1650.

79 Haraguchi, K. and Takehisa, T. (2002). Nanocomposite hydrogels: a unique organic–inorganic network structure with extraordinary mechanical, optical, and welling/de-swelling properties. *Advanced Materials* 14 (16): 1120–1124.

80 Zhang, Z.W., Li, B., Wang, C.Y. et al. (2003). Preparation of poly(*p*-xylene) thin films by vapor deposition. *Material Guide* 6: 83, 86, 87.

81 Hu, Y.M. (2012). Preparation of poly(*p*-xylene) thin films and their application in organic light-emitting diodes. Fudan University.

82 Kazem, N., Hellebrekers, T., and Majidi, C. (2017). Soft multifunctional composites and emulsions with liquid metals. *Advanced Materials* 29 (27): 1605985.

83 Gao, Y., Yu, L., Yeo, J.C., and Lim, C.T. (2020). Flexible hybrid sensors for health monitoring: materials and mechanisms to render wearability. *Advanced Materials* 32 (15): 1902133.

84 Lim, H., Kim, H.S., and Qazi, R. (2020). Wearable flexible hybrid electronics: advanced soft materials, sensor integrations, and applications of wearable flexible hybrid electronics in healthcare, energy, and environment. *Advanced Materials* 32 (15): 2070116.

85 Huang, Y., Huang, Y., Meng, W. et al. (2015). Enhanced tolerance to stretch-induced performance degradation of stretchable MnO_2-based supercapacitors. *ACS Applied Materials & Interfaces* 7 (4): 2569–2574.

86 Zhang, Y., Wang, S., Li, X. et al. (2014). Experimental and theoretical studies of serpentine microstructures bonded to prestrained elastomers for stretchable electronics. *Advanced Functional Materials* 24 (14): 2028–2037.

87 MacDonald, W.A., Looney, M.K., and MacKerron, D. (2007). Latest advances in substrates for flexible electronics. *Journal of the Society for Information Display* 15 (12): 1075–1083.

88 Gluschke, J.G., Richter, F., and Micolich, A.P. (2019). A parylene coating system specifically designed for producing ultra-thin films for nanoscale device applications. *Review of Scientific Instruments* 90 (8): 083901.

89 Crawford, G.P. (ed.) (2005). *Flexible Flat Panel Displays*. Chichester, West Sussex, England/Hoboken, NJ: Wiley.

90 Zhang, X., Zhang, X., Yan, Y. et al. (2015). Adhesion and thermal stability enhancement of IZO films by adding a primer layer on polycarbonate substrate: adhesion and thermal stability enhancement of IZO films. *Physica Status Solidi A* 212 (4): 831–839.

91 Jankowski, P., Ogończyk, D., Lisowski, W., and Garstecki, P. (2012). Polyethyleneimine coating renders polycarbonate resistant to organic solvents. *Lab on a Chip*, The Royal Society of Chemistry 12 (14): 2580–2584.

92 Choi, M.-C., Kim, Y., Ha, C.S. et al. (2008). Polymers for flexible displays: from material selection to device applications. *Progress in Polymer Science* 33 (6): 581–630.

93 MacDonald, W.A., Looney, M.K., MacKerron, D. et al. (2008). Designing and manufacturing substrates for flexible electronics. *Plastics, Rubber and Composites* 37 (2–4): 41–45.

94 Su, N.-C., Wang, S.-J., Huang, C.-C. et al. (2010). Low-voltage-driven flexible in GaZnO thin-film transistor with small subthreshold swing. *Electron Device Letters* 31 (7): 680–682.

95 Duan, S., Wang, Z., Zhang, L. et al. (2018). Strain sensors: a highly stretchable, sensitive, and transparent strain sensor based on binary hybrid network consisting of hierarchical multiscale metal nanowires. *Advanced Materials Technologies* 3 (6): 1870020.

96 Lv, Z., Luo, Y., Tang, Y. et al. (2018). Stretchable supercapacitors: editable supercapacitors with customizable stretchability based on mechanically strengthened ultralong MnO_2 nanowire composite. *Advanced Materials* 30 (2): 1870008.

97 Feng, P., Ji, H., Zhang, L. et al. (2019). Highly stretchable patternable conductive circuits and wearable strain sensors based on polydimethylsiloxane and silver nanoparticles. *Nanotechnology* 30 (18): 185501.

98 Lee, D., Kim, J., Kim, H. et al. (2018). High-performance transparent pressure sensors based on sea urchin shaped metal nanoparticles and polyurethane micro dome arrays for real-time monitoring. *Nanoscale* 10 (39): 18812–18820.

99 Garner, S.M., He, M., Lo, P.Y. et al. (2012). Electrophoretic displays fabricated on ultra-slim flexible glass substrates. *Journal of Display Technology* 8 (10): 590–595.

100 Kumar, K.A. (2012). Studies on properties of bio-composites from ecoflex/ramie fabric-mechanical and barrier properties. *Journal of Biomaterials and Nanobiotechnology*, Scientific Research Publishing 3 (3): 396–404.

101 Ling, S., Kaplan, D.L., and Buehler, M.J. (2018). Nanofibrils in nature and materials engineering. *Nature Reviews Materials* 3 (4): 18016.

102 Wang, C., Wang, S., Chen, G. et al. (2018). Flexible, bio-compatible nanofluidic ion conductor. *Chemistry of Materials* 30 (21): 7707–7713.

103 Kim, D.-H., Kim, Y.-S., Amsden, J. et al. (2009). Silicon electronics on silk as a path to bioresorbable, implantable devices. *Applied Physics Letters* 95 (13): 133701.

104 Kim, T., McCall, J.G., Jung, Y.H. et al. (2013). Injectable, cellular-scale optoelectronics with applications for wireless optogenetics. *Science* 340 (6129): 211–216.

105 Zyung, T., Kim, S.H., Chu, H.Y. et al. (2005). Flexible organic LED and organic thin-film transistor. *Proceedings of the IEEE* 93 (7): 1265–1272.

106 Perelaer, J., Jani, R., Grouchko, M. et al. (2012). Plasma and microwave flash sintering of a tailored silver nanoparticle ink, yielding 60% bulk conductivity on cost-effective polymer foils. *Advanced Materials* 24 (29): 3993–3998.

107 Yang, L.Y., Yin, S.G., Hua, Y.L. et al. (2006). Substrate material and packaging technology of flexible display device. *Functional Materials* 1: 10–13.

108 Sekitani, T., Nakajima, H., Maeda, H. et al. (2009). Stretchable active-matrix organic light-emitting diode display using printable elastic conductors. *Nature Materials* 8: 1–6.

109 Lee, J.H., Jeong, Y.R., Lee, G. et al. (2018). Highly conductive, stretchable and transparent PEDOT:PSS electrodes fabricated with triblock copolymer additives and acid treatment. *ACS Applied Materials & Interfaces* 1–33.

110 Wang, Y., Zhu, C., Pfattner, R. et al. (2017). A highly stretchable, transparent, and conductive polymer. *Science Advances* 3: 1–10.

111 Mannsfeld, S.C.B., Tee, B.C., Stoltenberg, R.M. et al. (2010). Highly sensitive flexible pressure sensors with micro structured rubber dielectric layers. *Nature Materials* 9: 859–864.

112 Wang, X., Gu, Y., Xiong, Z. et al. (2013). Silk molded flexible, ultrasensitive, and highly stable electronic skin for monitoring human physiological signals. *Advanced Materials* 26: 7.

113 Fan, F.-R., Lin, L., Zhu, G. et al. (2012). Transparent triboelectric nanogenerators and self-powered pressure sensors based on micropatterned plastic films. *Nano Letters* 12 (6): 3109–3114.

114 Liu, X., Lu, C., Wu, X., and Zhang, X. (2017). Self-healing strain sensors based on nanostructured supramolecular conductive elastomers. *Journal of Materials Chemistry A* 19 (4): 20.

115 Wang, Z.Y., Wang, T., Zhuang, M.D. et al. (2019). Research progress of functional polymer materials in flexible electronics. *Journal of China University of Science and Technology* 49 (11): 878–891.

116 Zardetto, V., Brown, T.M., Reale, A., and Di Carlo, A. (2011). Substrates for flexible electronics: a practical investigation on the electrical, film flexibility, optical, temperature, and solvent resistance properties. *Journal of Polymer Science Part B-Polymer Physics* 49: 638–648.

117 Lu, X. and Xia, Y. (2006). Buckling down for flexible electronics. *Electronic Materials* 1: 163–164.

118 Wang, T., Zhang, Y., Liu, Q. et al. (2018). A self-healable, highly stretchable, and solution processable conductive polymer composite for ultrasensitive strain and pressure sensing. *Advanced Functional Materials* 28 (7): 1705551.

119 Yang, Y. and Urban, M.W. (2013). Self-healing polymeric materials. *Chemical Society Reviews* 42: 7446–7467.

120 Hsu, E., Barmak, K., West, A.C., and Park, A.H.A. (2018). Advancements in the treatment and processing of electronic waste with sustainability: a review of metal extraction and recovery technologies. *Green Chemistry* 1 (3): 21.

121 Kiddee, P., Naidu, R., and Wong, M.H. (2013). Electronic waste management approaches: an overview. *Waste Management* 33: 1237–1250.

122 Fu, K., Wang, Z., Dai, J. et al. (2016). Transient electronics: materials and devices. *Chemistry of Materials* 28: 3527–3539.

123 Fang, Z., Zhu, H., Yuan, Y. et al. (2014). Novel nanostructured paper with ultrahigh transparency and ultrahigh haze for solar cells. *Nano Letters* 14 (2): 765–773.

124 Zhu, H., Li, Y., and Fang, Z. (2014). Highly thermally conductive papers with percolative layered boron nitride nanosheets. *ACS Nano* 8 (4): 3606–3613.

125 Jung, Y.H., Chang, T.-H., Zhang, H. et al. (2015). High-performance green flexible electronics based on biodegradable cellulose nanofibril paper. *Nature Communications* 6 (1): 7170.

126 Meredith, H.J. and Wilker, J. (2015). The interplay of modulus, strength, and ductility in adhesive design using biomimetic polymer chemistry. *Advanced Functional Materials* 25: 5057–5065.

127 Wirthl, D., Pichler, R., Drack, M. et al. (2017). Instant tough bonding of hydrogels for soft machines and electronics. *Science Advances* 3: 1–10.

128 Li, Z., Zhang, R., Moon, K.-S. et al. (2013). Highly conductive, flexible, polyurethane based adhesives for flexible and printed electronics. *Advanced Functional Materials* 23: 1459–1465.

129 Vernengo, A.J. (2016). Adhesive materials for biomedical applications. *Adhesive Materials for Biomedical Applications* 6: 111–133.

4

Test Methods

Junjie Yuan

Peking University Shenzhen Graduate School, Lishui Road, 518055, Shenzhen, China

4.1 Sealing Test

The hermeticity of thin-film packaging materials is critical; they must ensure that all electronics are completely impermeable to the surrounding circulating biofluids. Even small holes, exposed areas, point defects, or pinholes can allow water-based solutions to penetrate the interior and compromise functional behavior and performance, leading to short circuits. Therefore, qualified thin-film encapsulation materials must meet ultralow water permeability (25 °C, water-vapor transmis-sion rate (WVTR) $<10^{-6}$ g/m^2/day) as well as achieve a life span of decades at certain physiological conditions of temperature and pH.

Many authors have now developed different systems to assess the airtightness of TFEs, especially the water permeation rate, which is dominant compared to oxygen uptake in most cases [1, 2]. WVTR is the most widely available metric to assess barrier airtightness since water is the main routinely measurable environmental factor contributing to package degradation. This is particularly true for devices in contact with body fluids and can occur through delamination, dissolution, hydrolysis, blistering, mass loss, erosion, and loss of adhesion. In recent years, due to the rapid development of ultrahigh barrier materials, new concepts have emerged that are capable of detecting ultralow permeation rates, where most traditional methods fail in terms of sensitivity. Depending on the measurement method, the test environment and the type of flexible coatings that may be tested differ [3–5]. In the first classification, two types of test methods can be distinguished. Direct methods provide parameters that are direct indices of permeability (WVTR, OTR, etc.) and are usually based on diffusion cells. Indirect methods provide a quantity that can be related to the barrier permeability since water permeation affects other physical quantities.

Flexible Electronic Packaging and Encapsulation Technology, First Edition.
Edited by Hong Meng and Wei Huang.

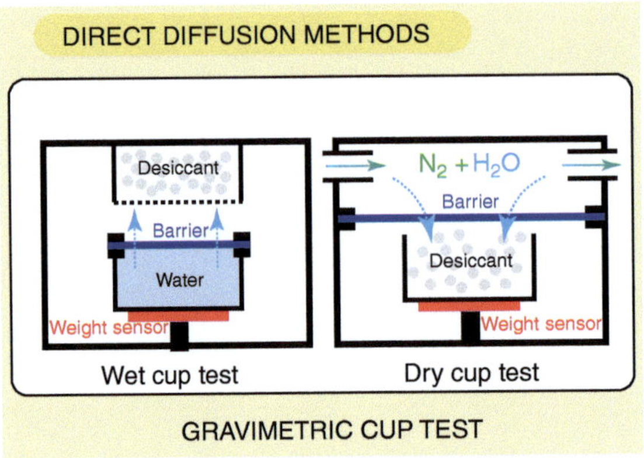

Figure 4.1 Schematic diagram of the weight cup test. Source: Herbst et al. [6] / John Wiley & Sons / CC BY 4.0.

4.1.1 Direct Diffusion Method

4.1.1.1 Weight Cup Test

The weight cup test (Figure 4.1) follows the ASTM E96 standard and is very effective in measuring the WVTR of polymer foils. An impermeable cup is filled with desiccant (or water) and covered with a barrier foil in order to quantify the permeation through the barrier based on the change in mass after a certain time interval [7–9]. In particular, for the desiccant method (or dry cup test), the specimen is sealed on the open side of the desiccant-filled test dish, and the assembly is kept in a controlled, humid environment. Periodic weighing determines the increase in water that passes through the barrier and is absorbed by the desiccant. Conversely, in the water method (or wet cup test), the test dish contains distilled water, so weighing determines the amount of water that has been reduced through the barrier into the surrounding environment. For both methods, the attachment of the specimen to the test dish is critical and has a direct impact on the accuracy of the test: usually performed using a wax seal.

4.1.1.2 Differential Pressure Method

The differential pressure (manometry) method is based on the direct pressure measurement caused by the diffusion of water through a barrier membrane; the vacuum vessel is divided into two chambers by the barrier membrane [10, 11]. One chamber (the feed chamber) is filled with the measured gas (water vapor), and the other chamber (the detection chamber) is under vacuum, so that over time the empty chamber gradually fills with permeated water vapor, which increases the absolute pressure (Figure 4.2). By measuring the pressure, the rate of permeation through the barrier can be determined.

4.1.1.3 Balancing Method

The equalization method (Figure 4.3), also known as the isobaric method, differs from the manometric method because the two chambers are filled to achieve the same pressure but with different gases: water vapor on one side and an inert gas on

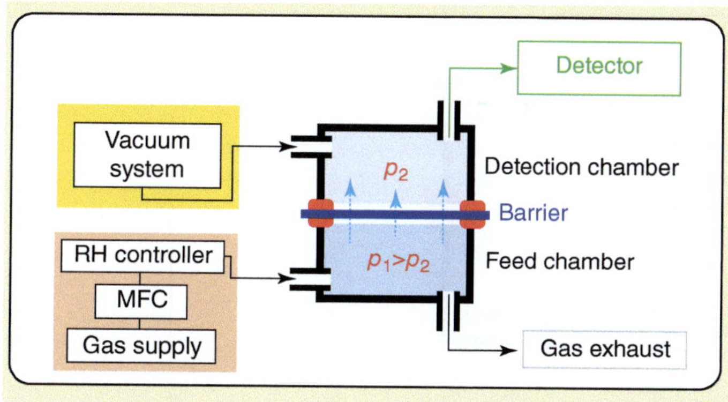

Figure 4.2 Schematic diagram of the differential pressure method of testing. Source: Herbst et al. [6] / John Wiley & Sons / CC BY 4.0.

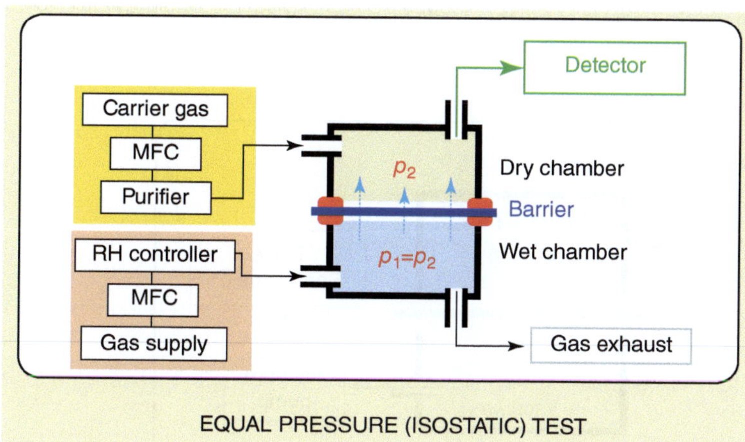

Figure 4.3 Equilibrium method test schematic. Source: Herbst et al. [6] / John Wiley & Sons / CC BY 4.0.

the other [12]. The inert carrier gas delivers the permeate water to a coulometric or IR sensor, which quantifies the amount of permeate water. The sensor is usually based on a phosphorus pentoxide (P_2O_5) desiccant, which electrolyzes the delivered water, and the resulting current is used to calculate the WVTR. This method is currently used in multiple machines, such as MOCON Aquatran [13] or Systech Illinois [14].

4.1.1.4 Tunable Diode Laser Absorption Spectrometry

An alternative method still based on a diffusion cell (with a dry nitrogen carrier gas stream) is the tunable diode laser absorption spectroscopy (TDLAS) method (Figure 4.4), which uses a laser diode spectroscopy gas sensor to evaluate WVTR [15, 16]. The detection principle is based on the measurement of the laser intensity decay caused by the excitation of permeable water molecules [5, 16].

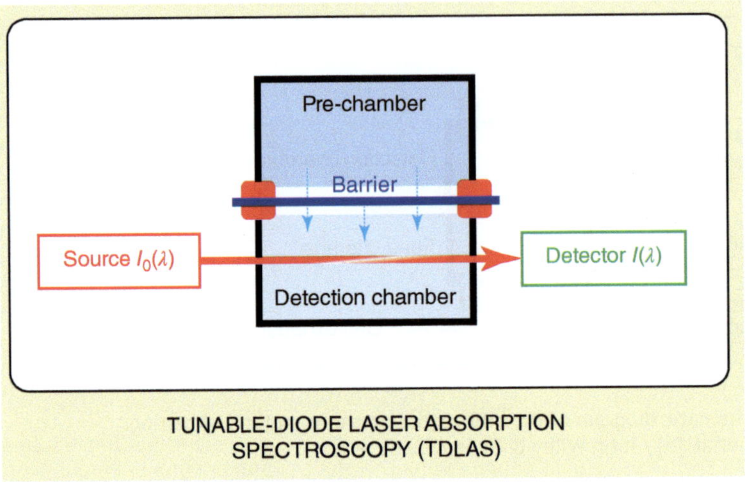

Figure 4.4 Schematic diagram of tunable diode laser absorption spectrometry testing. Source: Herbst et al. [6] / John Wiley & Sons / CC BY 4.0.

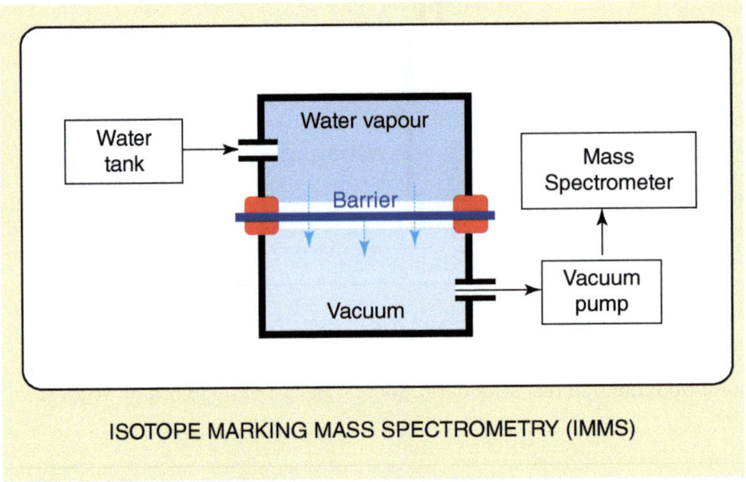

Figure 4.5 Schematic diagram of the mass spectrometry test. Source: Herbst et al. [6] / John Wiley & Sons / CC BY 4.0.

4.1.1.5 Isotope Labeling Mass Spectrometry

In the isotope-labeled mass spectrometry (IMMS) method (Figure 4.5), two gas chambers are separated by a barrier, one containing a high-pressure permeate gas and the other connected to a mass spectrometer under ultrahigh vacuum, with the aim of quantifying the mass of permeate water molecules [5, 17–20]. A constant concentration of the target gas is maintained in the first chamber and a quasi-zero concentration in the second chamber by means of a high-vacuum pumping system. Thanks to the mass spectrometer, several gases can be detected, including water and oxygen.

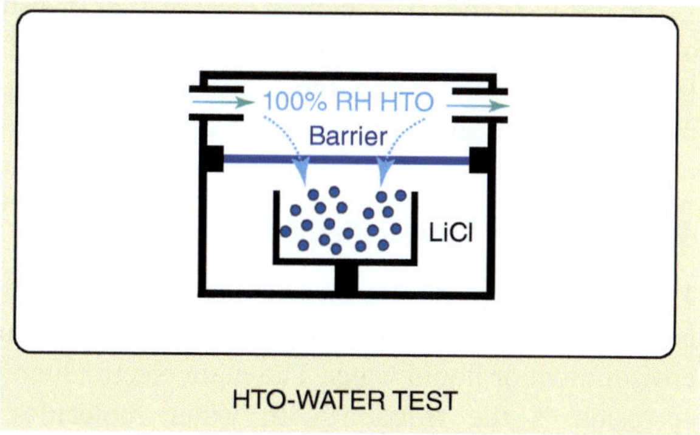

Figure 4.6 Schematic diagram of tritium water test. Source: Herbst et al. [6] / John Wiley & Sons / CC BY 4.0.

A special note must be made about the helium (He) leak detector method, a variant based on mass spectrometry that has proven to be an alternative to other more common methods for determining the moisture properties of barrier films [6, 21, 22]. One system for detecting helium leaks consists of two chambers separated by an isolation foil and evacuated with a vacuum pump at the same pressure level. Then, one chamber is filled with He, which permeates through the sample into the second chamber, where its pressure is detected by an appropriate device (leak detector) based on mass spectrometry. The WVTR can be derived from the helium transport rate (HeTR) by means of an appropriate correlation. The use of helium allows molecule-specific measurement systems to be faster than water vapor detection (due to the fact that helium atoms have a lighter and smaller diameter) and to be performed at lower pressure levels.

Another system is based on the radiotracer or tritium (HTO) test [23–25] (Figure 4.6). In fact, in this case, HTO-doped water is used as an indicator of permeation: two gas cells are separated by a barrier foil, one with 100% relative humidity (RH), including HTO water. The other cell purifies the permeated HTO water with a nonreactive, nonradioactive carrier gas up to the detector. The basic assumption is that the HTO atoms diffuse as molecules in HTO and that water and HTO permeate at the same rate. A vessel containing LiCl is used to absorb the HTO that permeates through the barrier membrane, and then the dissolved LiCl-based solution is placed in a liquid scintillation counter. The counts per minute (cpm) of the sample are determined by detecting the photons emitted by the phosphor present in the liquid scintillator. The energy of the emitted photons is provided by the beta-decay energy of the HTO droplet, whose activity is expressed in becquerels (Bq). The WVTR of the barrier film was extrapolated from cpm by a linear regression equation, yielding the following relationship [25]:

$$\mathrm{WVTR}_{\mathrm{HTO}} = \frac{D_{\mathrm{V}} \cdot (\mathrm{cpm} + C)}{1000 \cdot m \cdot A \cdot \Delta t \cdot a_0}\ (\mathrm{g/m^2/day}) \tag{4.1}$$

where D_{v} is the droplet volume (μl); cpm is counts per minute; C and m are the y-axis intercept and slope of the linear regression, respectively; A is the membrane surface area (m^2); Δt is the experiment duration (days); and a_0 is the initial droplet activity (kBq).

The results of the HDO method need further investigation because the absorption of HTO (HDO) may be influenced by water vapor from the external atmosphere. In addition, the test is hazardous due to the radioactivity of the radiotracer and its unavailability in some experimental situations.

4.1.2 Indirect Optical Method

The thin and opaque Ca layer is deposited directly on the barrier foil or placed between the barrier layer and the glass substrate, and when exposed to a humid environment or liquid water. The main electrochemical reaction occur-ring during corrosion is the reaction with water molecules: $Ca + 2H_2O \rightarrow Ca(OH)_2 + H_2$, $\Delta G° = -441$ kJ/mol. The corrosion reaction thus leads to the formation of transparent calcium hydroxide $Ca(OH)_2$. Therefore, optical methods calculate the WVTR based on the change in optical transmittance [26]. Optical transmittance curves can be obtained with a common UV/vis spectrophotometer; therefore, for a given wavelength, assuming a uniform decrease in Ca thickness, it is recommended to calculate WVTR using the following equation:

$$\mathrm{WVTR}_{\mathrm{o-Ca}} = -n\left(\frac{A_{\mathrm{Ca}}}{A_{\mathrm{B}}}\right)\left(\frac{M_{\mathrm{H_2O}}}{M_{\mathrm{Ca}}}\right) \delta_{\mathrm{Ca}}\frac{h_0}{T_0}\left(\frac{\mathrm{d}T}{\mathrm{d}t}\right)_{T,RH,\lambda} (\mathrm{g/m^2/day}) \tag{4.2}$$

where $n = 2$ is the number of molar equivalents in the corrosion reaction; A_{Ca} is the exposed Ca area; A_{B} is the permeation barrier area, $M_{\mathrm{H_2O}}$, M_{Ca} are the molecular weights of water and Ca, respectively; δ_{Ca} is the Ca mass density, h_0, T_0 is the initial thickness and transmittance of the Ca film; $(\mathrm{d}T/\mathrm{d}t)_{T,\mathrm{RH},\lambda}$ is the linearity of the transmittance-time curve for a fixed temperature, RH, and wavelength slope of the partial transmittance.

Another qualitative optical method uses a thin magnesium (Mg) pad: like calcium, Mg is sensitive to humidity and is subject to corrosion [27, 28]. The appearance of corrosion products on the surface of Mg after a period of immersion in PBS or after a period in a humid environment is a qualitative indicator of the lifetime of the coating. However, tests based on calcium and magnesium in optical methods are incomplete because corrosion is usually not uniform but localized in the form of pinholes. Furthermore, quantitative and reliable methods are lacking, and in fact, the main criterion used to determine the lifetime of an encapsulation consists of considering the time of appearance of the first pinhole or corrosion spot on the Mg (or Ca) surface. However, this can be misleading because it cannot distinguish between inhomogeneous corrosion due to the diffusion of water through the intact encapsulation and localized corrosion corresponding to pre-existing point defects in the coating. A convenient method often used by Ca sensor pass [29, 30] is to apply a large number ($\approx$100–150) of Ca micrometer square dot arrays to the encapsulation layer; the corrosion dots are then counted by optical inspection, and their number density is used as a quantitative indication of the corrosion rate, and thus for the water permeation rate. Novel quantitative electrical/optical/electrochemical

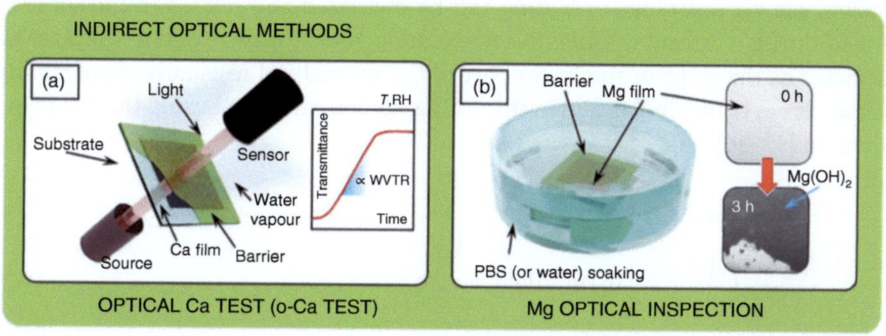

Figure 4.7 Schematic diagram of the optical method test ((a) Ca optical test; (b) Mg optical test). Source: Herbst et al. [6] / John Wiley & Sons / CC BY 4.0.

permeation rate measurements based on patterned Mg films are under development [31] (Figure 4.7).

4.1.3 Indirect Electrical Method

4.1.3.1 Calcium Electrical Test

A more refined method based on Ca sensors is the electrical Ca test (e-Ca test, Figure 4.8), which determines the rate of penetration based on the change in resistance of the Ca film (metal traces or pads) underwater corrosion (corrosion is related to material degradation and thickness reduction, changing resistance according to Ohm's law). As water penetrates through the barrier layer, it reacts with the conductive metal (Ca) to produce $Ca(OH)_2$, which acts as an insulator.

The amount of conductive material decreases and therefore its resistance increases (or its conductivity decreases). If the Ca test area and geometry are

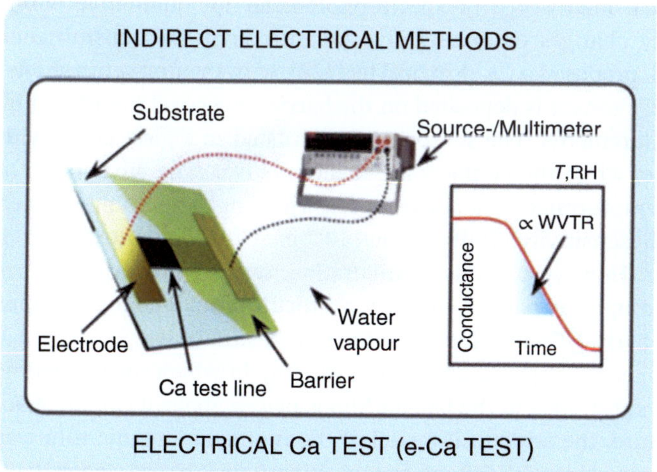

Figure 4.8 Schematic diagram of calcium electrical test. Source: Herbst et al. [6] / John Wiley & Sons / CC BY 4.0.

known, the Ca resistance over time can be measured and related to the Ca consumed during corrosion. Thus, the permeation rate at a steady state can also be extracted as the slope of the linear part of the normalized resistance/conductivity (versus time) curve; in particular, the WVTR can be calculated using the following equation [32–36]:

$$\mathrm{WVTR}_{\mathrm{e-Ca}} = -n\left(\frac{A_{\mathrm{Ca}}}{A_{\mathrm{B}}}\right)\left(\frac{M_{\mathrm{H_2O}}}{M_{\mathrm{Ca}}}\right)$$

$$\frac{\delta_{\mathrm{Ca}}\rho_{\mathrm{Ca}}l}{b}\left(\frac{\mathrm{d}G}{\mathrm{d}t}\right)_{T,\mathrm{RH}}\ (\mathrm{g/m^2/day}) \tag{4.3}$$

where ρ_{Ca} is the Ca resistivity; l is the distance between the measured electrical probes; b is the width of the Ca film (perpendicular to the dimension of the electron flow); and $(\mathrm{d}G/\mathrm{d}t)_{T,\mathrm{RH}}$ is the slope of the linear part of the conductance–time curve at a fixed temperature and RH. Other parameters have been introduced in (4.2).

Unlike the o-Ca method, which is able to detect local corrosion, the e-Ca test is a complete method that captures all processes leading to resistivity changes, regardless of local corrosion or not. Therefore, the two methods are complementary. While the optical method takes into account the change in transmittance/reflectance, which depends on the surface morphology and crystalline properties of the material over time, the electrical method is more sensitive to local corrosion spots and inhomogeneities, since the resistance is directly related to the electron mobility within the Ca film.

That is why the design of the Ca test setup has been improved and different methods have been defined during the last few years. Figure 4.9 reports the main configuration of the test, starting from the original o-Ca test patented by Bouten et al. (Figure 4.9a) [37]. It consists of an array of Ca spots or squares deposited on a blocking film and encapsulated in a glass cover (transparent). The sample is placed under an optical microscope or spectrophotometer for qualitative observation of transparency changes or quantitative measurements of transmittance changes. Paetzold et al. proposed a Ca electrical test [38], with the first setup shown in Figure 4.9b: a thin Ca sensor is deposited on the barrier layer and monitored by *in situ* resistance measurements. The sensor is encapsulated in a glass cover with edges sealed with glue, while the contacts are made of silver. The advantages of this technique were demonstrated on a polyester foil with single- and double-sided barrier coatings, with sensitivity limits below 10^{-6} g/m^2/day obtained with a reference glass barrier. However, in this configuration, water penetration occurs below the Ca layer through the barrier, making it difficult to understand whether corrosion starts at the barrier/Ca interface or at the top Ca surface.

For this reason, Choi et al. proposed an alternative setup in which the Ca layer is deposited on the glass substrate and the barrier film is placed above it (Figure 4.9c) [33]. Using this structure, the structural model of the inorganic/organic film was demonstrated: the steady-state WVTR of the multipotential barrier coated on both sides of the polyethersulfone (PES) substrate measured 5.17×10^{-3} g/m^2/day at 20 °C and 60% RH, with a test sensitivity as low as 10^{-6} g/m^2/day.

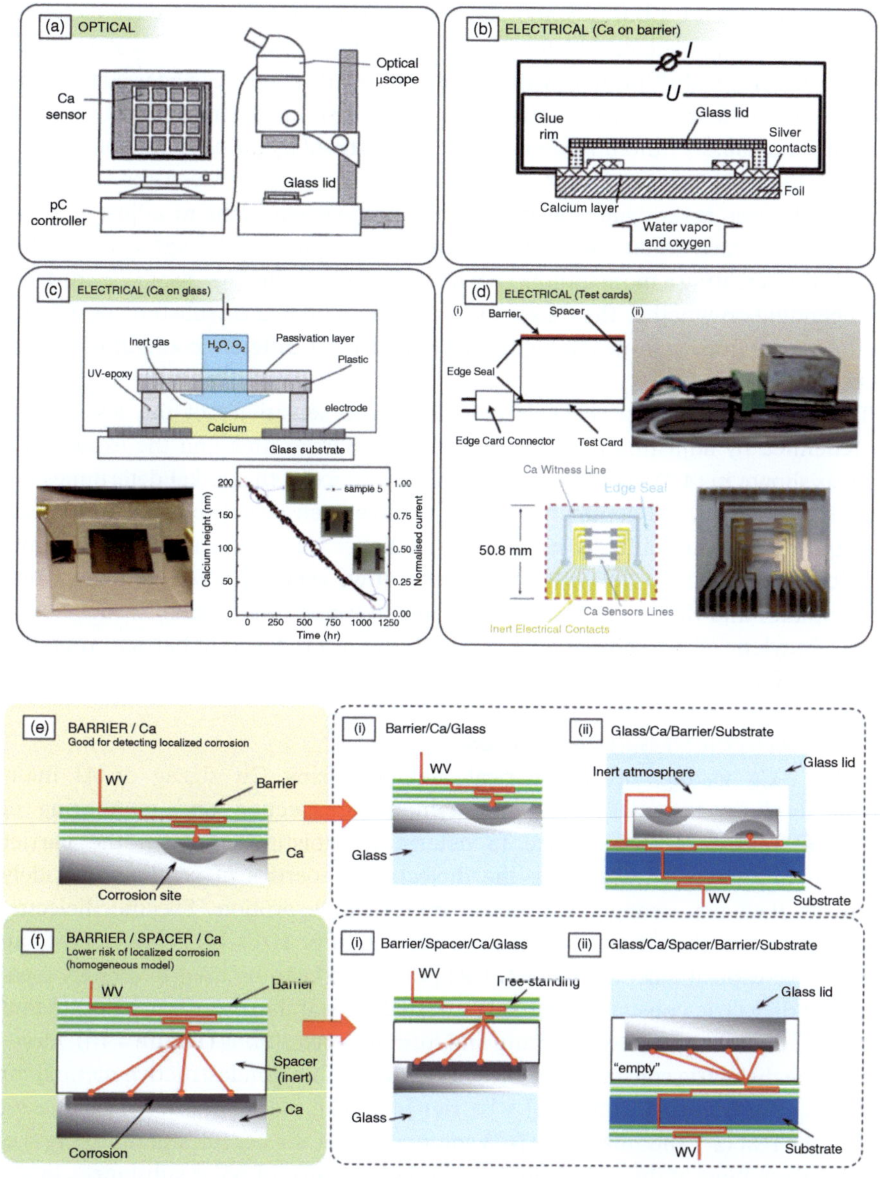

Figure 4.9 Evolution of the calcium electrical test setup. (a) Original optical calcium test. (b) Ca electrical test with a Ca film deposited on a barrier film and covered by a glass cover. (c) Ca electrical test with a Ca film deposited on a glass substrate below the barrier film. (d) Calcium electrical test based on the test card and top space optimization. (e, f) Schematic diagrams of several configurations of Ca test with barrier and Ca sensors with or without spacers, or with different relative positions of Ca and barriers: (e (i)) barrier/Ca/glass, (e (ii)) glass/Ca/barrier/substrate, (f (i)) barrier/spacer/Ca/, (f (ii)) glass/calcium/spacer/barrier/substrate. Source: Herbst et al. [6] / John Wiley & Sons / CC BY 4.0.

In the latest configuration, the e-Ca test is based on a test card made of a glass substrate in which Ca traces evaporate in an inert glove box and are contacted by evaporating Ti/Au contacts in a four-point arrangement as described by Reese et al. (Figure 4.9c) [39]. Multiple lines with different sizes and shapes can be used to obtain more accurate average WVTR measurements [32–35]. In addition, since the corrosion of Ca is usually inhomogeneous and characterized by pinholes and other local defects (Figure 4.9e), spacers or head spaces can be used between the Ca sensor and the barrier to mitigate the effects of inhomogeneity and to adjust the test sensitivity. In fact, water can penetrate the barrier and fill the space within the spacer, thus creating a more uniform environment above the Ca sensor (Figure 4.9f).

Depending on whether the Ca sensor is deposited on a glass substrate or directly on the barrier film, as shown in Figure 4.9e(i, ii), f(i, iii), different configurations can be used.

The design of the top space is crucial, as the sensitivity of the test should be redefined by adjusting the ratio between the barrier area and the exposed Ca area, as shown in (4.3). In addition, the sealing surface of the gasket determines the edge seal flow. Several commercial sealants have been tested for moisture lateral penetration and durability by using a simple linear Ca marker as a sensor. While poly(dimethylsiloxane) (PDMS) or ethylene vinyl acetate (EVA) had the worst performance and thus were highly permeable to water, other sealants, such as dry polyisobutylene (PIB), showed excellent performance in terms of barrier properties and ease of use [39].

4.1.3.2 Dielectric Measurement Method

Besides e-Ca testing, other electrical methods using Cu zigzag or Al interdigitated electrodes (IDE) rely on qualitative inspection and monitoring of electrical impedance/capacitance to determine moisture entry in the barrier [40]. In this context, measuring the dielectric properties of coatings is widely used, as they are strongly influenced by water absorption. Specific dielectric constant meters (based on parallel-plate capacitors or IDEs [41] connected to an inductance-capacitance-resistance (LCR) [42] meter) can be used to directly measure the dielectric constant of packages coated on standard substrates at different frequencies under given temperature and humidity conditions (Figure 4.10). Monitoring the dielectric constant (permittivity) is a suitable nondestructive method for assessing the water absorption of a barrier [43]. The main parameters of interest are (i) the real (ε') and imaginary (ε'') components of the complex permittivity (i.e. the measurement of the polarizability and the presence of polar substances in the material) [43], and (ii) the dielectric loss tangent or dissipation (the measurement of the energy dissipated in the material as the dipole tries to follow the change in the applied AC electric field), given by the following equation, respectively [43]:

$$\varepsilon = \varepsilon' - i\varepsilon'', \tan\delta = \frac{\varepsilon''}{\varepsilon'} \tag{4.4}$$

where δ is the dielectric loss angle.

The entry of water into the barrier leads to the following main consequences: (i) high dielectric constant of the absorbed water molecules, increasing the resistance

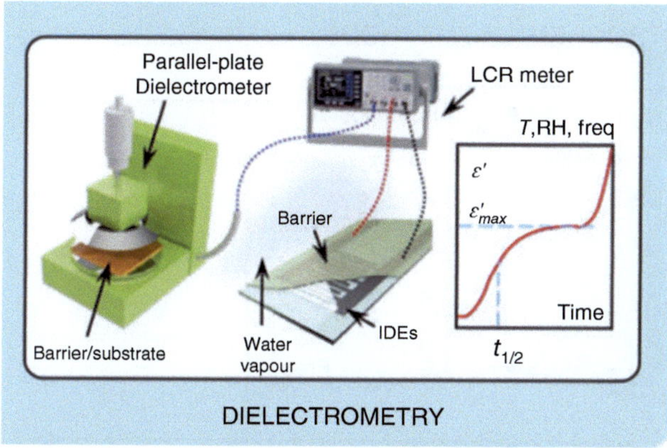

Figure 4.10 Schematic diagram of the dielectric measurement method. Source: Herbst et al. [6] / John Wiley & Sons / CC BY 4.0.

of the layer; (ii) this increase can also be attributed to the solvation response of the coating (depending on its material properties); (iii) dissociation of water molecules and penetration of solventized ions leading to a decrease in resistance [44–47]. It can be assumed that during isothermal experiments, the dielectric constant is given by the sum of the constant term and the water concentration-related contribution at each frequency [47]:

$$\varepsilon'(t) = \varepsilon_0{}' + \varepsilon_w{}'(t) \tag{4.5}$$

This time-dependent parameter increases during the exposure time due to water entry, reaching a constant value (maximum ε') corresponding to the maximum water concentration ($c_{\max}$) in the barrier. This increase is usually proportional to the concentration of water dipole (c) present in the system [48], and then the potential barrier diffusion coefficient and WVTR can be derived from the dielectric data [47]:

$$D = \frac{t_f{}^2}{t_{1/2}} \frac{\ln(\pi/8)}{\pi^2} \tag{4.6}$$

$$\mathrm{WVTR}_{\mathrm{dielec}} = \frac{Dc_{\max}}{t_f} \tag{4.7}$$

where t_f is the film thickness and $t_{1/2}$ is the time corresponding to half of the time at $(c - c_0)/(c_{\max} - c_0) = 0.5$, where c_0 is the initial water concentration in the system.

The sudden increase in the dielectric constant at longer times can be attributed to the formation of a charge layer at the interface between the electrode and the barrier material, resulting in an equivalent capacitance greater than that due to the water dipole alone [47].

4.1.4 Indirect Electrochemical Method

In addition, the corrosion protection performance of the package is measured by electrical and electrochemical tests; the most common methods are electrochemical

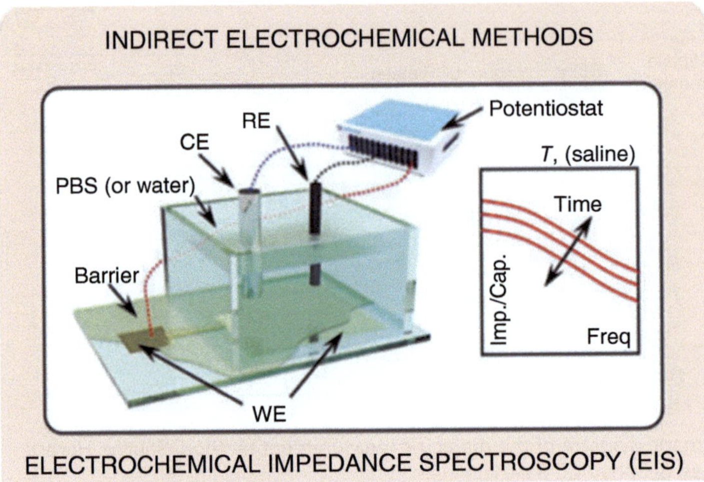

Figure 4.11 Schematic diagram of electrochemical impedance spectroscopy test. Source: Herbst et al. [6] / John Wiley & Sons / CC BY 4.0.

impedance spectroscopy (EIS), linear scanning voltammetry (LSV), and leakage current monitoring (LCM).

4.1.4.1 Electrochemical Impedance Spectroscopy (EIS)

The EIS method (Figure 4.11) consists of measuring the dielectric properties of a pair of electrodes coated with an external coating. These can be comb-shaped crossed-finger electrodes (IDE), which can increase the sensitivity and detection limits. IDEs are used as sensors for impedance and capacitance measurements in different technological applications, especially for biosensors [49]. When exposed to the environment, the coating is gradually permeated by water, leading to changes in capacitance, impedance, and phase of the electrodes, which can therefore be detected by frequency scanning, thus obtaining the corresponding Bode and Nyquist curves. The EIS data can then be fitted to an equivalent circuit model of the package, which is useful to study the failure mechanisms of the circuit components based on their variations.

4.1.4.2 Leakage Current Monitoring Method (LCM)

The LCM method (Figure 4.12) consists of monitoring the leakage current, i.e. the gradual transfer of electrical energy through the boundary represented by the insulating barrier coating when the coated device (electrode) is in contact with a wet or liquid environment under DC bias. The degradation of the barrier coating leads to the formation of conductive paths through its thickness, which increases the leakage current.

When this value is above the threshold (usually in the range of nA or μA, starting from a few tens of pA), the device can be considered as faulty or at the end of its life [50]. The same criteria can be applied by using complete devices (photoelectrodes, neural recording systems, PTs, etc.) as test tools to assess lifetime. In this case,

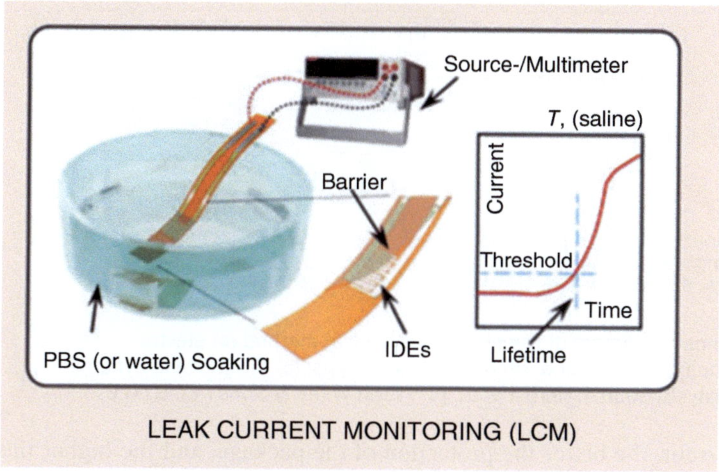

Figure 4.12 Schematic diagram of leakage current monitoring method. Source: Herbst et al. [6] / John Wiley & Sons / CC BY 4.0.

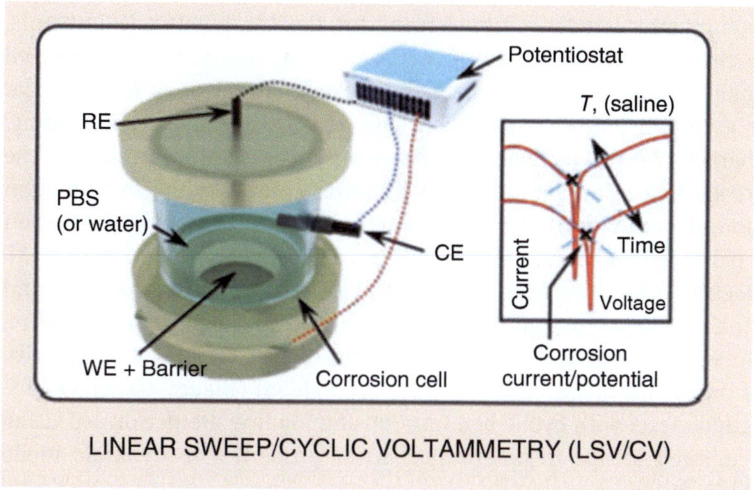

Figure 4.13 Schematic diagram of linear scanning voltammetry. Source: Herbst et al. [6] / John Wiley & Sons / CC BY 4.0.

physical quantities (e.g. light intensity, piezoelectric signals, etc.) are monitored, and their degradation over time indicates device failure.

4.1.4.3 Linear Scanning Voltammetry (LSV)

The LSV method (Figure 4.13) is based on the measurement of corrosion currents and voltages on a test metal plate covered with a coating and exposed to the environment.

The current is detected by linearly scanning the voltage. The resulting I, V diagram (Tafel diagram) provides information on the anode electron loss rate: the lower

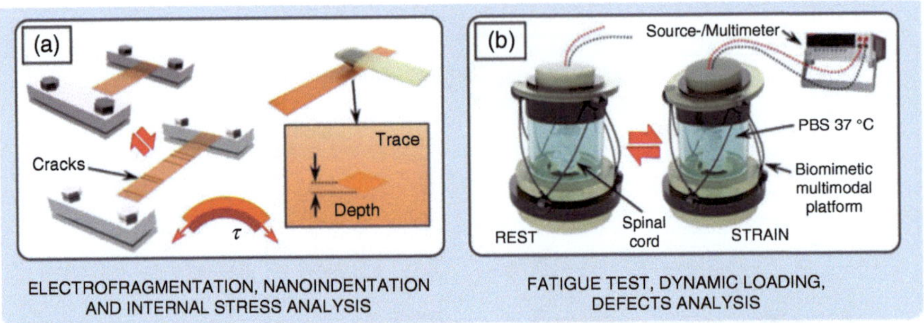

Figure 4.14 Schematic diagram of the electromechanical method (a) electro-fragmentation, nanoindentation and internal stress analysis; (b) fatigue test, dynamic loading, effects analysis. Source: Herbst et al. [6] / John Wiley & Sons / CC BY 4.0.

the corrosion current, the better the protection of the package, and the higher the seal [51].

4.1.5 Indirect Electromechanical Method

Mechanical compliance is a key requirement for flexible bioelectric implants. In this regard, the thermal absorption of barrier coatings should be characterized according to their mechanical reliability, e.g. the package should maintain its barrier quality under a certain amount of applied load and/or repeated cyclic loading. In order to characterize the mechanical influence on the coating properties, the standard approach is to evaluate mechanical failures caused by defect formation and propagation, delamination, and tearing effects, and to assess water diffusion in the case of partially cracked/degraded coatings, as described in reports [52, 53]. Therefore, (electrical) fracture tests associated with residual stress analysis and nanoindentation tests are commonly used to determine the critical strain for failure, the internal stress components, the Young's modulus of the nanofilm, and to derive the toughness properties of the film and interface [54–58] (Figure 4.14a). On the other hand, fatigue tests with cyclic bending/tension loading are performed using appropriately tailored setups to determine barrier properties in dynamic mode (Figure 4.14b) [59, 60].

4.2 Bending Test

Flexible electronics is a rapidly evolving technology for a variety of applications. Wearable devices and flexible displays are some examples of applications. Various technologies and processes are used to produce flexible electronics. An important aspect to consider when developing these systems is their reliability, especially in terms of repetitive bending. This section describes common methods for studying the bending reliability of flexible electronics.

Mature semiconductor technologies can be integrated into flexible electronics so that this combination has the ability to bend, deform into irregular shapes, or even stretch [61, 62]. Flexible electronics consist of electronic components such as surface-mounted devices (SMDs) or ultrathin chips on flexible substrates [63, 64]. The electronic components are not inherently flexible but are as thin as possible to provide the flexibility of a complete system when integrated on or in a flexible substrate [64, 65]. Bending strain is one of the main motions that cause cracks and lead to the failure of flexible structures [66]. The mechanical reliability of these flexible electronics is important to ensure the growing market demand for these systems. Ultrathin chips are very common in flexible electronics and therefore need to be robust enough to withstand the tensions generated during production and handling [67]. The integration of these ultrathin and bendable integrated circuits in a roll-to-roll process has enabled the development of inexpensive, mechanically flexible electronics. In order to make these systems reliable, the strength and deformation of the thin component layers must be studied in detail. Once these properties are known, the individual technological steps can be characterized and optimized in an optimal manner [68].

Flexible electronic devices must be able to deform, while functional characteristics and electronic parameters must be unaffected by bending during application [69]. In addition, with the growing reliability requirements for flexible systems, the need for characterizing the mechanical stability of components has increased [70]. Therefore, it is crucial to test their reliability under mechanical bending [71, 72]. The purpose of bending tests is to understand the behavior of flexible systems under bending and their fatigue stability. Bending-related failures seriously affect the reliability of components. Such failures are mainly caused by forces generated by bending or vibration of the circuit carrier [73].

4.2.1 Static Bending and Dynamic Bending

Static bending tests are the simplest and most reported tests to study the reliability of flexible electronic devices [74]. However, some applications can be bent not only once in their function but also dynamically. For these systems, static bending is not sufficient to understand the reliability conditions of the product in later work [75]. Therefore, it is necessary to consider dynamic bending.

Static bending tests can be performed to understand the system in terms of internal stresses. Depending on the application, static bending tests should be performed in concave or convex orientation. In the concave orientation, the components on the bottom layer are subjected to tensile stresses, while in the convex orientation, they are subjected to compressive stresses [76].

For the static bending test procedure, the flexible system will be bent on a rod or a tube with a specified radius (Figure 4.15). Depending on the system, two types are possible. In concave bending, the system bends in the internal area of the pipe at the location so that the system will be compressed and get compressive stress [77]. On the other hand, in convex bending, the system is in the rod; compared to concave bending, the system gets tensile stresses [74, 78, 79].

Depending on the type of structure and the material applied to the substrate, some structures are more sensitive to tension than compression. Bending tests should be

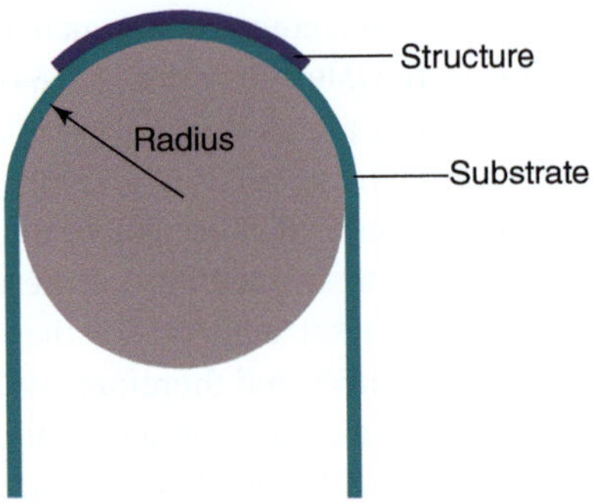

Figure 4.15 Schematic diagram of the system bending at the defined bending radius. Source: Adapted from Happonen et al. [77]; © 2006, American Chemical Society.

designed and selected according to the intended application. Both active and passive flexible electronic devices can be subjected to static and dynamic bending tests to study their stability under bending.

In addition, the critical bending ratio can be understood in this way. The bending ratio describes the ratio of the bending ratio to the thickness of the substrate. In applications where flexible electronics will undergo few bending cycles, static bending reliability studies are sufficient. In contrast, in some applications, flexible electronics are used to electrically connect to movable parts, such as mobile flip phones or displays in computers. Even connections to fast-moving components with very long-life cycles (e.g. the read head of an optical drive) can be achieved using flexible thin-film connections. Dynamic bending reliability studies of components are essential [80].

4.2.2 Three-Point Bending and Four-Point Bending

Three-point bending measurements are often used to study the stability and flexibility of ultrathin chips [67, 81]. The ball-and-ring test can be used to check stability. A force application system presses the ball on the chip and moves it slowly while a computer records data, such as force and distance, until the chip breaks. In another variant to study flexibility, the chip is placed on two rods. The force application system presses a third rod on the chip until it breaks [67].

Both three-point and four-point bending measurements are based on measuring chip curvature during the bending process and calculating the resulting bending stress based on the measured curvature [82]. Figure 4.16 shows the three-point bending setup.

In a three-point bending measurement, the maximum stress occurs in the center of the sample to be measured. On the one hand, this may lead to unexpected failures. On the other hand, the functional structure may be unevenly distributed over the sample and thus experience irregular stresses. Compared to three-point bending, four-point bending has many advantages. It ensures a uniform bending stress between the two supports, which is desirable in many cases. In addition, the errors caused by misalignment remain small, so good results are obtained. It should be noted that both methods are suitable for small displacements (a few microns) and

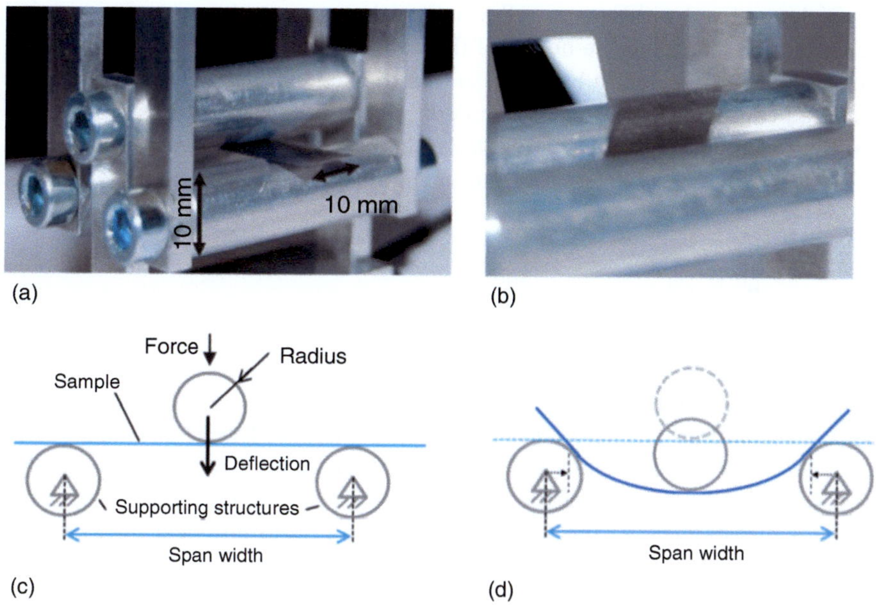

Figure 4.16 Three-point bending experimental setup. (a) Perspective view; (b) when the bending deformation is large; (c, d) Adapted from Van Den Ende et al. [82] deformation position of the sample. Source: Reprinted with permission from Van Den Ende et al. [82]; ©2014 / Reproduced with permission from Elsevier.

large radii of curvature. Therefore, these methods are more suitable for bending tests of single components and thick systems.

Four-point bending is also a common method for assessing the bond strength of multi-material stacks (called multi-flexible layers) [83]. In four-point bending, the upper support is pressed against the material with a certain force F, thus moving it to deflect z for bending. On the other hand, the lower support is fixed (Figure 4.17).

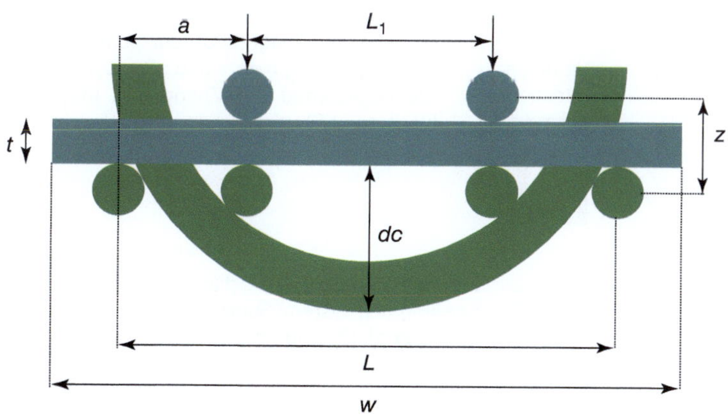

Figure 4.17 Diagram of four-point bending setup before and after bending. Source: Adapted from Van Den Ende et al. [82].

The thickness t and width w, as well as material parameters, such as the modulus of elasticity E, determine the height of the deflection [82]. If the thick member is bent to failure by four-point bending, so that the bending deflection z achieved is very small compared to the distance a between the inner and outer supports, the maximum bending stress σ occurring in the center of the member can be calculated by Eq. (4.8) using the following common bending equation for smaller deflections [82, 84]:

$$\sigma = \frac{3 \cdot F \cdot a}{w \cdot t^2} \tag{4.8}$$

4.2.3 Push Bending and Roll Bending

4.2.3.1 Push Bending

Figure 4.18 shows the setup of a bending cantilever under reverse load, etc. The results show that the deflection with curvature R is a function of the horizontal distance. During the deflection of the sample on this bending setup, the strain is not uniformly distributed over the sample but reaches a maximum near the fixed end and decreases sharply along the length of the sample. This suggests that the local fatigue life varies with the sample also due to the strain gradient [85].

Figure 4.19 shows another setup used to study the reliability of sample bending. Two plates are fixed at the ends and the sample is bent in the middle with a gap of $2R$. The upper plate is fixed and the lower plate is repeatedly moved linearly. This leads to fatigue damage in the sample. The strain in the bending region $2R$ is not uniform but has a gradient [86].

Due to the low thickness of the flexible electronics, it is necessary to use a support substrate to use push-bend [76]. Otherwise, the radius will be undefined. This is one of the disadvantages of using this system to study bending reliability. In addition, the connected active system becomes more rigid than usual due to the support system. Figure 4.20 shows a push–pull bending machine. In this bending machine, the flexible electronics are placed on a thicker support substrate.

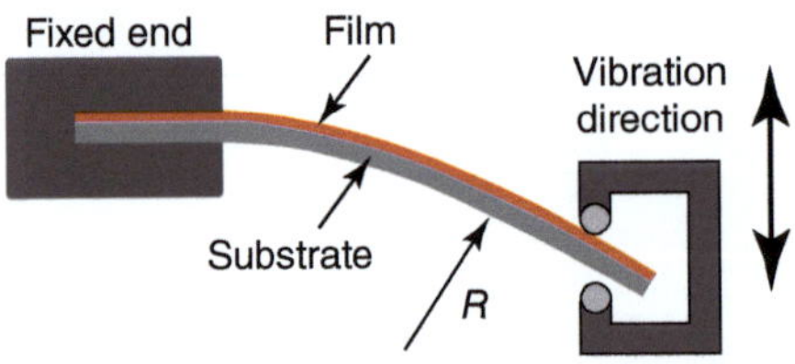

Figure 4.18 Diagram of four-point bending setting before and after bending. Source: Adapted from Zheng et al. [85] / Institute of Physics.

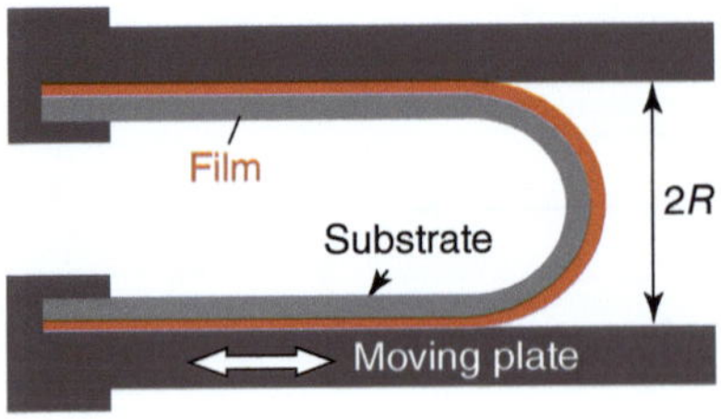

Figure 4.19 Schematic diagram of the cyclic bending test system. Source: Adapted from Zheng et al. [85] / Institute of Physics.

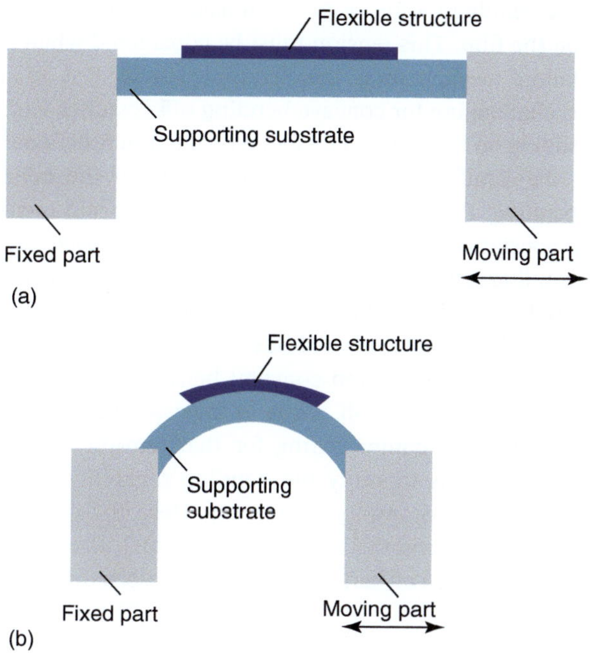

Figure 4.20 Schematic diagram of the dynamic push-bending setup: (a) in flat position, (b) in bending state. Source: Adapted from Fang et al. [27].

Thicker substrates govern the mechanical properties of the flexible system. Furthermore, the bending radius is not strictly defined. The strain in the substrate has a maximum in the region near the end of the clamped substrate [85]. One of the drawbacks of this bending measurement method is the undefined bending radius, called free bending. The sample to be tested does not experience a constant bending radius. In addition, each location on the sample has a different bending radius. In addition, it is difficult to achieve very small bending radii [74, 87].

The advantages of the push-bend measurement method include the possibility of using the same measurement device for both concave and convex bending tests since the sample is bent into the gap between the two ends and has no other contact with the part. In addition, the bending test setup can be upgraded to a torsional test setup by adding a rotary axis to the moving end. The push-bend setup is more suitable for applications with thicknesses above 100 μm [77, 88].

4.2.3.2 Rolling Bend

In this measurement method, the flexible electronics as well as the components mounted on the polymer foil are bent repeatedly on a cylinder, and the bending radius is kept constant throughout the measurement [75].

In contrast to the push-bend measurement method, with the rolling bend measurement method, the sample is bent onto a cylinder or roll, which means that a constant or even a smaller bend radius can be achieved. However, the sample must be loaded with some additional load to press the sample onto the curvature of the roll [74]. Although this load is necessary to provide tension to the system at the bending radius, it creates tension on the film and on the electronics mounted on the film. This tension must be considered when calculating and evaluating the failure mechanism of the electronics under test. In addition, this bending method is challenging for concave bending of substrates without protective layers because sliding on the roll surface can damage thin functional structures [74].

To significantly improve the quality of the bending investigation, using roll bending, three factors should be observed and considered. These factors include a constant bending radius, a constant roll speed, and a defined tension [87]. Roll bending setups are more suitable for thinner systems below 100 μm and are widely used in the field of foil-based systems [75, 89].

Ideally, bending tests of flexible electronics are performed without the influence of built-in tension and constant bending radius. This can be ensured by bending the sample onto a roll, while in the case of tension, the whole setup of the sample relaxes, thus compensating for that tension. Figure 4.21 shows two test stands that allow us to study the bending reliability without the influence of tension. Figure 4.21a shows the strap setup. Here, the sample is fixed on a flexible belt. The belt rolls around the bending roller, which slides over the spring structure and the track groove. The belt and the spring absorb the generated tension [74].

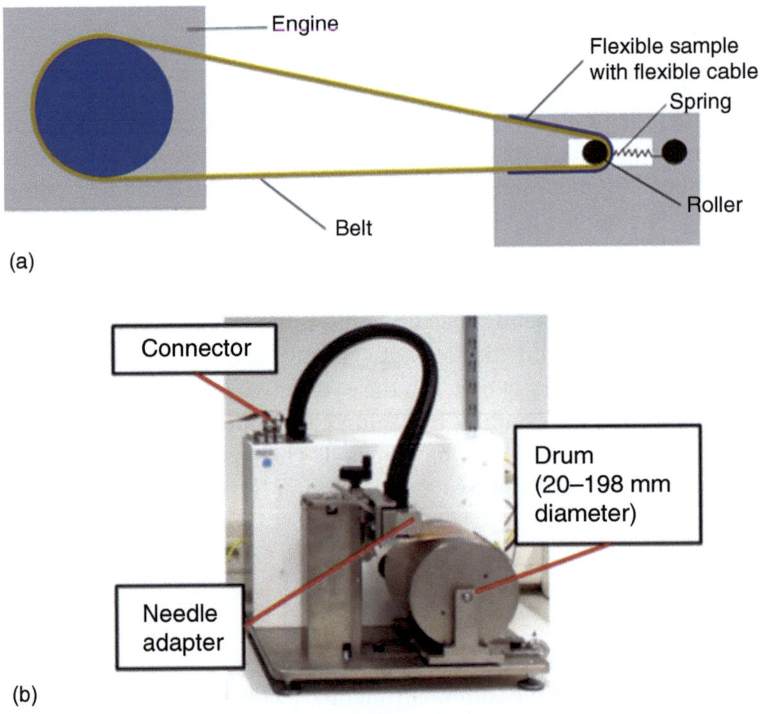

Figure 4.21 Examples of two bending test setups using rolling bending measurements and decoupling effect of tension on bent samples: (a) belt bending setup (a) Source: Adapted from Li and Jackson [74] and (b) bending test setup. Source: Li and Jackson [74] / Reproduced with permission from Elsevier.

Figure 4.21b shows another bending test setup. The sample is rolled onto the roller. The roller rotates and slides on the base plate so that the sample is not subjected to any tension [89, 90].

4.3 Mechanical Performance Testing

Due to the rapid development of flexible devices and sensors, the evaluation and analysis of the interfacial, mechanical, and electrical properties of thin films on flexible polymer substrates is becoming increasingly important. Metal and ceramic films are deposited on polymer substrates using not only traditional deposition methods but also advanced roll-to-roll [91, 92] and inkjet printing [93, 94] techniques. Due to the large differences in the elastic modulus of ceramics, metals, and polymers used to fabricate flexible devices, large strains (>10% or even up to 100%) can be achieved through proper device design [95]. In order to evaluate the failure mechanisms of flexible systems, electromechanical test methods are required. Uniaxial tensile testing, also known as fracture testing, is a common method used to investigate new materials or deposition processes used in the development of flexible electronics. During the fracture test, the film-substrate system is strained under tension, causing cracks to form perpendicular to the direction of strain, and cracks continue to form until a saturated crack spacing λ. When the crack spacing is reached, the film typically delaminates between cracked fragments. These flexural delaminations can be used to measure the adhesion at the film-substrate interface [96].

The fracture properties of hard films on pliable substrates can be assessed by measuring the interfacial shear strength (τ_{IFSS}) through the fragmentation test. τ_{IFSS} is a statistical analysis of the film or coating strength depending on the fracture geometry at saturation:

$$\tau_{\text{IFSS}} = K\frac{\sigma_{\text{f}} h}{\lambda} \tag{4.9}$$

where h is the film thickness, K is a constant that depends on the saturation crack spacing λ, and σ_{f} is the fracture stress. The fracture stress is usually determined according to Hooke's law ($\sigma_{\text{f}} = E\varepsilon_{\text{f}}$), where E is the elastic modulus of the film and ε_{f} is the fracture strain or crack initiation strain (COS) when the crack is first observed in an *in situ* experiment [97–99]. Equation (4.9) are based on shear hysteresis theory, where the shear stress transfer along the film-substrate interface is limited by τ_{IFSS} [100, 101].

For flexible electronics, the key reliability parameter to characterize a brittle film or layer on a polymer substrate is COS, which is the strain at which cracking is first observed. Assuming that cracking leads to immediate device failure, it is important that the mechanical strain that may occur during the fabrication of flexible electronics (e.g. in a roll-to-roll device) or during use (bending, stretching, or flexing) should be kept well below the COS. For fast and accurate measurement of COS, cracks that occur during tensile testing should be detected using *in situ*

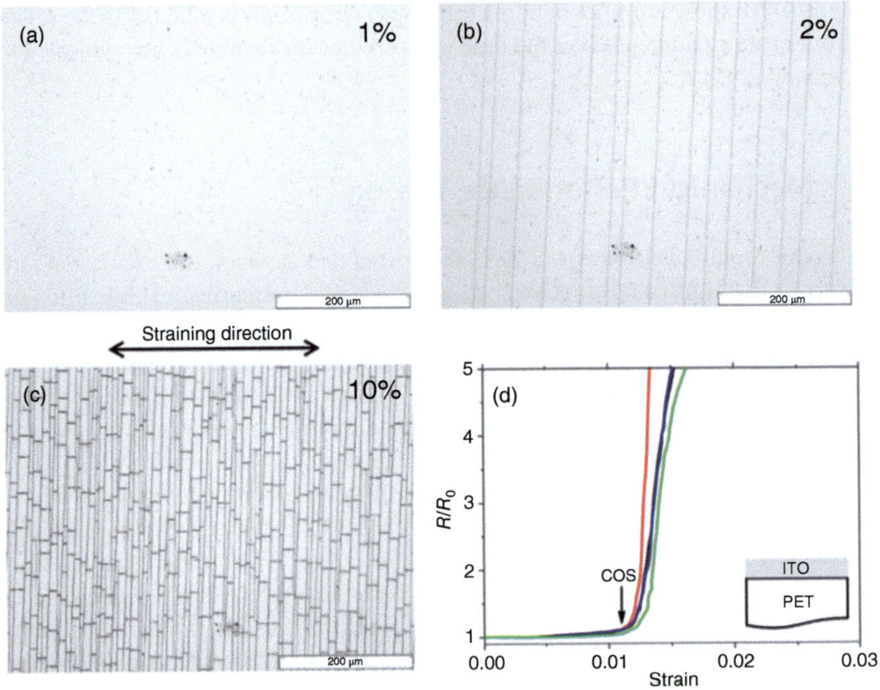

Figure 4.22 Stretching tests were performed on a 100 nm thick ITO film on PET. (a–c) Optical images of the same surface area taken *in situ* at different applied strain values. (d) *In situ* measurements of the resistance of three ITO samples during strain. Source: Glushko et al. [105] / Reproduced with permission from Springer Nature.

monitoring techniques. For example, tensile tests can be performed under an optical microscope (OLM) or within a scanning electron microscope (SEM) [102–104]. Strain is usually performed in small steps (e.g. 1% steps) with a holding time of two to five minutes for focusing and image acquisition. Figure 4.22a–c gives an *in situ* optical example showing the same area of 100 nm thick indium tin oxide (ITO) film on polyethylene terephthalate (PET) at different strains. It can be clearly seen that at 1% strain (Figure 4.22a), no cracks are observed in the observed area, while at 2% strain (Figure 4.22b), many cracks are observed. At 10% strain, a very dense network of cracks and flexures was formed in the film (Figure 4.22c). To accurately determine COS, very small steps between 1% and 2% strain need to be used, which is very time-consuming, or use continuous strain and fast video acquisition, which can be a costly solution. Another problem is the relaxation of the viscoelastic substrate during the holding time, which can lead to errors in COS determination [106]. COS is performed in a more reliable way if the resistance of the sample can be recorded during the strain. *In situ* four-point probe (4PP) resistance measurements are useful when studying thin films on polymer substrates [105, 107, 108]. The electrical properties as a function of strain can be studied, and the exact value of the critical strain at which crack initiation affects conductivity can be obtained more accurately by *in situ* 4PP resistance measurements. The electrical response as a function of strain is described theoretically in Eq. (4.10):

$$\left({}^{R}\!/_{R_0}\right) = \left({}^{L}\!/_{L_0}\right)^2 = (1+\varepsilon)^2, \tag{4.10}$$

where R is the measured instantaneous resistance, R_0 is the initial resistance, L is the measured instantaneous length between probes, L_0 is the original scale length, and ε is the engineering strain [109]. The sharp increase in relative resistance R/R_0 indicates that the critical crack density affects the measured resistance or COS. This may also be related to the extension of the first crack over the entire width of the sample. Figure 4.22d shows the dependence of the relative resistance ratio on the applied strain. The different curves depict the resistances of three samples with the same loading parameters. Using this method, the COS of 1.1% strain can be accurately measured, since the rapid increase in resistance is caused by crack extension. It is important that the 4PP experimental setup is robust and can repeat the same electrical conditions. Therefore, it is advisable to incorporate the 4PP into the fixture of the tensioning device and to correct for any parasitic resistance present in the fixture [105]. Another important parameter measured with *in situ* 4PP is the electrical response of the film during unloading. The no-load resistance is the "true" value of the device after it has been unfolded, stretched, or bent into the applied shape. It takes into account any grain growth, crack bridging, or viscoelastic recovery of the substrate [110, 111].

Using OLM or SEM, *in situ* methods provide information about COS, crack spacing evolution, and even brittle film flexural strain. Other techniques have also been used to examine the behavior of ductile and brittle films. For example, *in situ* atomic force microscopy (AFM), X-ray diffraction (XRD), or confocal laser scanning microscopy (CLSM) can provide further insight into the deformation mechanisms and buckling phenomena occurring during strain. *In situ* AFM studies have been used to better compare ductile deformation and brittle fracture by using deformation spacing and through-thickness crack spacing as well as flexural growth [112]. XRD studies have been fruitful in the study of microstructural effects, thickness, and even intercalation effects [113]. The use of *in situ* CLSM imaging is a relatively new *in situ* method that can provide similar 3D information to AFM without height or scan area limitations. This *in situ* imaging method is useful when studying interfacial adhesion in stretch-induced delamination.

As discussed, *in situ* fracture testing is a valuable method for conductive films. However, using only one *in situ* observation method provides only limited information on the electrical or mechanical behavior. By using multiple *in situ* techniques such as AFM and 4PP [114], more is learned about the combined electromechanical behavior and brings new insights into how mechanical damage affects the electrical response of ductile films. The new combination of *in situ* fragmentation methods and the valuable information that can be obtained about the conductive films will be the focus of this study. First, the combination of *in situ* XRD and 4PP provides accurate fracture and flexural stresses, as well as critical data on how stresses develop in multilayers. CLSM and 4PP can also be combined to study the

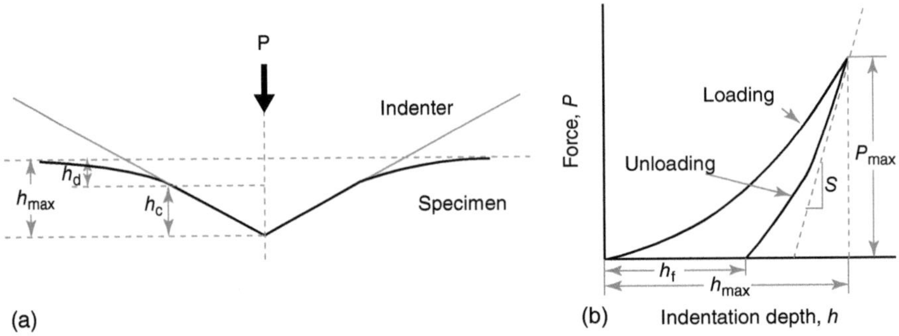

Figure 4.23 Schematic diagram of typical force indentation depth curve (a) and contact pattern between indenter and sample (b). Source: Adapted from Lee et al. [115].

relationship between crack extension and electrical behavior, as well as the study of interfacial adhesion.

In addition, nanoindentation can be used to measure the mechanical properties of small volumes and thin-film materials. The outstanding advantage of nanoindentation is that it has a simpler small-scale sample preparation procedure compared to other mechanical tests. As shown in Figure 4.23, force and indentation depth are measured continuously over a range of indentation depths from a few tens of nanometers to a few micrometers when indentation is performed on the sample surface using a pointed indenter. Hardness and elastic modulus can be calculated from the force indentation depth data using the Oliver Pharr method [116–118]:

$$H = \frac{P}{A_c} \tag{4.11}$$

The hardness H is given by Eq. (4.11), which relates the force to the indentation depth, where P is the applied force and A_c is the projected contact area. The reduced modulus is given by the following equation:

$$E_r = \frac{S}{2}\sqrt{\frac{\pi}{A_c}} \tag{4.12}$$

where the slope of the curve at unloading, dP/dh, indicates the stiffness of the contact, S. Based on the reduced modulus, the elastic modulus of the sample E can be calculated as:

$$\frac{1}{E_r}\frac{(1-\nu^2)}{E} + \frac{(1-\nu^2)}{E_i} \tag{4.13}$$

where E and ν are the elastic modulus and Poisson's ratio of the sample, respectively, and E_i and ν_i are the same for indenters usually made of diamond. In order to calculate the projected contact area A_c as a function of the contact depth h_c, the elastic deflection h_d of the sample surface can be calculated by subtracting from the maximum indentation depth h_{max}:

$$h_c = h_{max} - h_d = h_{max} - \varepsilon\frac{P}{S} \tag{4.14}$$

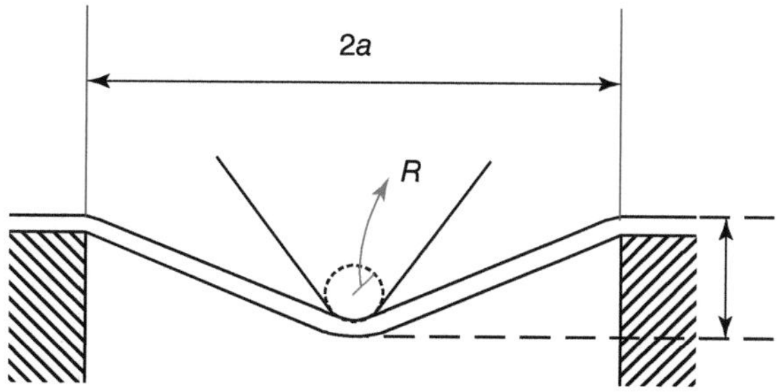

Figure 4.24 Schematic diagram of nanoindentation. Source: Adapted from Lee et al. [115].

where θ and ε are the equivalent angle of the indenter and a constant depending on the geometry of the indenter, respectively.

As mentioned above, nanoindentation can measure small-size materials, such as thin films, nanoparticles, etc. However, it is difficult to use nanoindentation when the sample thickness is below about 100 nm because the "1/10 rule" specifies the maximum indentation depth to be about 1/10 of the sample thickness to exclude substrates that adversely affect the experimental results. Graphene was nanoindented [115], as shown in Figure 4.24.

The mechanical properties of the membrane are calculated from the force penetration depth data based on the following nonlinear elastic response model:

$$F = \sigma_0^{2D}(\pi a)\left(\frac{\delta}{a}\right) + E^{2D}(q^3 a)\left(\frac{\delta}{a}\right)^3 \tag{4.15}$$

where F and δ are the force and deflection at the center, respectively; σ_0^{2D} and E^{2D} are the pretension and elastic stiffness, respectively; and q and a are functions of Poisson's ratio and the radius of the patterned hole, respectively. The maximum stress σ_{m}^{2D} under the indentation is defined by a simplc cquation as a function of the applied force:

$$\sigma_m^{2D} = \left(\frac{\mathrm{FE}^{2D}}{4\pi R}\right)^{\frac{1}{2}} \tag{4.16}$$

4.4 Stability Testing

Calcium titanite solar cells (PSCs) have shown great potential for next-generation photovoltaics. One of the main obstacles to their commercial use is their poor long-term stability under environmental conditions, especially their sensitivity to moisture and oxygen. Therefore, several encapsulation strategies are being developed in an attempt to improve the stability of PSCs in humid environments. The lack of a common test procedure makes the comparison of encapsulation strategies challenging. Mark Khenkin [119] investigated two common encapsulation strategies: laminated glass encapsulation for outdoor operation and commercial use (COM) and simple adhesive-based encapsulation (LAB) used mainly for laboratory research purposes (Figure 4.25).

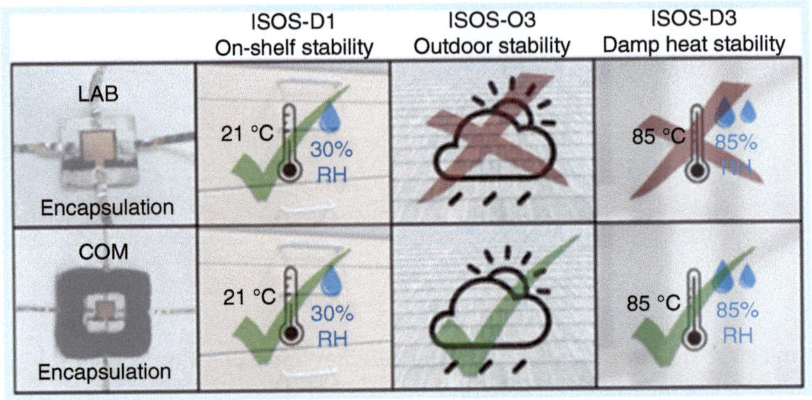

Figure 4.25 Schematic diagram of stability testing of adhesive-based and glass-encapsulated chalcogenide cells. Source: Ref. [119] / Reproduced with permission from American Chemical Society / CC BY 4.0.

They compared the two methods and evaluated their effectiveness in blocking moisture entry under three different test conditions: shelf storage at 21 °C and 30% RH (ISOS-D1), hygrothermal exposure at 85 °C and 85% RH (ISOS-D3), a rooftop test site in Berlin, Germany (ISOS-O3), and with maximum power point tracking. The device's performance was monitored for 10 consecutive months under maximum power point tracking. The LAB encapsulation of the chalcogenide devices consists of glue and cover glass and can be performed in an inert environment at ambient temperatures without the need for sophisticated equipment. This glue-based encapsulation procedure allows the PSC to maintain a conversion efficiency of more than 93% after 1566 hours of storage in an ambient atmosphere and is therefore adequate and suitable as a temporary encapsulation for cell transport or short-term experiments outside of an inert atmosphere. However, this simple encapsulation does not pass the IEC 61215 hygrothermal test, thus leading to a high potential for rapid cell degradation under outdoor conditions. The COM encapsulation procedure requires the use of a vacuum laminator with cells capable of withstanding short air exposure at high temperatures (in our case, 150 °C) for at least 20 minutes. This encapsulation method allows the cell to pass the IEC 61215 damp heat test and maintain more than 95% of its initial efficiency even after 1566 hours in a damp heat chamber. Most importantly, passing the damp heat test for COM-encapsulated devices means that the device maintains its initial efficiency completely throughout the entire outdoor test period (>10 months). This is one of the longest outdoor stability demonstrations of PSCs published to date. Both encapsulation methods described in this work are useful to the scientific community because they achieve different purposes: COM for prototyping for long-term real-world conditions validation, commercialization of calcium titanite solar cells, and LAB procedures to enable testing and experimentation of calcium titanite solar cells under noninvasive conditions (Table 4.1).

Table 4.1 Aging test specifications in accordance with IEC61215:2016 standard.

Test	Conditions	Hours/ Cycle time
Humidity and heat	85 °C/85% Relative Humidity (RH)	1400 h
Thermal cycle	−40 to 85 °C, Dwell time of 10 min for each condition	50 cycles
Humidity freeze	50 rounds of thermal cycling as a prerequisite, followed by −40 to 80 °C (30 min of dwell time) and 85% relative humidity (20 h of dwell time)	10 cycles
	Temperature rise rate of 100 °C per hour at 0 ⇋ 80 °C	
	Temperature rise rate of 200 °C per hour at 0 ⇋ −40 °C	

References

1 Nisato, G., Bouten, P.C.P., Slikkerveer, P.J. et al. (2001). *SID Conference Record of the International Display Research Conference*, 1435. USA: Society for Information Displays.

2 Schubert, S., Klumbies, H., Müller-Meskamp, L., and Leo, K. (2011). *The Review of Scientific Instruments* 82: 094101.

3 Song, E., Li, J., and Rogers, J.A. (2019). *APL Materials* 7: 050902.

4 Jeong, E.G., Kwon, J.H., Kang, K.S. et al. (2020). *Journal of Information Display* 21: 19.

5 Nisato, G., Klumbies, H., Fahlteich, J. et al. (2014). *Organic Electronics* 15: 3746.

6 Herbst, F., GroBer, S., With, P.C. et al. (2021). *Packaging Technology and Science* 34: 497.

7 Olaoye, T.S. and Dewsbury, M. (2019). In: *Proceedings of the 53rd International Conference of Architectural Science Association* (ed. A. Agrawal). India: Australian and New Zealand Architectural Science Association (ANZASA).

8 Mustapha, R., Zoughaib, A., Ghaddar, N., and Ghali, K. (2020). *Energy* 195: 117057. https://www.slide-share.net/Badanayak/moisture-management-and-wicking-behav-iour-of-textiles-210066584.

9 Badanayak, P. (2019). Moisture management and wicking behaviour of textiles. Department of Textile and Apparel Designing, College of Community Science, UAS, Dharawad (accessed January 2022).

10 Shingo, N. (2013). *Journal of Applied Physics* 114: 174302.

11 Suzuki, A., Takahagi, H., Uehigashi, A., and Hara, S. (2015). *AIP Advances* 5: 117204.

12 Nakano, Y., Yanase, T., Nagahama, T. et al. (2016). *Scientific Reports* 6: 35408. https://www.ametekmocon.com/products/www.ametekmocon.com/products/permeationanalyzers.

13 AMETEK MOCON (2021). OTR & WVTR permeation testing for films & packages (accessed January 2021). https://www.systechillinois.com/en/products/products/7000-water-vapor-per-meation-analyser.

14 Systech Illinois (2021). 7000 Water vapor permeation analyser (accessed January 2021).

15 Svensson, T., Lewander, M., and Svanberg, S. (2010). *Optics Express* 18: 16460. http://publica.fraunhofer.de/dokumente/N-244639.html, HiBarSens: Measurement of barrier properties by tunable diode laser absorption spectroscopy (TDLAS).

16 Beese, H., Grählert, W., Kaskel, S., et al. (2012). *Proceedings of the 2012 Web Coating and Handling Conference of the Association of Industrial Metallizers, Coaters and Laminators AIMCAL (AIMCAL, 2012)*, Prague, Czech Republic. p. 848.

17 Yoshida, H., Ebina, T., Arai, K. et al. (2017). *The Review of Scientific Instruments* 88: 043301.

18 Zhang, X.D., Lewis, J.S., Parker, C.B. et al. (2008). *Journal of Vacuum Science and Technology A* 26: 1128.

19 Morlier, A., Cros, S., Garandet, J.-P., and Alberola, N. (2014). *Thin Solid Films* 550: 85.

20 Firon, M., Cros, S., and Trouslard, P. (2009). Method and device for measurement of permeation. US7624621B2.

21 Jeong, E.W., Kim, D.Y., Choi, I.K. et al. (2014). *Micro & Nano Letters* 9: 866.

22 Schulze, S.H., Ehrich, C., Meitzner, R., and Pander, M. (2017). *Progress in Photovoltaics* 25: 1051.

23 Dameron, A.A., Davidson, S.D., Burton, B.B. et al. (2008). *Journal of Physical Chemistry C* 112: 4573.

24 Choi, B.I., Nham, H.S., Woo, S.B., and Kim, J.C. (2008). *Journal of the Korean Physical Society* 53: 2179.

25 Jarvis, K.L., Evans, P.J., Cooling, N.A. et al. (2017). *Surfaces and Interfaces* 9: 182.

26 Boldrighini, P., Fauveau, A., Thérias, S. et al. (2019). *The Review of Scientific Instruments* 90: 014710.

27 Fang, H., Zhao, J., Yu, K.J. et al. (2016). *Proceedings of the National Academy of Sciences of the United States of America* 113: 11682.

28 Li, J., Li, R., Du, H. et al. (2019). *ACS Nano* 13: 660.

29 Meyer, J., Schmidt, H., Kowalsky, W. et al. (2010). *Applied Physics Letters* 96: 243308.

30 Yoon, K.H., Kim, H., Lee, Y.-E. et al. (2017). *RSC Advances* 7: 5601.

31 Mariello, M., Kim, K., Van Gompel, M., et al. (2021). *Proceedings of International Conference on Micro & Nano Engineering*, MNE, Turin, Italy.

32 Zhang, S., Xue, W., and Yu, Z. (2015). *Thin Solid Films* C: 101.

33 Choi, J.H., Kim, Y.M., Park, Y.W. et al. (2007). *The Review of Scientific Instruments* 78: 064701.

34 Kempe, M.D., Reese, M.O., and Dameron, A.A. (2013). *The Review of Scientific Instruments* 84: 025109.

35 Dameron, A.A., Kempe, M.D., and Reese, M.O. (2014). *The Review of Scientific Instruments* 85: 075102.

36 Song, E.H., Park, Y.W., Choi, J.H. et al. (2011). *The Review of Scientific Instruments* 82: 054702.

37 Bouten, P.C.P., Nisato, G., Slikkerveer, P.J. (2006). Method for measuring a permeation rate, a test and an apparatus for measuring and testing. US6993956B2.

38 Paetzold, R., Winnacker, A., Henseler, D. et al. (2003). *The Review of Scientific Instruments* 74: 5147.

39 Reese, M.O., Dameron, A.A., and Kempe, M.D. (2011). *The Review of Scientific Instruments* 82: 085101.

40 Li, C., Cauwe, M., Mader, L. et al. (2020). *Coatings* 10: 19.

41 Tong, W. and Teng, A. (2000). Moisture diffusion monitoring in globtop encapsulant with microdielectric sensors. In: *International Symposium on Electronic Materials and Packaging (EMAP2000)* (Cat. No. 00EX458), pp. 464–469. Piscataway, NJ, USA: IEEE.

42 Choi, C.Y. (2008). Capacitive plate dielectrometer method and system for measuring dielectric properties. US20080111559A1.

43 Noel, G.T., Wood, V.E., McGinniss, V.D., et al. (1979). Measurement techniques and instruments suitable for life-prediction testing of photovoltaic arrays. Final Report *NASA-CR-157084*. Battelle Columbus Labs., OH, USA.

44 Gonon, P., Pham Hong, T., Lesaint, O. et al. (2005). *Polymer Testing* 24: 799.

45 Cotinaud, M., Bonniau, P., and Bunsell, A.R. (1982). *Journal of Materials Science* 17: 867.

46 Maxwell, I.D. and Pethrick, R.A. (1983). *Journal of Applied Polymer Science* 28: 2363.

47 Maffezzoli, A.M., Peterson, L., Seferis, J.C. et al. (1993). *Polymer Engineering and Science* 33: 75.

48 Aldrich, P.D., Thurow, S.K., McKennon, M.J., and Lyssy, M.E. (1987). *Polymer* 28: 2289.

49 Mazlan, N.S., Ramli, M.M., Abdullah, M.M.a.B. et al. (2017). *AIP Conf. Proc.* 1885: 020276.

50 Xie, X., Rieth, L., Merugu, S. et al. (2012). *Applied Physics Letters* 101: 093702.

51 Mirhashemihaghighi, S., Światowska, J., Maurice, V. et al. (2015). *Journal of the Electrochemical Society* 162: C377.

52 Chawla, V., Ruoho, M., Weber, M. et al. (2019). *Nanomaterials* 9: 88.

53 Bulusu, A., Singh, A., Wang, C.Y. et al. (2015). *Journal of Applied Physics* 118: 085501.

54 Kim, K., Pierron, O.N., and Graham, S. (2019). *Journal of Applied Physics* 125: 045301.

55 Kim, K., Luo, H., Zhu, T. et al. (2018). *Scientific Reports* 8: 4560.

56 Kim, K., Jia, X., Fuentes-Hernandez, C. et al. (2019). *ACS Applied Nano Materials* 2: 2525.

57 Singh, A.K., Kim, K., Chou, W.F. et al. (2020). *Thin Solid Films* 714: 138387.

58 Framil, D., Van Gompel, M., Bourgeois, F. et al. (2019). *Frontiers in Materials* 6: 319.

59 Top, M., Mulder, G., Prager, N. et al. (2018). *Surface and Coating Technology* 347: 46.

60 Tavares da Costa, M.V., Bolinsson, J., Neagu, R.C. et al. (2019). *Surface and Coating Technology* 370: 374.

61 Rogers, J.A., Someya, T., and Huang, Y. (2010). Materials and mechanics for stretchable electronics. *Science (80-)* 327: 1603–1607.

62 Su, Y., Liu, Z., Kim, S. et al. (2012). Mechanics of stretchable electronics with high fill factors. *International Journal of Solids and Structures* 49: 3416–3421.

63 Hamasha, M.M., Alzoubi, K., Switzer, J.C. et al. (2012). Reliability of sputtered aluminum thin film on flexible substrate under high cyclic bending fatigue conditions. *IEEE Transactions on Components, Packaging and Manufacturing Technology* 2: 2007–2016.

64 Janek, F., Saller, E., Müller, E. et al. (2020). Feasibility study of an automated assembly process for ultrathin chips. *Micromachines* 11: 654.

65 Akinwande, D., Petrone, N., and Hone, J. (2014). Two-dimensional flexible nanoelectronics. *Nature Communications* 5: 5678.

66 Dahiya, R.S. and Gennaro, S. (2013). Bendable ultra-thin chips on flexible foils. *IEEE Sensors Journal* 13: 4030–4037.

67 Kröninger, W.J. and Ossowski, L. (2004). Successful processing of thinned silicon chips thresholds and limits in mechanical properties. *Proceedings of the 2004 IEEE/SEMI Advanced Semiconductor Manufacturing Conference and Workshop* (IEEE Cat. No. 04CH37530), Boston, MA, USA (4–6 May 2004). pp. 232–236.

68 Schoenfelder, S., Ebert, M., Landesberger, C. et al. (2007). Investigations of the influence of dicing techniques on the strength properties of thin silicon. *Microelectronics and Reliability* 47: 168–178.

69 Harris, K.D., Elias, A.L., and Chung, H.J. (2016). Flexible electronics under strain: a review of mechanical characterization and durability enhancement strategies. *Journal of Materials Science* 51: 2771–2805.

70 Endler, S., Angelopoulos, E.A., Harendt, C. et al. (2011). Bestimmung der mechanischen Stabilität ultradünner Chips. In: *Mikrosystemtechnik Kongress*, 737–740. Berling, Germany: VDE Verlag GmbH.

71 Palavesam, N. and Landesberger, C. Bock K. (2014). Investigations of the fracture strength of thin silicon dies embedded in flexible foil substrates. *Proceedings of the 2014 IEEE 20th International Symposium for Design and Technology in Electronic Packaging (SIITME)*, Bucharest, Romania (23–26 October 2014), pp. 267–271. IEEE.

72 Endler, S., Hoang, T., Angelopoulos, E.A., et al. (2011). Mechanical characterisation of ultra-thin chips. *Proceedings of the IEEE—2011 Semiconductor Conference Dresden: Technology, Design, Packaging, Simulation and Test, SCD 2011—International Conference, Workshop and Table-Top Exhibition*, Dresden, Germany (27–28 September 2011), pp. 3–6. IEEE.

73 Al Ahmar, J., Wiss, E., and Wiese, S. (2018). Four-point-bending experiments on multilayer ceramic capacitors: microstructural details on crack initiation and propagation. *Proceedings of the 2018, 19th International Conference on Thermal, Mechanical and Multi-Physics Simulation and Experiments in*

Microelectronics and Microsystems (EuroSimE), Toulouse, France (15–18 April 2018), pp. 1–6. IEEE.

74 Li, H.U. and Jackson, T.N. (2016). Flexibility testing strategies and apparatus for flexible electronics. *IEEE Transactions on Electron Devices* 63: 1934–1939.

75 Wright, D.N., Vardoy, A.S.B., Belle, B.D., et al. (2017). Bending machine for testing reliability of flexible electronics. *Proceedings of the 2017 IMAPS Nordic Conference on Microelectronics Packaging (NordPac)*, Gothenburg, Sweden (18–20 June 2017), pp. 47–52. IEEE.

76 Kim, J.H., Lee, T.I., Shin, J.W. et al. (2016). Bending properties of anisotropic conductive films assembled chip-in-flex packages for wearable electronics applications. *IEEE Transactions on Components, Packaging and Manufacturing Technology* 6: 208–215.

77 Happonen, T., Hakkinen, J., and Fabritius, T. (2015). Cyclic bending reliability of silk screen printed silver traces on plastic and paper substrates. *IEEE Transactions on Device and Materials Reliability* 15: 394–401.

78 Gleskova, H., Cheng, I.C., Wagner, S. et al. (2006). Mechanics of thin-film transistors and solar cells on flexible substrates. *Solar Energy* 80: 687–693.

79 Lewis, J. (2006). Material challenge for outside of the active devices. *Materials Today* 9: 38–45.

80 Chen, Q., Xu, L., Jing, C., et al. (2008). Flexible device and component reliability study using simulations. *Proceedings of the EuroSimE 2008—International Conference on Thermal, Mechanical and Multi-Physics Simulation and Experiments in Microelectronics and Micro-Systems*, Freiburg im Breisgau, Germany (20–23 April 2008), pp. 1–5. IEEE.

81 Angelopoulos, E.A., Zimmermann, M., Appel W., et al. (2010). Ultra-thin chip technology for system-in-foil applications. *Proceedings of the 2010 International Electron Devices Meeting (IEDM)*, San Francisco, CA, USA (6–8 December 2010), pp. 2.5.1–2.5.4. IEEE.

82 Van Den Ende, D.A., Van De Wiel, H.J., Kusters, R.H.L. et al. (2014). Mechanical and electrical properties of ultra-thin chips and flexible electronics assemblies during bending. *Microelectronics and Reliability* 54: 2860–2870.

83 Zhu, X., Li, X., Nistala, R.R., et al. 4 Point bending characterization of interfacial adhesion strength of SiN/Cu film stack. *Proceedings of the 2017 IEEE 24th International Symposium on the Physical and Failure Analysis of Integrated Circuits (IPFA)*, Chengdu, China (4–7 July 2017), pp. 1–4. IEEE.

84 Lund, E. and Finstad, T.G. (2004). Design and construction of a four-point bending based set-up for measurement of piezoresistance in semiconductors. *The Review of Scientific Instruments* 75: 4960–4966.

85 Zheng, S.X., Luo, X.M., Wang, D. et al. (2018). A novel evaluation strategy for fatigue reliability of flexible nanoscale films. *Materials Research Express* 5: 035012.

86 Kim, B.J., Shin, H.A.S., Jung, S.Y. et al. (2013). Crack nucleation during mechanical fatigue in thin metal films on flexible substrates. *Acta Materialia* 61: 3473–3481.

87 Ernst, D., Zerna, T., and Wolter, K.J. (2011). Influences of organic materials on packaging technologies and their consideration for lifetime evaluation. *Proceedings of the 2011 34th International Spring Seminar on Electronics Technology (ISSE)*, Tratanska Lomnica, Slovakia (11–15 May 2011), pp. 288–293. IEEE.

88 Kim, J.H., Lee, T.I., Yoon, D.J., et al. (2017). Effects of anisotropic conductive films (ACFs) gap heights on the bending reliability of chip-in-flex (CIF) packages for wearable electronics applications. *Proceedings of the 2017 IEEE 67th Electronic Components and Technology Conference (ECTC)*, Orlando, FL, USA (30 May–2 June 2017), pp. 2161–2167. IEEE.

89 Harendt, C., Kostelnik, J., Kugler, A. et al. (2015). Hybrid Systems in Foil (HySiF) exploiting ultra-thin flexible chips. *Solid State Electronics* 113: 101–108.

90 Lorenz, E., Niemann, N., Koyuncu, M., et al. Reliability characterization of blind-hole vias for a System-in-Foil. *Proceedings of the 5th Electronics System-integration Technology Conference (ESTC)*, Helsinki, Finland (16–18 September 2014), IEEE.

91 Ahn, S.H. and Guo, L.J. (2008). High-speed roll-to-roll nanoimprint lithography on flexible plastic substrates. *Advanced Materials* 20: 2044–2049.

92 Dupont, S.R., Oliver, M., Krebs, F.C. et al. (2012). Interlayer adhesion in roll-to-roll processed flexible inverted polymer solar cells. *Solar Energy Materials & Solar Cells* 97: 171–175.

93 Forrest, S.R. (2004). The path to ubiquitous and low-cost organic electronic appliances on plastic. *Nature* 428: 911–918.

94 Gamerith, S., Klug, A., Scheiber, H. et al. (2007). Direct ink-jet printing of Ag-Cu nanoparticle and Ag-precursor based electrodes for OFET applications. *Advanced Functional Materials* 17: 3111–3118.

95 Wagner, S. and Bauer, S. (2012). Materials for stretchable electronics. *MRS Bulletin* 37: 207–213.

96 Cordill, M.J., Fischer, F.D., Rammerstorfer, F.G. et al. (2010). Adhesion energies of Cr thin films on polyimide determined from buckling: experiment and model. *Acta Materialia* 58: 5520–5531.

97 Agrawal, D.C. and Raj, R. (1989). Measurement of the ultimate shear strength of a metal-ceramic interface. *Acta Metallurgica* 37: 1265–1270.

98 Taylor, A.A., Edlmayr, V., Cordill, M.J. et al. (2011). The effect of film thickness variations in periodic cracking: analysis and experiments. *Surface and Coatings Technology* 206: 1830–1836.

99 Taylor, A.A., Cordill, M.J., and Dehm, G. (2012). On the limits of the interfacial yield model for fragmentation testing of brittle films on polymer substrates. *Philosophical Magazine* 92: 3363–3380.

100 Cox, H.L. (1952). The elasticity and strength of paper and other fibrous materials. *British Journal of Applied Physics* 3: 72–79.

101 Kelly Tyson, W.R. (1965). Tensile properties of fibre-reinforced metals: copper/tungsten and copper/molybdenum. *Journal of the Mechanics and Physics of Solids* 13: 329–350.

102 Cordill, M.J., Taylor, A.A., Berger, J. et al. (2012). Robust mechanical performance of chromium-coated polyethylene terephthalate over a broad range of conditions. *Philosophical Magazine* 92: 3346–3362.

103 Cordill, M.J., Taylor, A.A., and Schmidegg, K. (2011). Thickness effects on the fracture of chromium films on polyethylene terephthalate. *BHM Berg- und Hüttenmännische Monatshefte* 156: 434–437.

104 Vellinga, W.P., Hosson, J.T.M.D., and Bouten, P. (2011). Direct measurement of intrinsic critical strain and internal strain in barrier films. *Journal of Applied Physics* 110 (4): 1–120.

105 Glushko, O. and Cordill, M.J. (2016). Electrical resistance of metal films on polymer substrates under tension. *Experimental Techniques* 40 (1): 303–310.

106 Glushko, O., Marx, V.M., Kirchlechner, C. et al. (2014). Recovery of electrical resistance in copper films on polyethylene terephthalate subjected to a tensile strain. *Thin Solid Films* 552: 141–145.

107 Lu, N., Wang, X., Suo, Z., and Vlassak, J. (2009). Failure by simultaneous grain growth, strain localization, and interface debonding in metal films on polymer substrates. *Journal of Materials Research* 24: 379–385.

108 Peng, C., Jia, Z., Bianculli, D. et al. (2011). In situ electro-mechanical experiments and mechanics modeling of tensile cracking in indium tin oxide thin films on polyimide substrates. *Journal of Applied Physics* 109: 1–8.

109 Lu, N., Xi, W., Suo, Z. et al. (2007). Metal films on polymer substrates stretched beyond 50%. *Applied Physics Letters* 91 (22): 911–919.

110 Glushko, O. and Cordill, M.J. (2014). Electrical resistance decrease due to grain coarsening under cyclic deformation. *JOM* 66: 1–4.

111 Putz, B., Schoeppner, R.L., Glushko, O. et al. (2015). Improved electro-mechanical performance of gold films on polyimide without adhesion layers. *Scripta Materialia* 102: 23–26.

112 Jin, H., Lu, W.Y., Cordill, M.J. et al. (2011). In situ study of cracking and buckling of chromium films on PET substrates. *Experimental Mechanics* 51: 219–227.

113 Olliges, S., Gruber, P.A., Orso, S. et al. (2008). In situ observation of cracks in gold nano-interconnects on flexible substrates. *Scripta Materialia* 58: 175–178.

114 Cordill, M.J., Glushko, O., Kreith, J. et al. Measuring electro-mechanical properties of thin films on polymer substrates. *Microelectronic Engineering* 137: 96–100.

115 Lee, C., Wei, X., Kysar, J.W. et al. (2008). Measurement of the elastic properties and intrinsic strength of monolayer graphene. *Science* 321: 385–388.

116 Sharpe, W.N. (2003). Murray lecture tensile testing at the micrometer scale: opportunities in experimental mechanics. *Experimental Mechanics* 43: 228–237.

117 Oliver, W.C. and Pharr, G.M. (1992). An improved technique for determining hardness and elastic modulus using load and displacement sensing indentation experiments. *Journal of Materials Research* 7: 1564–1583.

118 Oliver, W.C. and Pharr, G.M. (2004). Measurement of hardness and elastic modulus by instrumented indentation: advances in understanding and refinements to methodology. *Journal of Materials Research* 19: 3–20.

119 Emery, Q., Remec, M., Paramasivam, G. et al. (2022). Encapsulation and outdoor testing of perovskite solar cells: comparing industrially relevant process with a simplified lab procedure. *ACS Applied Materials & Interfaces* 14: 5159–5167.

5

Flexible Electronic Encapsulation

Tao Yu

Science Center for Flexible Electronics (FSCFE) and Xi'an Institute of Flexible Electronics (IFE), Northwestern Polytechnical University, 127 West Youyi Road, Xi'an 710072, China

The development of encapsulation technology and materials mutually complements, constrains, and promotes each other. In recent years, the growth of encapsulating materials has been rapid and significant. The choice of materials for OLED encapsulation is critical, as their electrical, physical, and chemical properties impact the performance of OLED devices. A qualified encapsulating material should possess several characteristics, such as good thermal conductivity, a low coefficient of thermal expansion, or match the chip to prevent thermal stress damage, adequate stiffness and strength to support and protect the chip, and good mechanical and electrical properties. Moreover, a low dielectric constant improves electrical signal transmission characteristics, and excellent encapsulating processability helps to meet commercialization requirements [1]. Usually, flexible electronic encapsulating materials are classified as inorganic, organic, and organic–inorganic hybrid materials. Devices encapsulated with inorganic materials have better hermeticity, higher cost, and are suitable for use in applications with high reliability requirements. For example, metal encapsulating materials have the advantages of rich thermal conductivity, easy processing, and molding, but due to the difference between the coefficient of thermal expansion of metal and coincidental pieces, high density, high cost, and other disadvantages, they limit the application of metal encapsulating in some areas. Ceramic encapsulating materials are less dense, have higher thermal conductivity, and have expansion coefficient matching, making them comprehensively better encapsulating materials, but their high cost and more complex production and processing make them more suitable for advanced microencapsulating of electronic devices in aerospace and military fields. Organic materials have a lower dielectric constant, are light materials with ductility, are easy to thin and curl, and are easy to achieve high-density interconnections, so they have become the mainstream of optoelectronic encapsulating materials. More than 90% of the transistors and 70–80% of the integrated circuits (IC) use organic encapsulating materials. In addition, inorganic–organic hybrid materials are widely used in many applications due to their unique chemical and physical properties and the

Flexible Electronic Packaging and Encapsulation Technology, First Edition.
Edited by Hong Meng and Wei Huang.

advantages of both organic and inorganic materials. The last part of this chapter focuses on the application of inorganic–organic materials in the field of flexible electronics and their future development trends because of their good tunability and promising applications in the field of electronic encapsulating in particular.

5.1 Inorganic Encapsulating Material

Inorganic materials for electronic encapsulation can be divided into metal and ceramic encapsulating materials. Metal encapsulation utilizes metals such as copper, aluminum, tungsten, molybdenum, or copper–tungsten alloys as the main material for the tube shell, with the lead connecting internal and external circuits. This is advantageous due to its good mechanical strength, thermal conductivity, electromagnetic shielding function, and ease of mechanical processing. On the other hand, ceramic encapsulation materials provide an excellent hermetic seal and have advantages such as good moisture resistance, high mechanical strength, stable chemical properties, and good thermal properties like thermal expansion rate and thermal conductivity. These materials, including aluminum oxide, aluminum nitride, beryllium oxide, and boron nitride (BN), are often used for encapsulation shells and substrates. Overall, the choice of encapsulating material will depend on the specific requirements of the electronic components being used, with both metal and ceramic encapsulating materials providing unique advantages in different situations.

5.1.1 Metal Encapsulating Material

5.1.1.1 Copper, Aluminum

Copper's thermal conductivity of 401 W (m/K) can be applied to the high thermal or electrical conductivity of the encapsulation, making it the ideal encapsulation shell; however, its coefficient of thermal expansion is very high (16.5×10^{-6}/K) so that it will cause a lot of thermal stress on the substrate, generally by using several small substrates instead of large substrates to reduce thermal stress.

Aluminum and its alloys, because of their lightweight, low price, easy processing, and high thermal conductivity, are often used in microwave IC case. However, aluminum has a higher coefficient of thermal expansion (23.2×10^{-6}/K), which can produce greater thermal stress and failure. Considering the thermal adaptation of copper and aluminum to the substrate and the almost linear increase in the thickness of the copper–aluminum intermetallic compound layer at any time (Figure 5.1a) [2]. The diffusion coefficient of copper–aluminum metal compounds increases with increasing temperature (Figure 5.1b), so copper and aluminum are greatly influenced by time as well as temperature [3], leading to great difficulties for copper and aluminum in electronic encapsulating.

Molybdenum (Mo) has an average linear expansion coefficient of 3.21×10^{-6}/K between 300 and 4.2 K. Mo has high thermal conductivity and low specific heat capacity at low temperatures [4]. Mo is widely used as an all-around excellent electronic encapsulating material for lead substrates and cold stages in CdHdTe

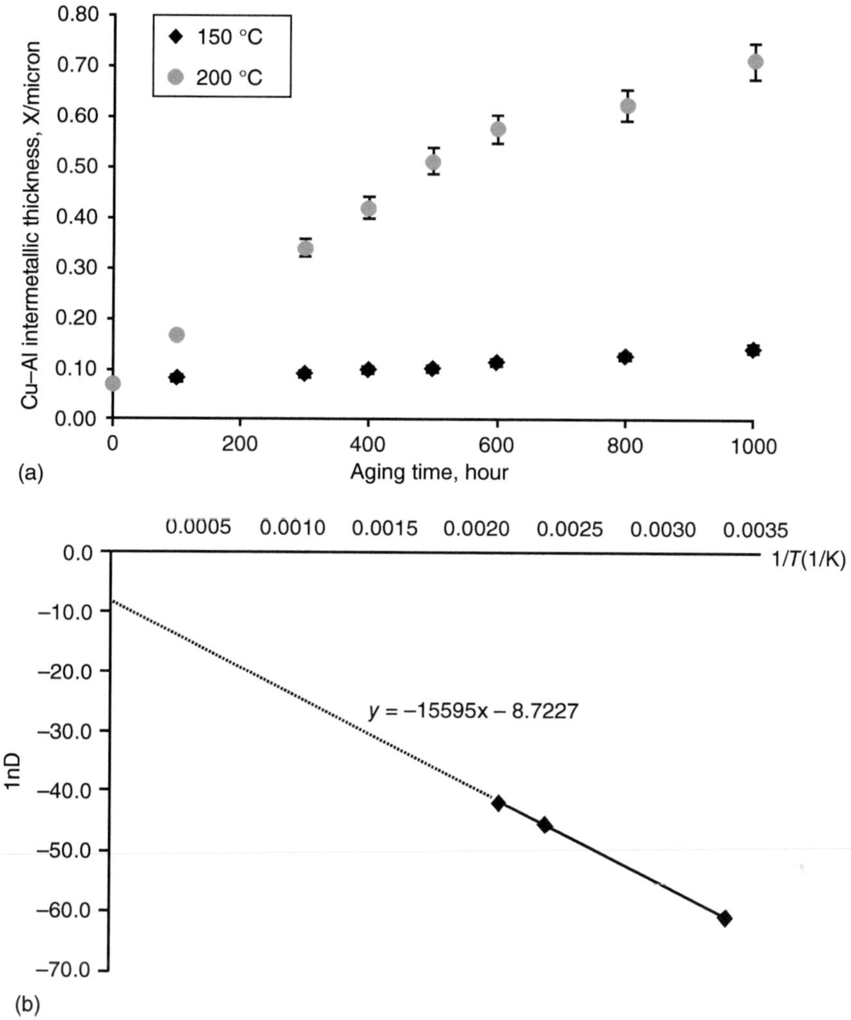

Figure 5.1 (a) Thickness of Cu–Al intermetallic compound versus aging time (h) at 150 °C; (b) diffusion coefficient (ln*D*) versus *temperature* (1/*T*) for Cu–Al interdiffusion system; (2) tungsten (W) and molybdenum (Mo). Source: Reproduced with permission from Chee Wei and Razak Daud [2]; © 2002, MCB UP Limited.

infrared focal plane detector assemblies (Figure 5.2) [5]. W has a coefficient of thermal expansion similar to that of Si and GaAs and has good thermal conductivity, which can be used as a support material for chips. Mo has a mixed tough-brittle fracture mode at 300 K based on the macroscopic and microscopic morphology of the tensile fracture mode, where the material changes from ductile at 300 K to brittle at 77 and 4.2 K. The fracture behaviors are crystal penetration fracture at 77 K and intergranular fracture at 4.2 K, respectively [5, 6]. Due to the high price of W and Mo, they are not suitable for mass use, and in addition, their high density makes them unsuitable for aviation and aerospace.

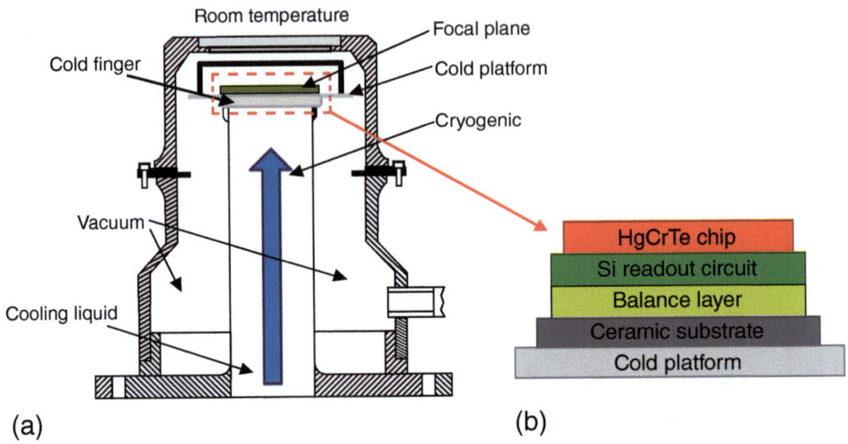

Figure 5.2 Schematic diagram of the CdHgTe IR focal plane detector encapsulation dewar structure; (a) cooled encapsulation dewar; (b) multilayer material structure. Source: Reproduced with permission from Su et al. [5]; © 2022, Elsevier B.V.

5.1.1.2 Favorable Alloys

Valuable alloys are also known as controlled expansion alloys or sealing alloys because they have a coefficient of thermal expansion similar to that of glass and ceramics, resulting in sealing requirements that match the sealing material [7]. Favorable materials have a low coefficient of thermal expansion, low density, high thermal conductivity, good thermomechanical stability, and compatibility with microelectronic assembly techniques [8]. They are most widely used in low-power-density metal encapsulations. However, its low thermal conductivity, high resistivity, and high density have greatly limited its wide application.

5.1.1.3 Copper–Tungsten Alloy (Cu–W)

Tungsten–copper (W–Cu) composites/pseudo-alloys with excellent properties are well suited for use as electrode materials, functional gradient materials, electronic encapsulating materials, electrical contacts, and other aerospace and military candidates. W–Cu alloys are considered good heat dissipation materials for high-power microelectronic devices [9]. With the increase in integrated products and the increase in current and voltage, encapsulating is an important factor in power electronics efficiency, which requires more stable mechanical support, higher heat dissipation, and greater electrical interconnections [10]. Compared to other materials, W–Cu alloys provide efficient thermal management to avoid warpage and effectively reduce thermal stress. In addition, the increase in W content reduces the thermal expansion of higher thermal output products. W–Cu alloys have the necessary properties of low coefficient of thermal expansion and high thermal conductivity for future three-dimensional encapsulations with better thermal management, smaller size, and more complex geometries. Challenges exist in W–Cu alloys due to the presence of large tungsten particles in the material as a result of traditional methods of preparing W–Cu alloys.

5.1.2 Ceramic Encapsulating Material

5.1.2.1 Al_2O_3 Ceramic Encapsulation Material

Pure alumina (Al_2O_3) is a ceramic that exhibits a high thermal conductivity of about 40 W/mK at 20 °C. However, its sintering temperature in the air is about 1700 °C, which is too high for LED encapsulating applications. Therefore, a small amount of glass material is usually added to Al_2O_3 to lower the sintering temperature. However, the thermal conductivity of Al_2O_3 glass composites is reduced to about 1.1, which is not sufficient to ensure suitable operating conditions for high-power LEDs [11]. Therefore, one of the main problems with Al_2O_3 glass composites currently used as encapsulation materials for LEDs (light-emitting diodes) is the deterioration of their thermal emission properties due to the low thermal conductivity of the glass.

5.1.2.2 AlN Ceramic Encapsulation Materials

Aluminum nitride (AlN) ceramic substrate is a new type of encapsulation material with excellent electrical and thermal properties and higher thermal conductivity compared to alumina (generally more than five times that of alumina ceramics) for increased power, high lead, and large-size chips. Reliability problems have arisen due to the mismatch in thermal expansion coefficients between metal, ceramic, and semiconductor materials in traditional power electronics encapsulations. To improve performance by eliminating DBCs and substrates [12] (Figure 5.3). The ability to bond AlN substrates directly to AlSiC heat sinks is generally achieved by using a Cu–Al binary system through transient liquid phase bonding [13]. Aluminum nitride is a hard material, so it can be made into very thin substrates for different encapsulating substrate applications. However, the preparation process of aluminum nitride is complicated and costly, thus limiting its mass production and application.

5.1.2.3 BeO Ceramic Encapsulation Material

Beryllium oxide (BeO) has a thermal conductivity close to that of metallic materials and is also a good insulating material. BeO has a low dielectric constant, low dielectric loss, and is highly adaptable to the encapsulation process. Compact optical encapsulations for PIN arrays are made of BeO, which can effectively remove the heat generated by the array. The design simplifies the testing and assembly of optical link systems, as each fiber strip terminated with MT collars can be easily connected/disconnected while maintaining good optical coupling efficiency [14]. However, BeO is highly toxic, which limits the promotion of both preparations as well as applications.

5.1.2.4 BN Ceramic Encapsulation Materials

BN is known as "white graphene" due to its high thermal conductivity, low dielectric constant, good insulation properties, etc. [15], and is often used in radar windows, tube holders, and housings of high-power transistors, heat sinks, and microwave output windows. However, cubic BN is expensive and not suitable for the

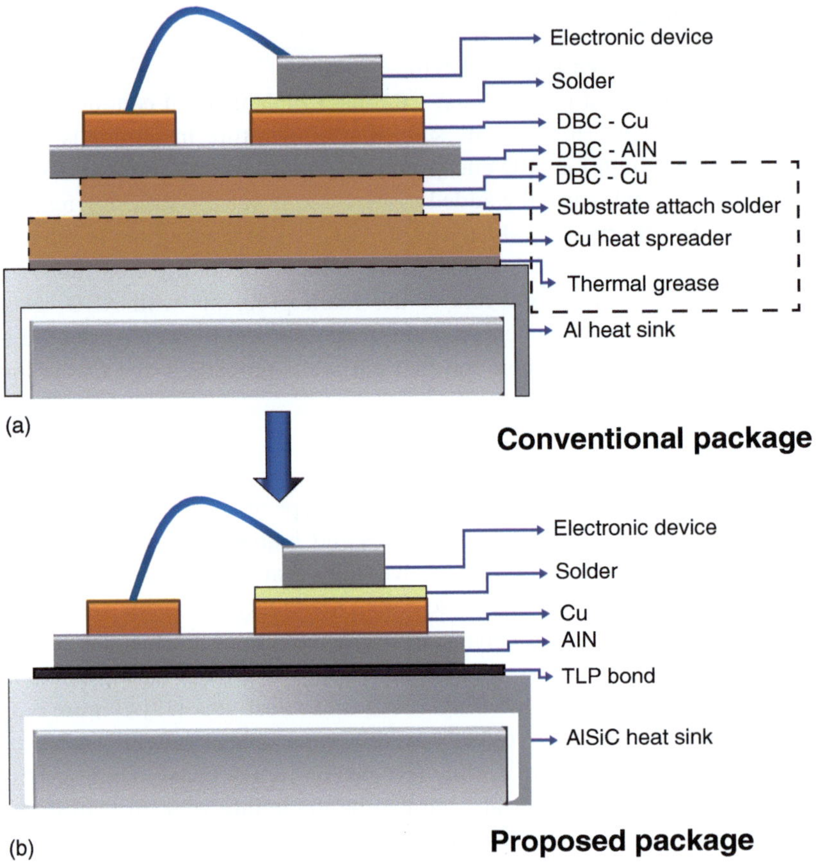

Figure 5.3 (a) Conventional electronic encapsulation display device, DBC, substrate, and heat sink; (b) encapsulation device without standard DBC and substrate. Source: Reproduced with permission from Pahinkar et al. [13]; © 2018, John Wiley and Sons.

production of ceramics with high thermal conductivity, and the mismatch between the coefficient of thermal expansion and silicon also limits its application.

5.1.3 New Trend in Inorganic Encapsulating Materials Combined with Flexible Electronic Technology

Flexible electronics technology is facing a challenge in electronic encapsulation, which involves seamlessly encapsulating components integrated on a flexible substrate to achieve the desired function. Inorganic encapsulating materials are being applied to flexible substrates using advanced encapsulating technologies such as wafer-level encapsulating and three-dimensional encapsulating systems (3-D SIP), which are rapidly penetrating the electronic component market. To reduce costs, a new approach to panel-level encapsulating technology called (PLP) has been proposed by Schein et al. This approach uses printed circuit board (PCB) laminate cores for chip embedding while avoiding high bumps to form chip through-holes.

With the development of IoT applications, the thermocompression bonding (TCB) process with nonconductive film (NCF) is considered an effective solution for fine-pitch applications. However, production volumes were limited to a small level due to its expensive assembly cost. To mitigate the cost impact, many participants introduced multimode bonding methods. Okaniwa et al. presented the development of a new NCF with a new resin combination. The properties of the developed NCF met the requirements of the collective bonding process and other multimode bonding processes. TCB experiments using the newly developed NCF confirmed that reliable solder joints were formed without abnormalities during the bonding stage, even after thermal exposure at 130 °C for 180 minutes.

The subject of intermetallic compounds has received increasing attention in recent years due to their growing influence on the reliability of small solder joints and on all-metal interconnects manufactured by solid–liquid interdiffusion. The formation of the Cu_6Sn_5 phase at the interface between tin-based solder and copper solder disc metallization marks the actual joining process of solder interconnects for electronic components. Mueller et al. discussed solidification experiments without interfacial reactions on SnCu solder balls (Ø270 μm) with copper content ranging from 0.25 to 3.0 wt% [16]. Microstructural analysis showed that incipient Cu_6Sn_5 intermetallic compounds were formed only in samples with a Cu mass fraction of 1.5% or higher.

With the increasing requirement for high I/O density, the encapsulation size is shrinking, and the requirement for fine-pitch solder joints is increasing to achieve miniaturized 3D interconnects. With the trend toward miniaturization of solder size, the consumption of under bump metallization (UBM) and the formation of Kirkendall voids at the Cu-UBM interface using tin-based solder during the product lifetime may create reliability issues. Hou et al. proposed to study the solid-state interfacial response of Ni-UBM and Cu/Ni-UBM alternative metals with Sn by in situ resistance measurements [17]. The resistivity of the formed IMC phase was estimated from the measured resistance changes with known reduction rates of UBM and Sn layer thicknesses. Kinetic parameters such as power factor and activation energy were also extracted from the measured resistance changes with time. In order to study the reaction and phase formation between UBM and solder materials in three-dimensional applications, the method has the advantage of fast and accurate measurements.

Ultrathin sensors have a wide range of applications. The next paper is titled Ernst et al. proposed active magnetic bearings (AMBs) for ultrathin sensors [18]. The focus of this paper is to optimize the encapsulating process of AMB polyimide (PI) Hall sensors. Currently, the chip soldering method is preferred, but bonding is also considered as a promising technique. The next step is to describe the electromagnetic properties of bonded sensors and evaluate them in an AMB device.

Current electronic encapsulations typically use die compounds, glues, and lead frames or laminated substrates. The resulting encapsulation families such as ball grid arrays (BGA), quad planar leadless (QFN), and quad planar encapsulations (QFP) are well known and have been in use for many years. Chip embedding on copper lead frames is a new encapsulating technology that is particularly suitable for power devices. There is limited available knowledge about long-term reliability compared to conventional encapsulating. The last paper was presented by Fruehauf et al. on the reliability of chip-embedded encapsulations [19]. In addition to traditional reliability tests, new test methods have been developed to meet the specific requirements of typical applications (e.g. solar inverters).

5.2 Organic Encapsulating Material

In addition to inorganic systems, encapsulating materials include organic systems and composite systems. Organic materials are often used as substrates and sealing materials. Generally, inorganic encapsulating materials have a lower coefficient of thermal expansion and higher thermal conductivity, with high reliability, but the dielectric constant of these materials is relatively high, usually not suitable for use in high-frequency circuits; in contrast, organic encapsulating materials, although the thermal expansion rate is slightly higher, are not conducive to heat dissipation, but have a lower dielectric constant, and the material is light, easy to process, has ductility, and is easy to thin. From the perspective of development trends, in recent years, with thin and light flexible structures and flexible curling and folding characteristics, easy-to-achieve high-density interconnection of flexible organic cncapsulating materials demand is growing rapidly, and the proportion of the overall encapsulating materials market is increasing. To date, the market size of encapsulating materials exceeds \$20 billion, of which encapsulating substrates account for more than 60%, and organic substrates are expanding year by year with a broad market prospect.

At present, the widely used organic encapsulating materials are polymers, such as epoxy resin (EP), bismaleimide (BMI), bismaleimide triazine (BT), PI, and silicone. In the following subsections, we will introduce the characteristics of each material and the application of organic encapsulating materials in the preparation of flexible electronic devices and analyze the development trend and outlook.

5.2.1 Polymer Encapsulating Material

Before systematically introducing polymer-based encapsulation materials, I would like to explain that in order to prevent the characteristics of encapsulated products from deteriorating under the action of time and environmental conditions, encapsulation materials are required to have a low coefficient of thermal expansion, low water vapor transmission, good adhesion of lead pins, and no impurities affecting the performance of the components within the material. From the current process and material development status, a single resin is difficult to fully meet the above requirements, so most industrial polymer encapsulating materials are composite materials composed of resins, accelerators, anti-flame agents, coupling agents, fillers and pigments, lubricants, and other components. The next section introduces several types of resin materials commonly used for electronic encapsulation and briefly presents a few cases of modification to inspire the readers [20].

5.2.1.1 Epoxy Resins

Polymer-based encapsulation materials, also known as plastic encapsulation materials, are the most extensively researched and commonly used organic encapsulation materials. Currently, over 90% of electronic devices and ICs are encapsulated in plastics, with EP materials comprising more than 90% of plastic encapsulation materials due to their exceptional advantages. These advantages include low shrinkage, high adhesive strength, good corrosion resistance, excellent electrical properties, easy curing and molding, easy production and processing, and low cost [21].

EP is a polymer with the molecular formula $(C_{11}H_{12}O_3)n$. As long as the polymer molecular chain contains more than two epoxy groups, it can be called EP. This type of resin can be obtained by the condensation reaction of epichlorohydrin with bisphenol A or polyol. As a thermosetting resin, it cures and crosslinks when subjected to heat, pressure, or UV light, gradually hardens and molds, and does not soften when subjected to further heat. EP materials are widely used in electronics and electrical appliances because of their many outstanding properties. However, after curing unmodified EP, its texture is hard and brittle, easy to fracture, has poor impact resistance, fatigue resistance, moisture resistance, etc. Therefore, with the increasing integration of ICs, increasingly fine wiring, miniaturization of chip size, and increase in encapsulating speed, in order to broaden the application of EP to meet the demand for IC materials, researchers at home and abroad have been committed to the In order to broaden the application of EPs to meet the demand of IC materials, researchers at home and abroad have been working on the modification of EPs in order to obtain EPs with excellent heat and humidity resistance, high purity, low stress, and low linear expansion coefficient. For example, by using phenolic epoxy curing agents and their catalysts, the reaction activity can be greatly improved, thus effectively shortening the curing time while still maintaining the electrical insulation and reliability of the material [22]; by increasing the content of fillers in EPs to reduce their water absorption; and by improving the low molecular weight and multifunctional group of EPs, as well as introducing rigid groups in the flexible chain segments to improve the resin's high thermal conductivity and comprehensive performance [23].

The current composition of commercial resin sealing materials is EP (base resin and curing agent), filler (silica), curing accelerator, coupling agent (for improving wettability and adhesion with filler), flame retardant, rapture-giving agent, coloring agent, ion trapping agent (curing of corrosive ions), release agent, etc.

5.2.1.2 Polyimide Resins

PI resin is the most commonly used material in flexible encapsulating substrates. PI refers to a class of polymers containing an imide ring on the main chain, among which the polymer containing a phthalimide structure is particularly important, mainly synthesized by diamine and dianhydride. However, its high thermal expansion coefficient and low transparency limit its application in flexible electronics. In order to improve the performance of PI resins as flexible electronic encapsulating materials, researchers have linked fluorine atoms and alicyclic units with PI from the perspective of molecular structure design to increase their transparency and give them other excellent properties [24]. For example, by introducing dicyclopentadiene and cyclohexane structures into the PI molecular chain, Fukukawa et al. significantly improved its transparency while still maintaining excellent thermal stability of the material and could make the thermal expansion coefficient of PI significantly lower, with the lowest value being less than that of pure copper (17×10^{-6}/K) to 8×10^{-6}/K [25].

Currently, PI resin materials are widely used in flexible PCBs and IC carrier boards, especially in strip IC carrier boards. Flexible film substrates are mainly divided into three-layer adhesive substrates and two-layer nonadhesive substrates, as shown in Figure 5.4. The three-layer adhesive board is mainly composed of PI film, copper foil, and binder, initially mainly used in military electronics such as launch vehicles, cruise missiles, and space satellites, and later also extended to a variety of civilian electronic products; the nonadhesive board is made of PI film and copper foil directly bonded; its thickness is smaller, suitable for high-density wiring; heat resistance; thin line; and thinning has obvious advantages; the product is widely used in consumer electronics, automotive electronics, and other fields; this is the main development direction of the future flexible encapsulating substrate.

Driven by the demand for line miniaturization, lightweight, thinness, and high heat dissipation, PI resins are mainly used for high-density, multi-I/O-count ICs encapsulating for portable electronics.

5.2.1.3 Organic Silicon

Silicone materials are a class of polymers with Si—O—Si bond as the main chain structure and side chains connected with various organic groups through silicon atoms, mainly used in semiconductor chip coatings and LED encapsulating adhesives. They have high bond energy (452 kJ/mol), a large bond length (0.165 nm), and bond angle (109°) of Si—O bond, low rotation resistance of Si—O bond, and very good flexibility of chain segments.

These intrinsic properties make silicone materials have a series of excellent performances: excellent high and low temperature resistance, can work in the temperature range of −50–250 °C; excellent aging resistance (heat resistance, UV resistance, ozone aging resistance), with decades of service life in the natural environment; good water repellent properties; surface energy as low as 21–22 mN/m^2; very high light transmission and very low internal stress; good electrical insulation properties, etc.

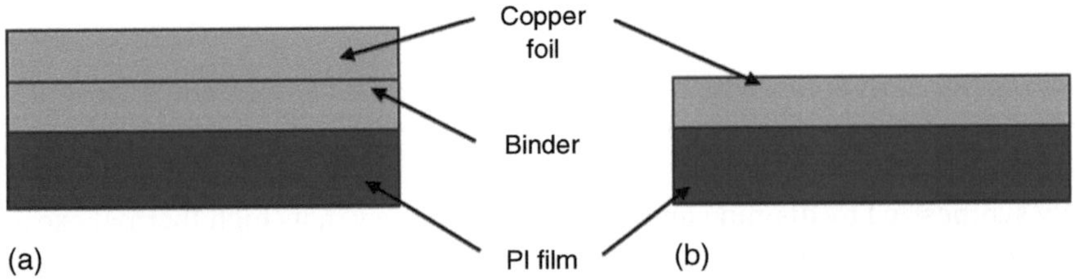

Figure 5.4 Schematic diagram of the structure of flexible copper-clad substrates. (a) Adhesive type three-layer structure; (b) Non-adhesive two-layer construction.

Silicone encapsulation materials can be classified into two main categories: low refractive index and high refractive index. In some cases, medium refractive index products are also used. The refractive index of low-refractive index silicone encapsulation materials typically ranges from 1.40 to 1.45, and they are primarily composed of methyl silicone. High-refractive index silicone encapsulation materials, on the other hand, typically have a refractive index ranging from 1.50 to 1.55, and they are mostly made of phenyl silicone.

When the refractive index of the encapsulant is significantly different from that of the chip, some of the light is reflected back into the chip, which can negatively impact the light output. This is particularly true for low-refractive index silicone encapsulants. Therefore, it is crucial to select the appropriate encapsulation material to ensure optimal light output.

The current research on OLED encapsulation materials is the use of silicone-modified EP, so that the material combines the excellent UV aging resistance and heat resistance of silicone resin and many characteristics of EP. For example, 4-vinyl ethylene oxide is mixed with a variety of phenylsilane reactions, made of excellent cold and heat shock resistance and UV aging resistance, high light transmission, high thermal expansion coefficient, and chip similar to the silicone-modified epoxy encapsulation materials for LED. In order to simplify and shorten the process, the silicone anhydride curing agent can also be mixed with EP directly, and the resulting silicone-modified EP encapsulation material has 88% light transmittance (the light transmittance remains 90% after 40 hours of irradiation by a 400 nm UV lamp). The refractive index of the material can be effectively improved by adding metal compounds to the silicone resin, and Basin et al. effectively improved the device's luminous efficiency by adding 2.5–5.0% TiO_2 and ZrO_2 to GaN-type LED devices [26]. The raw material of silicone encapsulation is a two-component, colorless, transparent liquid-like substance, and the A component generally includes base adhesive, catalyst, and tackifier, and the B component generally includes base adhesive, crosslinker, and inhibitor. A small amount of reinforcing filler and pigment can be added optionally, according to the usage.

5.2.1.4 Bismaleimide

BMI is an important class of bifunctional compounds derived from PI resin system, which can react with a variety of compounds with maleimide as the end group and is a high temperature-resistant thermosetting resin with excellent performance, which can maintain high physical and mechanical properties in a wide temperature range. Since the molecular chain of BMI resin has high rigidity and glass transition temperature, its dipole loss is small in a wide temperature range, so the electrical properties are very good, with the advantages of low dielectric constant, small dielectric loss, and large volume resistance. These properties can still be maintained at a high level over a wide range of temperatures and frequencies. Since BMI resin can meet the requirements of manufacturing high-speed digital and high-frequency PCB substrates, it is often used to make high-performance multilayer boards. Pure BMI resins are very brittle and have no adhesion. It must be toughened and viscosity modified to retain its heat resistance and electrical properties but also to improve its toughness and adhesion. At present, the main ways to modify BMI resin for PCB are diamine modification, allyl phenyl compound modification, cyanate modification EP modification, and so on. For example, through the introduction of allyl phenyl compounds to significantly improve the toughness of BMI, and for example, through cyanate-modified BMI resin, which, due to its excellent mechanical properties, stability, and other advantages, has become one of the important substrate materials for encapsulating substrates.

5.2.1.5 Bismaleimide Triazine Resin

BT resin is a high-performance resin matrix that is obtained by copolymerization of cyanate resin with –OCN group and BMI at 170–240 °C, and the reaction formula is as follows Figure 5.5.

BT resin is a unique material that combines the excellent properties of BMI and cyanate ester (CE). It offers exceptional heat resistance, processability, and environmental resistance, as well as good mechanical properties, low moisture absorption (less than 1.5%), and high heat resistance. By complementing the strengths of both materials, BT resin can improve the impact resistance, dielectric properties, and processability of BMI while also enhancing the hydrolysis resistance and brittleness of CE. This makes it an ideal choice for demanding environments that require excellent dielectric, thermal, and mechanical properties. Developed and commercialized by Mitsubishi Gas Chemical Corporation of Japan, BT resin is widely used in high-frequency and high-temperature PCB resin matrices. It has a large market in the field and has shown great potential in the electronics industry and aerospace field. However, the production of BT resin is costly and requires harsh production conditions, making it difficult to obtain high-quality products. Additionally, when used in electronic components, BT resin is a poor heat conductor, which can hinder heat removal from electronic parts.

To address these challenges, researchers have explored various methods to modify BT resins. These resins contain many heterocyclic structures, such as triazine rings and imide rings, with high reactivity in their molecules. Copolymerization with allyl compounds, EPs, benzoxazine, and polyphenylene ethers can change the internal crosslinking structure to obtain modified resins that meet different needs. For example, reactive modifiers containing CE and alkenyl groups have been shown to improve the thermomechanical and water absorption properties of cured BMI/CE blends.

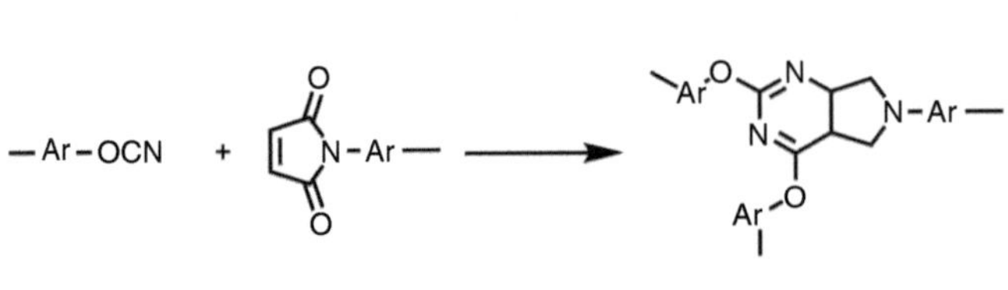

Figure 5.5 Reaction formula of bismaleimide triazine resin.

5.2.2 Development Trend of Organic Encapsulating Materials in Flexible Electronic Devices

Electronic encapsulating materials are essential components in the manufacturing of electronic devices. These materials include substrates, wiring, interlayer media, and sealing materials. Organic encapsulating materials are commonly used in substrates and sealing materials, and sometimes in interlayer media. In the coming decades, the microelectronics encapsulating industry is expected to experience rapid growth and become a high-tech, high-efficiency industrial field with significant potential for development. Plastic-based materials will continue to dominate the electronic encapsulating industry for the foreseeable future. However, with the increasing miniaturization of electronic components and the explosive development of flexible wearable devices and portable electronic products, there is a growing demand for miniaturization, high efficiency, high frequency, and high-speed electronic components. To meet these demands, there is a need to modify substrate materials to improve their performance. Ceramic substrates are brittle and complicated to prepare, making the use of composite and organic substrates a more viable option. This trend toward using composite and organic substrates is expected to continue in the future. In conclusion, the electronic encapsulating industry is poised for significant growth in the coming years. The use of composite and organic substrates will become increasingly important as the demand for miniaturization and high-performance electronic components continues to grow.

In the future, with high density, higher glass transition temperature and modulus of elasticity, lower coefficient of thermal expansion, better electrical properties, and physical stability of the material is bound to be the future development direction of organic substrate materials. I speculate that the new epoxy molding plastic will be popular in the market; silicone or PI class has a very promising development. At present, the electronic encapsulating substrate materials mainly exist in the following two scientific issues: (i) for composite substrates, how to improve their thermal conductivity and reduce thermal expansion performance while at the same time both maintaining high transparency and excellent mechanical properties. (ii) For organic substrate materials, how to improve their thermal conductivity, maintain a low coefficient of thermal expansion and high thermal stability while maintaining high flexibility. It is believed that these problems will eventually be solved with the advancement of research and technology.

In addition, green electronic encapsulating is also the future development direction of encapsulating materials, which refers to the entire encapsulating process in electronic products, must take into account resource and energy consumption, environmental impact, and other issues, and take into account technical and economic factors, so that the economic and social benefits of electronic encapsulating enterprises can achieve coordinated optimization of the concept of electronic encapsulating. Specifically, in the production process, the use of lead, halogen, and antimony-containing materials is prohibited.

5.3 Organic–Inorganic Hybrid Encapsulating Material

Inorganic–organic hybrid materials have been widely used in many applications due to their unique chemical and physical properties and the advantages of both organic and inorganic materials. The composite of inorganic–organic hybrid materials can form a variety of functional materials such as optical materials, high-temperature resistant materials, and mechanical materials. The research has become one of the fastest-growing fields in materials chemistry. Hybrid materials are obtained by combining at least two components (usually inorganic and organic) at the nanoscale. Methods for synthesizing inorganic–organic hybrid materials are usually based on physical and chemical methods: physical crushing, mechanical ball milling, and chemical methods: sol-gel processes, intercalation, exchange, or grafting. Considering the diversity of combinations of components (and properties), inorganic–organic hybrids represent an interesting class of materials with a wide range of applications. The field of organic–inorganic hybrids has developed rapidly in the last 20 years and is undoubtedly one of the major areas of research and one of the most exciting areas today. The rise of new characterization techniques and knowledge at the nanoscale and even at the molecular scale has been a driving force for inorganic–organic hybrid materials in many major fields such as sensors, electronics, optics, lightning, medicine, catalysis, energy storage, energy conversion, and many others. Moreover, the need to realize complex devices requiring better materials with better performance and multifunctional response is a constant need that can often be met by the use of composites. This section summarizes the applications of inorganic–organic hybrids in flexible electronics and lists the future trends of organic–inorganic-like hybrids.

5.3.1 Application of Organic–Inorganic Hybrid Materials in Flexible Electronics

In recent years, the development of flexible electronic devices has attracted great attention. The application of flexible electronics in strategic emerging areas such as electronic skin, medical electronics, environmental detection, and wearable devices is an important research direction. However, the development of flexible electronic devices still faces considerable challenges. Most importantly, the devices should have stable performance and be durable under complex stress conditions such as bending, stretching, folding, and twisting. Therefore, the selection of materials with excellent electronic and mechanical properties plays a crucial role in the application of flexible sensors. Currently, there are various materials used to fabricate flexible sensors, including metallic materials, inorganic semiconductor materials, and organic materials. Among them, inorganic semiconductor materials have excellent optoelectronic properties, but their bending ability is limited. Meanwhile, organic materials also face some limitations: (i) less sensitive to signals that respond to changes in piezoresistance and capacitance, and (ii) low carrier mobility and conductivity. These inherent properties of single materials limit the further development of flexible sensors. Therefore, composite hybrid materials are receiving increasing attention in the development of flexible sensors.

Hybrid materials are composite materials consisting of two or more components mixed at the molecular level. They are commonly found in natural organisms such as plants and bones. Organic–inorganic hybrids have undergone many years of evolution and synthesis strategies regarding artificial hybrid materials. Based on structural differences, hybrid materials were divided into two main categories: organic molecules or polymers embedded in inorganic matrices, and organic components grafted into a network of inorganic compounds by chemical bonding. Many studies have demonstrated that hybridized materials extend and enhance the properties of a single material. For example, the epoxy polymer material was improved due to the covalent integration of nanotubes with the epoxy matrix in nanotube-enhanced epoxy polymer hybrid material systems. In hybrid material systems, the structures of organic and inorganic components can interact with each other. Similarly, macromolecules from the calcite shell layer can control the precipitation of mineralized biotite or calcite. And metal organic backbones (MOFs) can influence the polymer structure in MOF polymer hybrid materials. In addition, multiple components are often integrated into hybrid materials to seek synergistic advantages from a collection of individual physicochemical properties. By adjusting the ratio of organic and inorganic phase components in hybrid material, the functionality of hybrid material can be "cut", "assembled" to meet the requirements. Thus, the development of hybrid materials opens up the possibility of increasing the flexibility and functionality of flexible electronic devices.

Currently, flexible electronic devices based on organic–inorganic hybrid materials have been widely used in various applications such as smart devices [27], optical displays [28, 29], healthcare [30], corrosion protection [31], and functional coatings [32]. Depending on the application, a variety of devices are prepared, including pressure/strain sensors, temperature sensors, humidity sensors, optical sensors, gas sensors, and biosensors. In this section, we focus on the practical applications based on organic–inorganic hybrid materials in the field of flexible electronics.

5.3.1.1 Strain and Pressure Sensors

Elastic strain and pressure sensors are widely used in smart clothing, smart sports, and electronic skin. In addition, strain and pressure sensors have the potential to be used in the fields of aerospace, autonomous driving, and three-dimensional global positioning using atmospheric pressure information. Sensitivity is the most critical superior value for calibrating the performance of flexible pressure and strain sensors, which determines the efficiency and measurement accuracy of the sensor. Currently, there are two ways to improve the sensitivity of tactile sensors. In addition to employing micro-nano or other structures to amplify external stimuli, hybrid materials can be used. For example, strain sensors based on hybrid ZnO-NW/polystyrene (PS) NF structures can strain up to 50% at high strain rates, as shown in Figure 5.6a, and the response curves of strain sensors reveal their potential applications in electronic skin or health monitoring [33]. Carbon nanotube/silicone rubber nanocomposite membranes were also used as strain sensors, and Ecoflex

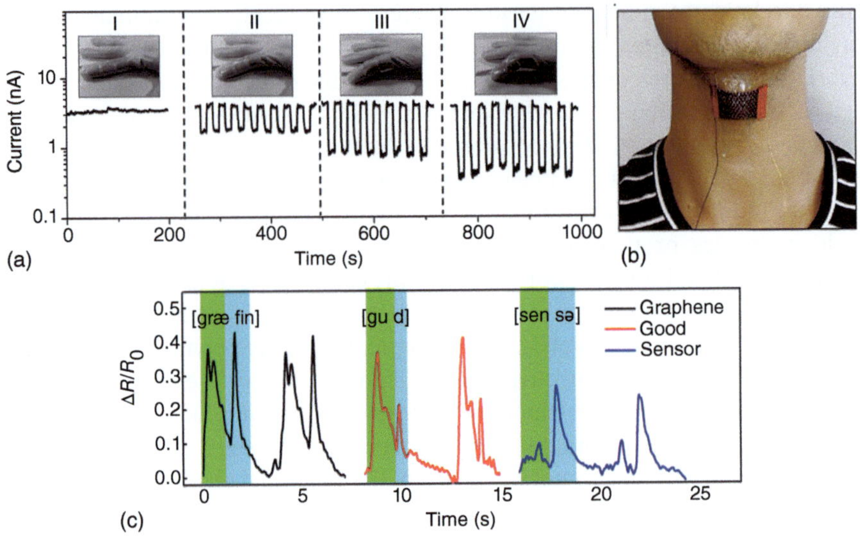

Figure 5.6 (a) Response curves of the ZnO NW/PS NF strain sensor at different levels of stretching; inset: photos from the experiment. (b, c) Application of the NR/ChNCs-CB strain to sensors to detect minute human motion. Source: (a) Reproduced with permission from Xiao et al. [33]; © 2011, John Wiley and Sons; (b, c) Reproduced with permission from Ma et al. [34]; © 2018, American Chemical Society.

was chosen as the flexible substrate. Skin motion is detected by attaching the sensor to different parts of the body. The sensitivity of the material can be controlled by the number density of the carbon nanotube permeation network. In another example, the natural rubber (NR)/chitin nanocrystals (ChNCs)–carbon black (CB) composite was designed as a helical structure by a solution mixing-spraying-rolling method. The NR/ChNCs CB strain sensor can detect even small body movements such as fingers or vocal cords, as shown in Figure 5.6b, and also three different articulation signals, with the results shown in Figure 5.6c. They can be used to monitor heart rate or respiratory rate and thus the future health of the body [34].

5.3.1.2 Temperature Sensor

Temperature changes can cause changes in resistance or conductivity, and temperature sensors based on pure nanoparticles have excellent performance. However, the formation of pure nanoparticles requires a suitable flexible substrate material with low-temperature thermoplasticity. Organic–inorganic hybrid materials can satisfy both flexibility and temperature sensing. Sensors based on hybrid materials have good electrical conductivity and thermal motion and can show reproducible and stable thermal responses. For example, sensors based on scaled graphite/carbon nanotube/polydimethylsiloxane composites have high temperature sensitivity and excellent linearity. Most importantly, the resistance of the flexible sensor on the soft prosthetic hand is stable when the hand exhibits different postures, as shown in Figure 5.7. Flexible temperature sensors have potential applications in diverse

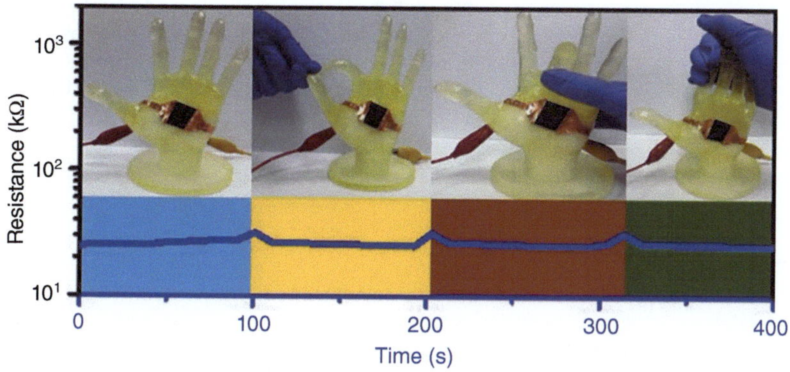

Figure 5.7 Resistance variation of the temperature sensor on the soft prosthetic hand. Source: Reproduced with permission from Yang et al. [35], © 2016, John Wiley and Sons.

fields such as medical health monitoring, sports, communications, aerospace, and firefighting [35].

5.3.1.3 Humidity Sensor

Flexible humidity sensors have numerous advantages, including low cost, low energy consumption, easy manufacturing, and easy integration. They can be widely used in various fields, such as meteorological monitoring, pharmaceuticals, and agricultural production. The wearable humidity analysis system based on molybdenum trioxide nanosheets has high sensitivity and can noninvasively monitor human respiration. The structural stability of the moisture-sensitive material can be improved by ensuring good dispersion in the hybrid material and increasing the number of chemical bonds between different materials in the system. The electrical parameters of the humidity sensor are related to the ambient humidity and can be influenced by the choice of hygroscopic material. For example, resistive humidity sensors based on polypyrrole (PPy)/zinc oxide hybrids have a good linear response. To fabricate stretchable humidity sensors with a wide range of mechanical deformations, Trung et al. discovered a flexible humidity sensor based on r-Go/polyurethane composites. This sensor can monitor the humidity of the environment, objects, human skin, and breath. The response speed and response time of the humidity sensor remain consistent under different conditions. The electrical properties of the device can be affected by the catalyst, applied potential, and AC frequency. For example, flexible humidity sensors based on new surface-modified carbon nanotubes containing thiol-ene have better electrical properties due to improved dispersion and interfacial interactions. The addition of MOF can also improve the water-molecule trapping performance of humidity sensors.

5.3.1.4 Optical Sensors

Light can distinguish the concentration of deoxyhemoglobin and oxygenated hemoglobin and thus indirectly calculate the oxygen saturation in tissues.

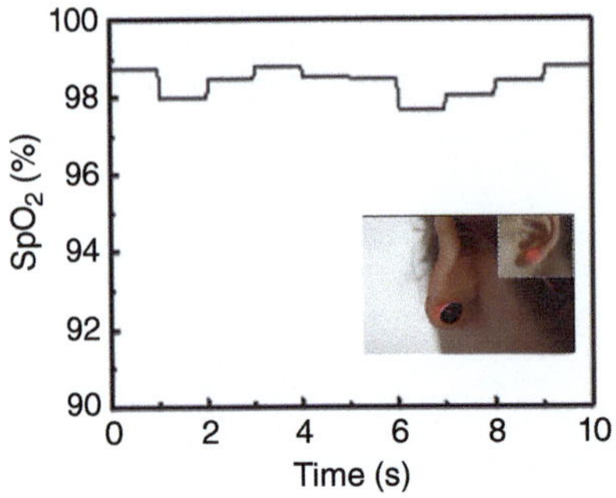

Figure 5.8 Optical sensor-based earlobe SpO_2 measurements; the inset shows the earlobe during device operation. Source: Reproduced with permission from Kim et al. [36], © 2017, John Wiley and Sons.

Therefore, optical sensors can be used to measure clinical information. Certain types of optical sensors are available that allow noninvasive measurement of hemoglobin oxygen saturation. Flexible optical sensors attached to the finger or wrist can measure oxygen levels in the blood and are not affected by tissue deformation caused by arm movements (Figure 5.8). In addition, optical sensors can be used in digital cameras, industrial and automotive systems, and wearable electronics, including electronic eyes, multifunctional robots, and artificial skin [36].

5.3.1.5 Other Types of Sensing Devices

In addition to the previously mentioned strain and pressure sensors, temperature sensors, humidity sensors, and optical sensors, chemical sensors are also a category of sensing devices with a wide range of applications. Flexible electronic sensing elements prepared on the basis of organic–inorganic hybrid materials can be used to measure (bio)chemical signals and thus for medical detection and environmental pollution. Some materials with framework structures, such as MOF, can further improve the adsorption capacity due to their large porosity, high specific surface area, periodic channels, and more unsaturated metal sites of these framework materials of materials [37]. Besides, organic–inorganic hybrid materials can also be used to prepare self-healing flexible devices, self-powered sensor devices, multichannel sensing flexible devices, gas-sensing flexible devices, and pH-responsive flexible devices.

5.3.2 Development Trends of Organic–Inorganic Hybrid Materials

With the rapid development of modern technology, single-performance materials can no longer meet the diverse needs of people. Currently, composites with excellent performance can be prepared by functionally combining two or more materials with complementary and optimized properties. Inorganic–organic hybrid materials are composites in which inorganic and organic materials are combined at the nanoscale, and strong forces exist between the two phases or form interpenetrating network structures. Organic–inorganic hybrid materials have excellent ductility and flexibility and have made significant progress in the development of flexible sensors, flexible electronics, and other fields. They are very convenient and can accurately and quickly measure unique environments and signals. Organic–inorganic hybrids are expected to be the leading technology in the field of photoelectric, thermoelectric, or electroluminescent devices due to their excellent properties. They offer better mechanical flexibility and are more energy-efficient than pure organic materials or single inorganic materials. The use of organic–inorganic hybrids often results in improved performance due to synergistic effects. Flexible sensors allow for miniaturization, integration, and creative development and have an important role in electronic skin, biomedical, wearable electronics, and aerospace applications. With the huge demand for flexible sensors, organic–inorganic hybrids are the main material of choice. By mixing inorganic conductive materials into organic matrices as fillers, hybrid material-based sensors offer flexibility and high conductivity compared to comparable sensors. The research and development of flexible electronics using hybrid materials will greatly accelerate the practical application of flexible sensors.

Despite significant progress in the field of flexible sensors based on organic–inorganic hybrids, challenges remain in practical applications. Firstly, improved material preparation is required to obtain large-area and high-quality benefits. Additionally, material development must be optimized before materials can be used on a large scale for fabrication. Furthermore, these materials must be effectively combined with soft substrates. Some high-performance sensors are affected by external noise signals, making device design a challenge. From a practical point of view, these sensors should be highly durable, accurate, enduring, consistent, and reliable while maintaining relatively low material and processing costs. The combination of hybrid materials is critical for accelerating the practical application of sensors in the coming years, and many materials remain to be investigated in the future. These materials can improve stability, sensitivity, and longevity while reducing costs. Each material has its own unique physical properties, and understanding these properties can improve utilization and facilitate the rational design of hybrid material structures. Secondly, by understanding the structure–property relationships associated with the materials and applying them in practice, the performance of sensor applications can be improved. Material design, synthesis, and selection are among the most important solutions to improve device performance.

References

1 Du, Z.Y. (2010). *Electronics Manufacturing and Packaging*. Electronic Industry Press.

2 Chee Wei, T. and Razak Daud, A. (2002). The effects of aged Cu-Al intermetallics to electrical resistance in microelectronics encapsulating. *Microelectronics International* 19 (2): 38–43.

3 Braunovic, M. and Alexandrov, N. (1994). Intermetallic compounds at aluminum-to-copper electrical interfaces: effect of temperature and electric current. *IEEE Transactions on Components, Encapsulating, and Manufacturing Technology, Part A* 17: 78–85.

4 Cao, Y.-k., Liu, Y., Liu, B. et al. (2019). Effects of Al and Mo on high temperature oxidation behavior of refractory high entropy alloys. *Transactions of Nonferrous Metals Society of China* 29 (7): 1476–1483.

5 Su, W., Pan, M., Mo, D. et al. (2022). Mechanical and thermal properties of Mo for encapsulating HgCdTe IRFPA detector at cryogenic temperatures. *Infrared Physics & Technology* 127: 104373.

6 Majumdar, S., Kapoor, R., Raveendra, S. et al. (2009). A study of hot deformation behavior and microstructural characterization of Mo–TZM alloy. *Journal of Nuclear Materials* 385 (3): 545–551.

7 Chern, T.-S. and Tsai, H.-L. (2007). Wetting and sealing of interface between 7056 Glass and Kovar alloy. *Materials Chemistry and Physics* 104 (2): 472–478.

8 Gao, Z., Ba, X., Yang, H. et al. (2020). Joining of silicon particle-reinforced aluminum matrix composites to Kovar alloys using active melt-spun ribbons in vacuum conditions. *Materials* 13 (13): 2965.

9 Krauss, W., Lorenz, J., Konys, J. et al. (2017). Mechanical characterization of electrochemically based W–Cu joints for low-temperature heat sink application. *Fusion Engineering and Design* 124: 220–225.

10 Wang, Y., Zhuo, L., and Yin, E. (2021). Progress, challenges and potentials/trends of tungsten-copper (WCu) composites/pseudo-alloys: fabrication, regulation and application. *International Journal of Refractory Metals and Hard Materials* 100: 105648.

11 Kang, M. and Kang, S. (2012). Effects of Al_2O_3 addition on physical properties of diopside based glass–ceramics for LED encapsulations [J]. *Ceramics International* 38: S551–S555.

12 Pietranico, S., Pommier, S., Lefebvre, S. et al. (2009). Thermal fatigue and failure of electronic power device substrates. *International Journal of Fatigue* 31 (11): 1911–1920.

13 Pahinkar, D.G., Puckett, W., Graham, S. et al. (2018). Transient liquid phase bonding of AlN to AlSiC for durable power electronic encapsulations. *Advanced Engineering Materials* 20 (10): 1800039.

14 Gan, K.K. (2009). An MT-style optical encapsulation for VCSEL and PIN arrays. *Nuclear Instruments and Methods in Physics Research Section A: Accelerators, Spectrometers, Detectors and Associated Equipment* 607 (3): 527–529.

15 Fang, H., Bai, S.-L., and Wong, C.P. (2017). Thermal, mechanical and dielectric properties of flexible BN foam and BN nanosheets reinforced polymer composites for electronic encapsulating application. *Composites Part A: Applied Science and Manufacturing* 100: 71–80.

16 Mueller, M., Panchenko, I., Wiese, S. et al. (2020). Morphologies of primary Cu_6Sn_5 and Ag3Sn intermetallics in Sn–Ag–Cu solder balls. *IEEE Transactions on Components, Encapsulating and Manufacturing Technology* 10 (1): 18–29.

17 Hou, L., Derakhshandeh, J., Beyne, E. et al. (2020). A novel resistance measurement methodology for in-situ UBM/solder interfacial reaction monitoring. *IEEE Transactions on Components, Encapsulating and Manufacturing Technology* 10 (1): 30–38.

18 Ernst, D., Faghih, M., Liebfried, R. et al. (2020). Encapsulating of ultrathin flexible magnetic field sensors with polyimide interposer and integration in an active magnetic bearing. *IEEE Transactions on Components, Encapsulating and Manufacturing Technology* 10 (1): 39–43.

19 Fruehauf, P., Munding, A., Pressel, K. et al. (2020). Chip-encapsulation-board reliability of system-in-encapsulation using laminate chip embedding technology based on Cu leadframe. *IEEE Transactions on Components, Encapsulating and Manufacturing Technology* 10 (1): 44–56.

20 Zhan, Y. and Zhang, X.M. (2011). *Packaging Technology of Semiconductor Optoelectronic Devices*. Electronic Industry Press.

21 Fu, Z.L. and Wu, B.Y. (2003). Progress of epoxy resins for electronic plastic sealing materials in semiconductor optoelectronic device packaging process. *Journal of Wuhan Institute of Chemical Technology* 25 (4): 47–50.

22 Zhou Q.F. A fast-curing epoxy powder composition. 201310005605, 23 December 2015.

23 Cheng, X.M. (2004). Properties and development trend of epoxy molding plastics. *Semiconductor Technology* 29 (1): 40–45.

24 Damaceanu, M.D., Constantin, C.P., Nicolescu, A. et al. (2014). Highly transparent and hydrophobic fluorinated polyimide films with ortho-kink structure. *European Polymer Journal* 50: 200–213.

25 Fukukawa, K., Okazaki, M., Sakata, Y. et al. (2013). Synthesis and properties of multi-block semi-alicyclic polyimides for thermally stable transparent and low CTE film. *Polymer* 54 (3): 1053–1063.

26 Basin G, Haque A.S, Chen C.H, et al. LED with particles in encapsulant for increased light extraction and non-yellow off-state color. *US7791093 B2*. 7 September 2010.

27 Yang, W.S., Non, J.H., Jeon, N.J. et al. (2015). High-performance photovoltaic perovskite layers fabricated through intramolecular exchange. *Science* 348 (6240): 1234–1237.

28 Kim, Y.H., Cho, H.C., Heo, J.H. et al. (2015). Multicolored organic/inorganic hybrid perovskite light-emitting diodes. *Advanced Materials* 27 (7): 1248–1254.

29 Zhang, B.H. *Preparation of Organic and Inorganic Hybrid Perovskite Materials and Their Application in Optoelectronic devices*. Nanjing University.

30 Ng, K.K. and Zheng, G. (2015). Molecular interactions in organic nanoparticles for phototheranostic applications. *Chemical Reviews* 115 (19): 11012–11042.

31 Li, H., Lv, Y., Zhou, Z. et al. (2022). Coordinated anionic inorganic module—an efficient approach towards highly efficient blue-emitting copper halide ionic hybrid structures. *Angewandte Chemie International Edition* 61 (8): e202115225.

32 Tong, H., Li, H.B., Li, H.J. et al. (2021). Incorporation of an emissive Cu_4I_4 core into cross-linked networks: an effective strategy for luminescent organic–inorganic hybrid coatings. *Inorganic Chemistry* 60 (20): 15049–15054.

33 Xiao, X., Yuan, L.Y., Zhong, J.W. et al. (2011). High-strain sensors based on ZnO nanowire/polystyrene hybridized flexible films. *Advanced Materials* 23 (45): 5440–5444.

34 Ma, J.H., Wang, P., Chen, H.Y. et al. (2019). Highly sensitive and large-range strain sensor with a self-compensated two-order structure for human motion detection. *ACS Applied Materials & Interfaces* 11 (8): 8527–8536.

35 Yang, H., Qi, D.P., Liu, Z.Y. et al. (2016). Soft thermal sensor with mechanical adaptability. *Advanced Materials* 28 (41): 9175–9181.

36 Kim, J., Gutruf, P., Chiarelli, A.M. et al. (2017). Miniaturized battery-free wireless systems for wearable pulse oximetry. *Advanced Functional Materials* 27 (1): 1604373.

37 Lu, Y. and Yan, B. (2014). Lanthanide organic–inorganic hybrids based on functionalized metal–organic frameworks (MOFs) for a near-UV white LED. *Chemical Communications* 50 (97): 15443–15446.

6

Development of Flexible Electronics Packaging Technology

Qiushi Rao

Peking University Shenzhen Graduate School, 2199 Lishui Road, Nanshan, Shenzhen, Guangdong Province, Shenzhen, China

6.1 Flexible Electronics Packaging

In general, the advancement of encapsulation technology is crucial for enhancing the lifespan and dependability of organic electronic devices. Various encapsulation techniques have been developed, typically involving the use of inert gases such as nitrogen or argon to fill the sealing container. The container is then sealed with a UV-curable epoxy resin and covered with a metal or glass lid. An absorbent material, such as calcium or barium, can also be added to remove any residual water or water vapor that may have diffused through the epoxy sealant. However, these rigid materials are not suitable for use in flexible electronic devices. Several flexible packaging methods have been developed, including the use of ultrathin glass and vacuum deposition of thin films. Of these flexible packaging methods, thin films have garnered the most attention due to their lightweight, transparency, and high mechanical flexibility.

6.1.1 Single-Layer Thin-Film Packaging

Thin-film encapsulation technology refers to a method of protecting devices by preparing a dense film on the device surface to isolate water and oxygen. This method has the advantage of effectively blocking water and oxygen while maintaining the flexibility of the device and is generally used for the encapsulation of flexible devices [1]. Thin-film encapsulation includes single-layer film encapsulation and multilayer film encapsulation. Single-layer films can simplify the fabrication process, but the prepared films usually have defects that provide a pathway for the transmission of water vapor and oxygen. Therefore, the main research goal for developing single-layer encapsulation films is to reduce defects and improve barrier performance. Currently, the main approaches for developing monolayer encapsulation films are based on plasma-enhanced chemical vapor deposition (PECVD) and atomic layer deposition (ALD), which typically use inorganic oxides such as Si–O/Si–N and Al–O/Al–N for encapsulation. PECVD is a more attractive

Flexible Electronic Packaging and Encapsulation Technology, First Edition.
Edited by Hong Meng and Wei Huang.

fabrication technology than ALD due to its high deposition rates (approximately 2 orders of magnitude faster than ALD) and its ability to fabricate monolayer barrier films for organic electronics at low deposition temperatures. Current SiO_x and SiN_x films deposited using PECVD can be used for thin-film packaging, but the defect density of these films is too high, leading to actual WVTR values for devices greater than 0.01 g/m^2/day, which is several orders of magnitude higher than required for packaging targets in organic electronics. The use of multilayer film packaging technology can effectively address the problem of high water and oxygen transmission rates in single-layer film packaging to meet the packaging requirements of flexible electronic devices.

6.1.2 Multi-Layer Thin-Film Packaging

Multilayer films consist of inorganic and organic films. Organic films are very flexible but have poor water and oxygen barrier capabilities, making it difficult to apply them directly to device packaging. Inorganic films have relatively good water and oxygen barrier capabilities but are prone to defects during the preparation process. The size of these defects greatly impacts the device. Generally, when the defect hole is less than 0.3 nm, the water and oxygen penetration path is more tortuous, and the penetration rate is slow. However, when the defect hole is larger than 0.3 nm, the water and oxygen penetration path is reduced and the penetration rate accelerates. When the defect hole increases to 1 nm, water and oxygen can penetrate directly into the device, greatly affecting its lifespan [2]. Increasing the thickness of the inorganic layer can improve water–oxygen barrier capacity to a certain extent, but as thickness increases, film stress also increases and defects increase. Inorganic materials are more brittle and prone to cracking when bent, which affects encapsulation. Combining inorganic and organic films using multilayer film encapsulation technology can reduce defect points in inorganic films and increase the water–oxygen permeation path. Organic films can also be used to adjust stress mismatch during film bending and increase film flexibility, making it a promising encapsulation method [3] (Figure 6.1).

6.1.2.1 Barix Multilayer Thin-Film Packaging

The Barix encapsulation layer is composed of alternating layers of organic and inorganic films with a total thickness of about 3 μm, as shown in Figure 6.2.

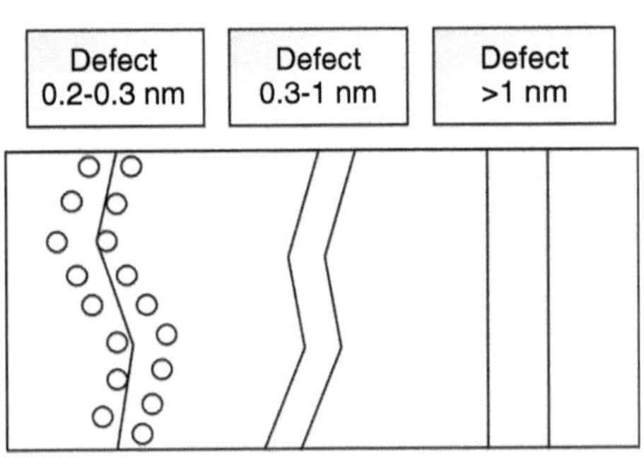

Figure 6.1 Water and oxygen permeation pathways in thin-film packages. Source: Reproduced with permission from Park et al. [3]; © 2011, The Institute of Physics.

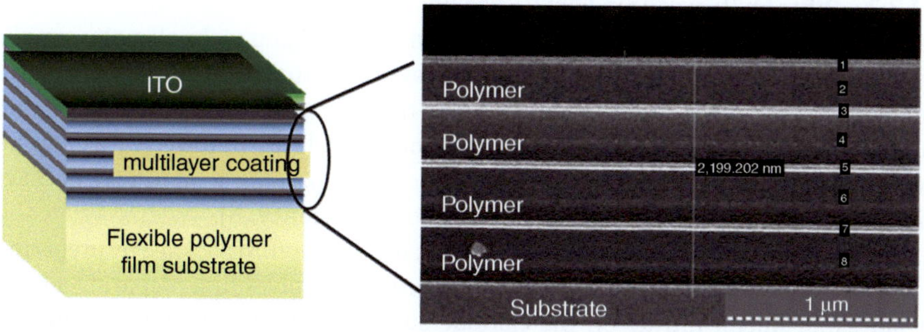

Figure 6.2 Multilayer thin-film package structure. Source: Reproduced with permission from Park et al. [3]; © 2011, The Institute of Physics.

A polymer organic layer is formed on the substrate using an organic material such as polyacrylate. This is prepared by evaporating the polymer monomer master batch and introducing it into the vacuum chamber with airflow. The monomer coalesces in liquid form on the substrate in the vacuum chamber and is then transferred to a UV light source to cure the polymer film. An inorganic layer is then prepared on the organic layer by magnetron sputtering, chemical vapor deposition (CVD), or PECVD. This layer is usually made of aluminum oxide but can also be made of silicon nitride or silicon oxide. Since the preparation process involves coalescing the monomer in liquid form on the substrate, the top surface has atomic-level smoothness. Polymerization can also form an atomic-level smooth polymer film, and the inorganic layer deposited on it will have few defects. This can effectively enhance the water and oxygen barrier capability of this film, which is a unique advantage of Barix [4]. However, these two layers alone have limited water–oxygen barrier capacity, so it is necessary to repeat the preparation of a total of seven organic–inorganic stacked layers on top of this. Each inorganic layer has excellent water–oxygen barrier capacity. By repeating the alternating process to deposit multiple layers of polymer films, the polymer films can decouple defects in the inorganic layers, preventing defect propagation through the multilayer structure. When the thickness of the organic layer is smaller than the average size of the pinholes (defects) in the inorganic barrier layer (Al_2O_3), the permeation process is controlled by a "zigzag path" mechanism. The WVTR is strongly dependent on the thickness of the organic layer, providing a longer permeation diffusion path and lower pinhole density through the multilayer structure to improve barrier performance. Barix films prepared with this alternating organic–inorganic multilayer structure can maintain water permeability at 10^{-4}–10^{-6} g/m^2/day and oxygen permeability at 0.005 cc/cm^2/day. These films have excellent water and oxygen barrier capabilities and can meet the packaging requirements of organic electronic devices [5].

In practical applications, due to the intrinsic cohesive stress of inorganic materials, the inorganic layer is susceptible to holes through which water and oxygen can penetrate into the organic layer, affecting material performance.

6.1.2.2 Other Multilayer Thin-Film Packaging

Samsung Display has optimized Barix's thin-film packaging technology by simplifying it from a seven-layer structure to a five-layer structure. The initial three layers are still prepared as an alternating structure of inorganic, organic, and inorganic layers. A resin coating is then applied to the inorganic layer, followed by a complementary inorganic layer. This simplifies the process of preparing multilayer packaging films and greatly improves film encapsulation efficiency [6].

Applied Materials has improved on the Barix multilayer film package by preparing a three-layer film structure. The preparation process involves first preparing a layer of inorganic silicon nitride on the cathode surface. The inorganic silicon nitride layer has excellent water and oxygen barrier properties but can form pinholes. To address this, an organic hexamethyldisiloxane (HMDSO) layer can be prepared on top of the inorganic silicon nitride layer to smooth the interface, followed by another inorganic silicon nitride layer. The organic layer in the middle of the film is a hybrid polymer with a highly crosslinked amorphous structure, strong bonding, heat, and corrosion resistance. It can cover the inorganic layer well and buffer the release of stress accumulated on the film layer. This can reduce defects in the barrier film species and increase the length of the water–oxygen permeation channel. The combination with the silicon nitride layer can effectively reduce water–oxygen permeation [7]. This three-layer film preparation process can be produced using roll-to-roll equipment, simplifying the process while providing the possibility of large-scale film preparation with good application prospects.

General Electric has developed a gradient barrier layer that can be prepared by PECVD deposition at 55 °C. It consists of alternating organic and inorganic layers with a total thickness of less than 1 μm. The organic layer is prepared using a combination of silicone precursors and Ar gas, allowing for the preparation of organic materials (SiO_xC_y) with a modulus of less than 10 GPa. Inorganic layers are prepared using a combination of silane (2% silane diluted in He), ammonia, and oxygen to produce SiO_xN_y layers with a modulus of 100–500 GPa. This alternating organic/inorganic layer has a very low water–oxygen transmission rate.

Jeon-Min Han of Yonsei University in Korea used a multilayer film encapsulation technique by preparing a SiO_2 layer followed by a polydimethylsiloxane layer on a polycarbonate (PC) film. The film was cured at 80 °C for 20 minutes and then at 60 °C for 1 hour to obtain organic, inorganic, and organic three-layer encapsulation films. The prepared films can be directly applied to device encapsulation [8].

Today, devices can be fabricated on almost any substrate and then encapsulated directly with thin films. While TFE technology has the disadvantage of potentially damaging devices through direct deposition and limited processing conditions, it also has the advantage of preventing moisture and oxygen from penetrating from the edges. When using glass or barrier foil as an encapsulation film, adhesives are inevitably used, which can lead to parasitic effects and contamination. However, since TFE does not require the use of adhesives, this result can be avoided. Most importantly, TFE is the most promising display encapsulation technology because it is flexible and easy to apply to any device.

6.2 Thin-Film Packaging Technology

6.2.1 PECVD Atomic Layer Deposition Packaging Technology

6.2.1.1 Introduction to PECVD Technology

CVD is a commonly used technique for preparing thin-film materials developed in the last decade or so. It involves preparing a high-quality thin film by chemically reacting gases at the solid–gas interface or in the gas phase to produce a solid material for deposition on the substrate surface. The substrate is heated to a certain temperature, and a reaction gas containing the atoms of the film to be prepared is introduced. The reaction gas forms the desired film on the substrate through a series of reactions. CVD can prepare many kinds of thin films, including metal, nonmetal, metal alloy, ceramic, and other chemicals. The deposited films have few defects and high crystallinity. The deposition process parameters can be adjusted to obtain films with different morphologies and crystal structures.

Several CVD methods have been developed to deposit thin films. These include atmospheric pressure chemical vapor deposition (APCVD), PECVD, and low-pressure chemical vapor deposition (LPCVD), which are used to produce thin films in high volume [9, 10]. The APCVD technique has the advantages of fewer steps, good coverage, uniformity, and a relatively low cost. However, this deposition process works at high temperatures. Since organic or organic–inorganic hybrid materials used in flexible devices may degrade at high temperatures or during encapsulation processes such as high-energy electron beams, the direct application of these techniques needs to be avoided, and encapsulation needs to be performed using a gentler process. To reduce the deposition temperature, the PECVD technique is used. This is a plasma-assisted deposition technique that allows for the industrial deposition of high-density, high-quality insulating barrier films with good adhesion. Along with the advantages of low deposition temperature, low ion damage, fast deposition rate, and large deposition area, PECVD is commonly used to deposit silicon nitride and silicon oxide films, among others [11].

While ordinary thermochemical vapor deposition methods usually require deposition at temperatures of 700–900 °C, deposition of similar films by PECVD requires only 250 °C or lower. Due to the inherent characteristics of organic materials and polymer substrates used in flexible electronics, deposition temperatures are mostly limited to less than 100 °C. Therefore, the deposition of thin films at low temperatures by PECVD is common for organic device fabrication. The plasma is usually a mixed state of matter in which the outermost electrons of some atoms in the gas are detached by means of discharge or heating, forming equal amounts of charged electrons and positive ions and also containing some neutral particles. The properties of plasma are very different from those of the original state. The plasma contains a large number of high-energy electrons, which can provide the activation energy required in the deposition process. The collision of electrons with gas-phase molecules can promote the excitation, decomposition, and chemistry of gas molecules to generate highly reactive chemical groups. This can significantly

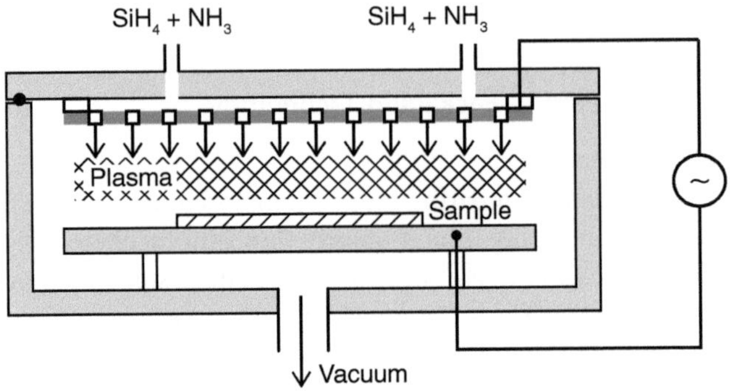

Figure 6.3 Plasma-enhanced chemical vapor deposition (PECVD).

reduce the temperature required for the reaction, allowing CVD to be carried out at lower temperatures [9, 12]. As shown in Figure 6.3, the preparation of thin films consists of three main processes:

(1) Gas excitation process: The plasma reacts with the gas in a primary reaction, causing the decomposition of the reacting gas and generating more reactive groups.
(2) Group diffusion process: The highly reactive groups diffuse and transport toward the substrate, while secondary reactions occur between the reactants.
(3) Film growth process: Each reactant and primary or secondary reaction product arriving at the substrate is adsorbed by the substrate and reacts with its surface to generate a thin film. This process is generally accompanied by the re-emission of gas-phase substances.

6.2.1.2 Development of PECVD Technology

PECVD has a long history of development and was first used for photovoltaic (PV) devices, such as amorphous silicon solar cells, in the mid-1970s. In 1981, Hezeland SchoKrner transferred the manufacture of SiN_x by PECVD from the microelectronics industry to the crystalline silicon PV industry [5]. Later, metal-insulator-semiconductor inversion layer (MIS-IL) cells were developed using plasma SiN_x. In 1999, Erlat et al. deposited SiO_x gas barrier coatings on polymer substrates by PECVD. They showed that the films deposited by PECVD have good water vapor barrier properties and can be deposited on polymer surfaces to make them flexible. The deposited SiO_x or SiN_x films have promising applications in biomedical devices and food packaging [7]. In 2005, Wuu et al. used PECVD-deposited SiO_x and SiN_x films to encapsulate OLEDs and further used a multilayer structure with multiple polymer/inorganic layers to improve barrier properties [13]. In their study, SiO_x (50 nm)/SiN_x (50 nm) barrier layers were prepared on PC substrates at 80 °C to reduce their WVTR and OTR to 0.01 and 0.01 g/m^2/day, respectively. This broadens the prospect of PECVD packaging and has a significant impact on packaging technology.

SiN_x films deposited by PECVD show weak barrier properties due to pinholes and low-density films at low-temperature deposition. In fact, the PECVD deposition process requires a complex vacuum system and is expensive. However, PECVD is performed at relatively low deposition temperatures below 250 °C compared to the original hot CVD of 300–800 °C and has good step coverage for industrial-scale applications [3]. However, pinholes are more likely to occur at low temperatures during the deposition of these single inorganic layers. Deposition temperatures above 100 °C remain critical for many functional devices, especially OLEDs, OPVs, and PSCs, where the mechanical properties of the glass are compromised by the glass transition temperature in the device, which is typically higher than 100 °C. Therefore, the deposition temperature during the deposition process is a critical factor for the performance of the thin-film barrier layer and the device [14].

OLED flexible barrier films require optimal film stress, high optical transparency, low surface roughness, good mechanical properties, and low deposition temperature. PECVD can meet these requirements well. PECVD can effectively achieve low-temperature deposition of thin films and is easy to operate with simple equipment. Deposition does not require a high-temperature environment, which can save energy, reduce costs, and avoid the effect of a high-temperature environment on some materials. The high activity of reactive groups and fast deposition rate in PECVD can effectively increase production capacity. The film growth temperature is much lower than the melting point of the film material, and the prepared films usually have high purity, crystallinity, and good density. They also have good adhesion to substrates and good coverage of complex-shaped substrates, which makes PECVD technology increasingly attractive [15]. However, PECVD has the disadvantages of high investment cost for equipment and high purity requirement for reaction gas. The production process may generate noise or dust that is harmful to the human body. Nowadays, PECVD technology is mainly used in the field of preparing high-quality thin films. Further development of more advanced PECVD technology is still needed in the future to make it have more extensive applications.

6.2.2 ALD Atomic Layer Deposition Packaging Technology

6.2.2.1 Introduction to ALD Technology

ALD is a method of forming thin films on a substrate by alternating pulses of gas-phase precursors into the reactor and chemically adsorbing and reacting with them. ALD is chemically similar to CVD, except that the ALD reaction splits the CVD reaction into two halves, keeping the precursor materials separated during the reaction. This is known as "purification" [5]. Separation of precursors is accomplished by purging the gas (usually nitrogen or argon) after each reactant pulse to remove excess precursors from the process chamber and prevent parasitic CVD deposition on the substrate. ALD is a continuous, self-limiting surface reaction that deposits thin-film materials onto substrates of different compositions. Most ALD processes are based on binary reaction sequences in which two surface reactions occur and binary compound films are deposited [7].

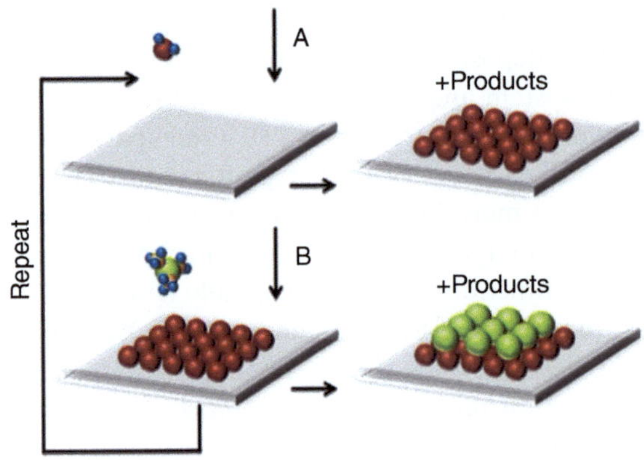

Figure 6.4 Schematic diagram of ALD with AB as an organometallic precursor reacting sequentially with an oxidant. Source: Reproduced with permission from Yu et al. [5]; © 2016, Elsevier.

The ALD preparation film is grown in a reactor, and the principle of the preparation method is shown in Figure 6.4. A gas-phase precursor is first passed into the reactor and chemisorbed with the substrate, undergoing a surface reaction. After enough time for the reaction, an atomic layer is formed. The reactor is then washed by passing an inert gas, and a second gas-phase precursor is passed into the reactor. The second precursor reacts with the already adsorbed first precursor until the first precursor is consumed. The reaction stops automatically, and a dense film with the desired atomic layer growth is formed. Since ALD films are prepared by an AB binary sequential reaction, the reaction can only deposit a limited number of reactant "B" molecules because there are only a limited number of reactant "A" chemisorption sites. Each of the two surface reactions is self-limiting, which is the key to ALD. The two reactions can be sequentially deposited to produce thin films with atomic levels [5]. It is worth noting that, basically, all gas-phase substances can be physisorbed on the substrate, but a certain activation energy is required to produce chemisorption on the material surface. The ability to produce chemisorption is the key to ALD, so it is important to choose the right precursor material. This property makes ALD a very promising technology for thin-film packaging with promising applications in the organic electronics industry [16].

The advantage of ALD is that the thickness can be precisely controlled by controlling the number of reaction cycles. Unlike CVD, there is less need for uniformity in the reaction process, providing large-area uniformity in film deposition with excellent reproducibility. In addition, the self-limiting nature of ALD produces extremely smooth, very continuous, and pinhole-free films because no surface sites are left behind during the film growth process. This is extremely important for improving diffusion barrier performance. However, the main limitation of ALD is its slow growth rate (low throughput). Typically, only a small fraction of monolayers are deposited in a single cycle. Researchers have improved the ALD technique by developing multilayer structures to improve ALD coating times [14, 17].

6.2.2.2 Development of ALD Technology

ALD is a technique widely used today, especially in the field of thin-film packaging. ALD, first known as atomic layer epitaxy (ALE), was reported by Suntola et al. in

1977 and has been widely developed and used worldwide [18]. The first application of ALE was thin-film electroluminescent displays, which could be prepared on large-area substrates with high-quality dielectric and luminescent films. With the development of silicon-based microelectronics, there is a need to prepare high-quality ultrathin films. ALD is considered to be one of the most promising deposition methods for producing ultrathin films that can control the thickness and composition of the film at the atomic level. These advantages enable scaled-down microelectronic devices, so ALD has received widespread attention [19]. ALD can be used to deposit many types of films, including various oxides, metal nitrides, metals, metal sulfides, etc.

In flexible device packaging, ALD is commonly used to deposit Al_2O_3 films. These have been used to passivate or encapsulate OLEDs and organic solar cells due to their excellent barrier properties to prevent water and oxygen penetration. To avoid the effect of high temperature on the substrate, plasma-enhanced atomic layer deposition (PEALD) was developed to prepare Al_2O_3 thin films. Similar to the principle of PECVD, these can be prepared below 100 °C due to the ability of energetic radicals to be deposited at low temperatures [20]. Al_2O_3 thin films have been reported using ALD technique at lower deposition temperatures using trimethylaluminum (TMA) and H_2O. Elemental analysis of the films showed that the hydrogen content increased with decreasing growth temperature. Also, the density of the Al_2O_3 films decreased with decreasing temperature and was 3.0 and 2.5 g/cm^3 at 177 and 33 °C, respectively. It is noteworthy that the reaction time became longer with decreasing growth temperature and therefore the reaction rate became slower and the purging time became longer [5]. Sarkar et al. [21] used O_3 instead of H_2O as an ALD oxidant and compared it with Yang et al.'s study. They also compared H_2O and O_3-based Al_2O_3 films and showed that the barrier properties of O_3-based films were better than those of H2O-based Al_2O_3 films. The prepared films had a lower WVTR value of 8.7×10^{-6} g/m^2/day [22].

Low-temperature ALD allows us to fabricate unique inorganic/organic polymer composites and to deposit a thin barrier film on thermosensitive materials such as organic polymers. Georges Groups [23] reported thin layers of Al_2O_3 prepared by the ALD method coated on polymeric substrates. They showed that the coated Al_2O_3 films could reduce the WVTR of the substrates at room temperature from 1 to 10^{-5} g/m^2/day with excellent barrier properties. ALD diffusion barrier films are attractive as thin film encapsulation (TFE) layers to meet the demanding packaging needs of high-sensitivity OLEDs for emerging flexible display and lighting technologies.

Monolayer Al_2O_3 films deposited by ALD and monolayer SiN_x films deposited by PECVD are typical packaging films for flexible OLEDs and OPVs. The successful application of these barrier films in flexible OLEDs and OPVs makes them very promising for encapsulating PSCs. ALD-deposited Al_2O_3 films are known for their ultrathin, conformal, pinhole-free, low characteristic microstructure, low stress, and good film integrity. The quality of ALD-deposited films can be precisely controlled due to their atomic-level film growth advantages. ALD is capable of depositing films on complex surfaces and is substrate-independent, making it suitable for devices with different structures [16]. In addition, ALD can be used to prepare nanostructured packaging materials and novel materials. All these properties are favorable for the development of ALD-based single-film packaging technology. However, ALD deposition processes usually require long processing times. This means that ALD still needs to be developed and improved if it is to be used for large-scale commercial fabrication.

Although ALD deposition can be deposited on large-area substrates, ALD and/or MLD processes should meet low-cost and high-yield process requirements for TFE processes. The cost of ALD is largely related to the cost of reactants and equipment. Most ALDs are costly with vacuum pumps used to move reactants and products through the reactor and maintain a clean environment in the reactor. To solve this problem, ALD deposition can be achieved at atmospheric pressure without a vacuum pump if the reactant and product gases in the reactor are pushed at a pressure higher than atmospheric pressure [18]. Recently, some groups have developed ALD based on a coating head type located on the substrate at atmospheric pressure, as shown in Figure 6.5. The ALD reactants are delivered through channels in the coating head, which are separated by an inert gas stream to prevent the gas phase reaction of the precursor. The coating head can be moved in a linear manner or back and forth to overlap the surface area and achieve the ALD process. Therefore, this novel ALD system can be utilized to solve the problem of high cost and slow preparation with low throughput for large-area substrate preparation [24, 25].

On the other hand, a self-limiting surface reaction similar to ALD can be used for the growth of organic polymers. This film growth is described as molecular layer deposition (MLD), where a molecular fragment is deposited in each MLD cycle and can be used to prepare organic polymer films. Nowadays, many polymer films are prepared, such as polyamides and polyimides. Poly(terephthalamide) (PPTA), polyurea, polyurethane, and poly(azomethine) have also been grown using the MLD technique [3].

Hybrid inorganic–organic polymer multilayer films have been prepared using inorganic precursors from ALDs and MLDs and various organic precursors. One class of hybrid inorganic MLD polymers is alucones. These hybrid polymers are produced by using alkyl aluminum precursors such as TMA and various organic diols such as ethylene glycol (EG). Figure 6.6 shows a multilayer structure with Al_2O_3

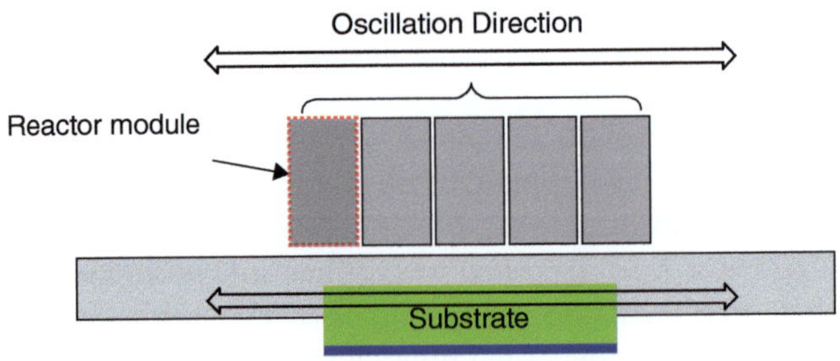

Figure 6.5 Schematic diagram of the atmospheric pressure ALD system. Source: Reproduced with permission from Park et al. [3]; © 2011, The Institute of Physics.

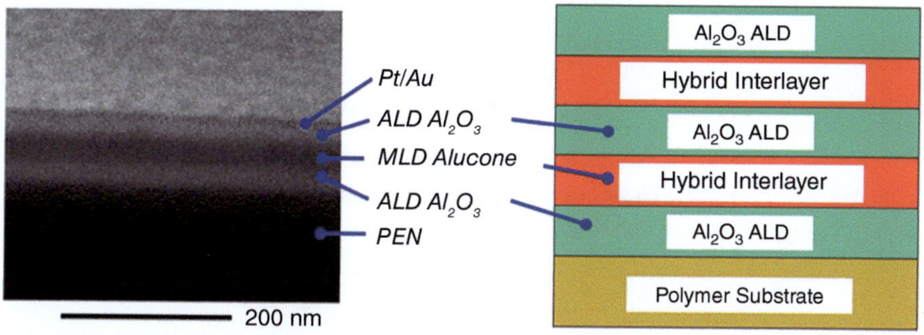

Figure 6.6 Al_2O_3 and alucone films prepared by ALD and MLD. Source: Reproduced with permission from George et al. [23]; © 2009, American Chemical Society.

and alucone layers deposited using TMA, EG, and water. In fact, MLD deposition of alucones can produce linear growth and is very stable and efficient. However, the MLD process still has difficulty solving the problem of deposition of heat-sensitive organic precursors and naturally porous organic films. For example, it still needs to be developed [23].

6.2.3 Inkjet Packaging Technology

6.2.3.1 Introduction to Inkjet Encapsulation Technology

Inkjet printing is a noncontact printing technique in which a digital image is recreated by jetting droplets of ink onto a substrate [26]. The technology is mainly used in the publishing industry, but it is becoming increasingly popular in the thin-film device industry due to its many advantages such as fast printing speed, cost-effectiveness, and low manufacturing temperature. Inkjet printing technology involves spraying tens of trillionths of a liter of solution onto the surface of a sample at a frequency of more than a few hundred times per second. The solvent is then removed to form a film. Inkjet printing technology finely sprays the material in its solution state with an inkjet nozzle where it is needed. In contrast to depositing films in a vacuum chamber, inkjet printers do not have to be in a vacuum environment. The printers are typically isolated in a nitrogen atmosphere before being sprayed and encapsulated. In addition, it has first-class compatibility with organic and inorganic materials and is commonly used to print highly conductive lines and patterns. Inkjet printing can usually be divided into two types according to the method of operation: continuous inkjet printing (CIJ) and drop-on-demand (DoD) inkjet printing [27].

6.2.3.2 Continuous Inkjet Printing

In the CIJ system (Figure 6.7), piezoelectric vibrations generate a continuous stream of ink, which is harmonically modulated and breaks up into droplets at the printhead nozzle. The droplets are typically between 80 and 100 μm in size. To create a pattern, individual droplets are navigated by binary or variable electrostatic deflection and move toward the substrate at a speed of approximately 20 m/s. On

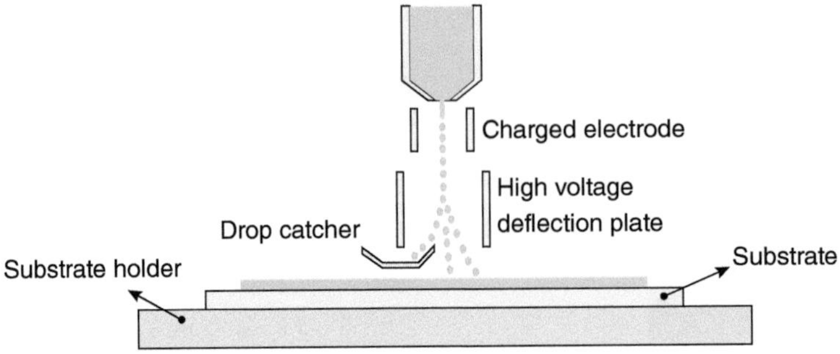

Figure 6.7 Continuous inkjet printing (CIJ). Source: Reproduced with permission from Ref. [26]; © 2021, Elsevier.

the substrate, the droplet diameter is approximately twice the size of the droplet before release. The CIJ method has some advantages over the DoD method, such as the use of lower-viscosity inks, higher droplet velocities, and thus faster printing speeds. However, it is not economically advantageous because more ink must be used, and individual droplets can be difficult to control [28].

6.2.3.3 Drop-on-Demand Inkjet Printing

In the DoD method (Figure 6.8), droplets are driven thermally, piezoelectrically, electrostatically, or acoustically. Jetting on demand by applying only (digital) signals or waveforms allows control of ink droplet jetting through each nozzle, thus avoiding ink waste. This method has been widely promoted for its advantages of high precision, controllability, and cost-effectiveness. Thermally activated by current pulses, the heating element inside each chamber of the cartridge causes rapid evaporation of the ink and the formation of ink vapor bubbles that force the ink droplets away from the nozzle. Ink droplets can be as small as 0.5 μm for methanol and 1 μm for toluene [29]. The temperature of the process, which can reach up to 300 °C, causes the evaporation of water from the nozzle surface and can easily clog the nozzle. This makes it unsuitable for printing biological materials and polymer inks. Thermal DoD is mainly used in applications such as desktop printing with water-based inks. Piezoelectric activation is achieved by applying a voltage to a piezoelectric element placed on the back of each ink container. The element changes shape and thus generates a pressure wave that forces smaller droplets of 10–20 pL to be ejected from the nozzle. Piezoelectric DoD is often used in a wide range of research and industrial applications because of its ability to print aqueous and UV-based inks on a variety of substrates. It is well suited for printing conjugated polymeric aqueous inks because the process does not require high temperatures.

However, both thermal and piezoelectric drive systems have some limitations in terms of nozzle array density (and thermal drive system size) and jet frequency due to thermal issues and physical limitations. In contrast, the liquid is directly manipulated by an electric field in an electrostatic drive system, which can be used to address these limitations. The liquid is placed within an electric field, and an electric charge is induced on the surface of the liquid-bending moon [27]. This creates

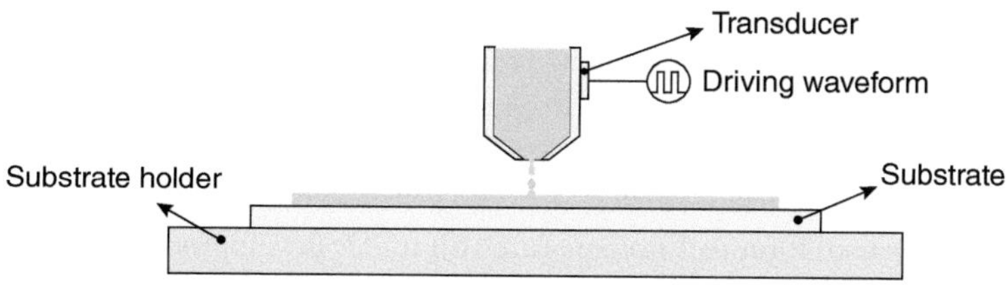

Figure 6.8 Drop-on-demand (DOD) inkjet printing technology. Source: Reproduced with permission from Verboven et al. [26]; © 2021, Elsevier.

an electrical stress that stretches the curved liquid surface in the direction of the electric field and forms a sharp cone under the influence of the electrostatic field. When the electrostatic force exceeds the surface tension, a uniform droplet of charged liquid is ejected from the nozzle. The sprayed droplets can be smaller than the nozzle size, resulting in higher resolution than piezoelectric inkjet printers and the ability to print more viscous inks. However, this method is not currently widely adopted commercially. A recently developed method is acoustic inkjet printing, which can use acoustic energy to generate droplets. Droplets are produced by focusing an ultrasonic beam with an acoustic lens onto a fluid surface. The droplet size can be dynamically changed by adjusting the distance from the fluid to the sensor. This printing method has the advantage that the nozzles do not clog and can print a wider range of ink viscosities [30].

Commercially available printers can print lines as wide as 50 μm with a minimum pitch of 20 μm. However, more advanced inkjet technology can reduce line width and pitch to between 3 and 20 μm. The ability to print very sharp patterns and images with high precision is one of the most attractive features of inkjet printing. Inkjet printing offers many other advantages, including fast print speeds, extremely low ink waste (primarily for DoD), and roll-to-roll compatibility. However, inkjet printing technology also has certain disadvantages. It is limited to printing on flat substrates and films of a certain thickness. Additionally, it requires a rather complex ink formulation, and the ink needs to have a certain solvent evaporation rate and a dynamic viscosity that can be easily adjusted. Inkjet printing technology is well suited for printing almost all OLED device layers. Printing on transparent layers is still challenging but can be solved by printing very narrow arrays or grids [26, 27].

6.2.3.4 Development of Inkjet Printing Technology

After acquiring South Korea's OLEDPlus in 2014, Kateeva, Inc. proposed changing the inkjet printing technology from printing the OLED luminescence layer to printing the OLED luminescence layer after the vapor deposition of the sealing layer. This was done for inkjet process packaging in flexible large-panel OLED displays. Sumitomo Chemical developed the encapsulation material solution for the inkjet printing process, while Samsung integrated Kateeva's inkjet print head and control software into Canon's vapor deposition equipment for printing the sealed encapsulation layer. Inkjet printing film preparation technology can reduce material waste compared to traditional vacuum film preparation technology. Additionally, the cost of investment in equipment is much lower than that of vacuum equipment. However, the inkjet process of encapsulation involves an organic, low-molecular solution. Its barrier rate, lifetime, and uniformity are lower than those of traditional inorganic barrier materials. The solution process development and inkjet process encapsulation technologies are still under development [26].

References

1 Channa, I.A., Distler, A., Zaiser, M. et al. (2019). Thin film encapsulation of organic solar cells by direct deposition of polysilazanes from solution. *Advanced Energy Materials* 9 (26): 1900598.

2 Behrendt, A., Meyer, J., van de Weijer, P. et al. (2016). Stress management in thin-film gas-permeation barriers. *ACS Applied Materials & Interfaces* 8: 4056.

3 Park, J.S., Chae, H., Chung, H.K., and Lee, S.I. (2011). Thin film encapsulation for flexible AM-OLED: a review. *Semiconductor Science and Technology* 26: 034001.

4 Yin, Q.W., Lian, W.D., Li, J.C. et al. (2022). A study of thin film encapsulation on improving electrical characteristics and reliability for flexible OLEDs. *Materials Science Forum* 1058: 93–98.

5 Yu, D., Yang, Y.Q., Chen, Z. et al. (2016). Recent progress on thin-film encapsulation technologies for organic electronic devices. *Optics Communication* 362: 43.

6 Guo, T., Wang, K., Lu, R. et al. (2019). 74–1: A new flexible thin film encapsulation structure with high reliability and wide color-viewing-angle. *SID International Symposium Digest of Technical Papers* 50: 1052.

7 Jeong, E.G., Kwon, J.H., Kang, K.S. et al. (2020). A review of highly reliable flexible encapsulation technologies towards rollable and foldable OLEDs. *Journal of Information Display* 21: 19.

8 Kwon, B.H., Lee, H., Kim, M. et al. (2021). A systematic study of the interactions in the top electrode/capping layer/thin film encapsulation of transparent OLEDs. *Journal of Industrial and Engineering Chemistry* 93: 237.

9 Park, K.W., Lee, S., Lee, H. et al. (2019). High-performance thin H:SiON OLED encapsulation layer deposited by PECVD at low temperature. *RSC Advances* 9: 58.

10 Prepelita, P., Filipescu, M., Stavarache, I. et al. (2017). Transparent thin films of indium tin oxide: morphology–optical investigations, inter dependence analyzes. *Applied Surface Science* 424: 368.

11 Greener, J., Ng, K.C., Vaeth, K.M., and Smith, T.M. (2007). Moisture permeability through multilayered barrier films as applied to flexible OLED display. *Journal of Applied Polymer Science* 106: 3534.

12 Zhou, A., Wang, H., Zhang, Y., and Hu, C. (2022). Recent progress of flexible OLEDs for wearable devices. *Fiber and Integrated Optics* 1–25.

13 Wu, J., Fei, F., Wei, C. et al. (2018). Efficient multi-barrier thin film encapsulation of OLED using alternating Al_2O_3 and polymer layers. *RSC Advances* 8: 5721.

14 Kim, B.J., Park, H., Seong, H. et al. (2017). A single-chamber system of initiated chemical vapor deposition and atomic layer deposition for fabrication of organic/inorganic multilayer films. *Advanced Engineering Materials* 19: 1600819-1.

15 Lu, Q., Yang, Z., Meng, X. et al. (2021). A review on encapsulation technology from organic light emitting diodes to organic and perovskite solar cells. *Advanced Functional Materials* 31 (23): 2100151.

16 Wang, L., Ruan, C., Li, M. et al. (2017). Enhanced moisture barrier performance for ALD-encapsulated OLEDs by introducing an organic protective layer. *Journal of Materials Chemistry C* 5: 4017.

17 Jen, S.H., Bertrand, J.A., and George, S.M. (2011). Critical tensile and compressive strains for cracking of Al_2O_3 films grown by atomic layer deposition. *Journal of Applied Physics* 109: 084305–084301.

18 Meyer, J., Görrn, P., Bertram, F. et al. (2009). Al_2O_3/ZrO_2 nanolaminates as ultrahigh gas-diffusion barriers—a strategy for reliable encapsulation of organic electronics. *Advanced Materials* 21: 1845.

19 Chen, G., Weng, Y., Sun, F. et al. (2019). Low-temperature atomic layer deposition of Al2O3-alucone nanolaminates for OLED encapsulation. *RSC Advances* 9: 20884.

20 Kwon, B.-H., Joo, C.W., Cho, H. et al. (2021). Organic/inorganic hybrid thin-film encapsulation using inkjet printing and PEALD for industrial large-area process suitability and flexible OLED application. *ACS Applied Materials & Interfaces* 13 (46): 55391–55402.

21 Sarkar, S., Culp, J.H., Whyland, J.T. et al. (2010). Encapsulation of organic solar cells with ultrathin barrier layers deposited by ozone-based atomic layer deposition. *Organic Electronics* 11 (12): 1896–1900.

22 Yang, Q., Li, H., Li, M. et al. (2017). Rayleigh instability-assisted satellite droplets elimination in inkjet printing. *ACS Applied Materials & Interfaces* 9: 41521.

23 George, S.M., Yoon, B., and Dameron, A.A. (2009). Surface chemistry for molecular layer deposition of organic and hybrid organic inorganic polymers. *Accounts of Chemical Research* 42 (4): 498–508.

24 Levy, D.H., Freeman, D., Nelson, S.F. et al. (2008). Stable ZnO thin film transistors by fast open air atomic layer deposition. *Applied Physics Letters* 92 (19): 192101.

25 Sun, J., Mourey, D.A., Zhao, D. et al. (2008). ZnO thin-film transistor ring oscillators with 31-ns propagation delay. *IEEE Electron Device Letters* 29 (7): 721–723.

26 Verboven, I. and Deferme, W. (2021). Printing of flexible light emitting devices: a review on different technologies and devices, printing technologies and state-of-the-art applications and future prospects. *Progress in Materials Science* 118: https://doi.org/10.1016/j.pmatsci.2020.100760.

27 Singh, M., Haverinen, H.M., Dhagat, P., and Jabbour, G.E. et al. (2010). Inkjet printing—process and its applications. *Advanced Materials* 22 (6): 673–685.

28 De Gans, B.J., Duineveld, P.C., and Schubert, U.S. (2004). Inkjet printing of polymers: state of the art and future developments. *Advanced Materials* 16 (3): 203–213.

29 Sirringhaus, H. and Shimoda, T. (2003). Inkjet printing of functional materials. *MRS Bulletin* 28 (11): 802–806.

30 Li, J., Ye, F., Vaziri, S. et al. (2013). Efficient inkjet printing of graphene. *Advanced Materials* 25 (29): 3985–3992.

7

Application of Flexible Electronics Packaging

Yuezhou Zhang

Northwestern Polytechnical University, Institute of Flexible Electronics, Dongxiang Road, Xi'an 710129, China

7.1 Industry Chain Analysis of Flexible Electronics Packaging

Flexible electronics is a rapidly emerging field encompassing the science and technology of using organic, inorganic, or organic–inorganic composite (hybrid) materials deposited on flexible substrates to create electronic, photonic, and optoelectronic components. These components, which go beyond traditional circuits, exhibit unique features such as deformability, portability, lightweightness, and wearability. The flexible electronics industry is experiencing significant growth and holds promising prospects for further development.

7.1.1 Upstream, Midstream, and Downstream of the Flexible Electronics Industry Chain

The flexible electronics industry is comprised of distinct segments along its supply chain. The upstream sector involves equipment manufacturing and the production of raw materials for flexible screens, such as organic light-emitting materials, film materials, and chemicals. Additionally, it includes the manufacturing of flexible electronic materials like driver chips, flexible sensors, and electronic components. The midstream sector focuses on the manufacturing and assembly of fully flexible displays, module assembly, and the manufacturing of fully flexible sensors and modules, as well as other aspects like flexible integrated circuit production and assembly. These stages require specialized technology and equipment, resulting in high barriers to entry. The downstream sector involves the utilization of flexible electronics in various end-application products. These include intelligent mobile terminals, intelligent transportation systems, entertainment and media devices, smart home technologies, sports and fashion products, and office and education applications. Each segment of the supply chain has clear divisions of labor, and there are significant technological and equipment barriers between them. Notably, encapsulation technology plays a critical role in the middle and upper levels of the flexible electronics

Flexible Electronic Packaging and Encapsulation Technology, First Edition.
Edited by Hong Meng and Wei Huang.

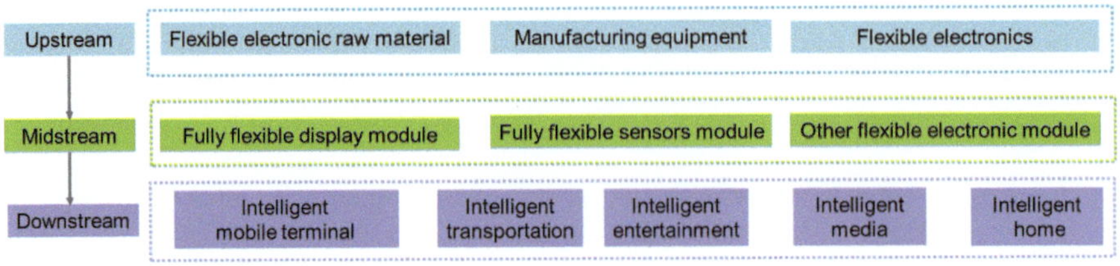

Figure 7.1 Diagram of the industry chain of the flexible electronics industry.

industry chain. For a visual representation of the flexible electronics industry chain, please refer to Figure 7.1.

The flexible electronics industry is poised for significant growth and has a promising market outlook in the future. On one hand, the emergence of fully flexible displays and sensors will liberate users from the constraints of rigid physical forms. These technologies offer advantages in terms of thickness, weight, durability, color, and energy efficiency. The introduction of flexible electronics will provide users with unprecedented smart experiences, driving the demand from downstream customers and fueling rapid industry growth. On the other hand, the flexible electronics industry is witnessing an increasing number of participants across the entire supply chain. Established flexible electronics solution providers, key component suppliers, and end-use product companies are making substantial investments in terms of finances and human resources to advance flexible electronics technology. These solution providers closely monitor consumer preferences and needs, collaborating with customers to develop products tailored to different industry sectors, thereby expanding the downstream market. According to projections, the global market size of the flexible electronics industry is expected to reach US$ 304.940 billion by 2025, exhibiting a compound annual growth rate (CAGR) of approximately 144.71% from 2019 to 2025 [1]. The demand for flexible electronics technology is rapidly increasing across various sectors, including smart mobile terminals, intelligent transportation, entertainment and media, smart home technologies, sports and fashion, and office education.

7.1.2 Overview of the Development of Flexible Packaging Materials

The development of the encapsulation materials market is in line with the electronic end-product market and the semiconductor encapsulation industry market and is closely related to downstream encapsulation demand and changes in the widely used encapsulation process. In general, foreign flexible encapsulation materials have been developed earlier and the technology is more mature, thus occupying the main market position.

At present, the main global suppliers of encapsulation materials are Samsung SDI and LG Chem from Korea, 3M from the United States, and Mitsui Chem from Japan. In terms of major producing countries and regions, the major global producing countries and regions of encapsulants are Japan, Korea, China, Taiwan, Europe, and America. Japan has long been the largest supplier of encapsulants and is an industry leader in terms of technology and product quality. In recent years, Japan has gradually outsourced a large number of low-profit, low-grade encapsulant production to other countries in order to achieve production cost advantages, improve production efficiency, or supply downstream customers nearby while maintaining domestic production of high-grade encapsulants. South Korea is the second largest producer of encapsulation materials, mainly producing IC carrier boards and interconnection leads, and in recent years, mold sealing materials and tin balls have also gradually emerged, with each material being mainly supplied to the local Korean market. Taiwan, China, is the third-largest encapsulant producer, mainly producing IC carrier boards, and the scale of IC carrier board production is only slightly behind that of Japan. The main suppliers are Changchun Petrochemical, Changxing Chemical, etc. New products in mold sealing materials are also worth watching. Domestic companies involved in high-barrier materials are mainly focused on food and pharmaceutical encapsulation, while the water vapor barrier film for IC carrier boards is still in its infancy.

7.2 Packaging Applications of Flexible OLED Devices

The mesmerizing display industry stands tall as a fundamental pillar within the realm of electronic information. Its vast landscape encompasses an array of subdivisions, from substrates and display materials to polarizers, optical films, and the remarkable panel industry. Embarking on a journey that traces back to the 1980s, the evolution of panel display principles has witnessed a captivating transformation, gracefully transitioning from the cathode ray tube (CRT) to the plasma display (PDP), and finally finding its zenith in the liquid crystal display (LCD) and organic light-emitting diode (OLED) technologies. Among these marvels, OLED reigns supreme as a cutting-edge current-type organic light-emitting device, distinguished by its luminous principles that deviate significantly from LCD display technology. Unlike its LCD counterpart, the brilliance of OLED lies in its individual pixel illumination, casting a harmonious veil of uniformity upon the screen while bestowing upon it an unparalleled contrast that seems to possess infinite depth. Astonishingly, the need for a liquid crystal box and backlight is rendered obsolete within the OLED realm, defying conventional expectations and slimming down the screen thickness with an elegant finesse. By harnessing the power of carrier injection and compound-induced light emission, OLED's required drive voltage plunges to new lows, resulting in lightning-fast response times that captivate the senses. But the wondrous qualities of OLED extend far beyond its technical prowess. Indeed, its remarkable flexibility and bendable nature redefine the boundaries of possibility, ushering in a new era of innovation. Initially, OLED technology found solace in the realm of high-end smartphones, gracing rigid screens and fixed curved displays with its vibrant hues and immersive experiences. With a color saturation and brightness that can only be described as breathtaking, OLED mesmerizes spectators, offering a visual feast that captivates the soul. While OLED technology indisputably outshines its traditional inorganic LED counterparts in terms of efficiency, lining ratio, response time, flexibility, and wide field of view, its journey toward dominance in the realm of display technology has not been without obstacles. The ever-present specter of OLED's short lifespan threatens to hinder its progress. However, the intrepid pursuit of encapsulation techniques offers a glimmer of hope, promising to extend the life of these remarkable devices and thus propel the OLED revolution forward.

7.2.1 Stability Issues of Flexible OLED Devices

OLEDs, those captivating tunneling diode devices, possess an insatiable hunger for electrode materials that perfectly match the HOMO–LUMO energy of the emitting material. Only through this meticulous alignment can the magic of efficient direct electron injection into the emitting polymer truly unfold [2]. As time unfolded, OLEDs gracefully ascended to the summit of display technology, assuming the mantle of its steadfast backbone. With a well-honed working mechanism in their arsenal, a multitude of academic research endeavors and industrious R&D teams embarked on a relentless quest to conquer the challenges that plagued early OLEDs – namely, their low external quantum efficiencies (EQEs) and demanding drive voltages. Armed with a ceaseless spirit of innovation, they delved into a diverse range of perspectives to unearth the secrets behind these obstacles. The result? A breathtaking symphony of complexity, as functional material layers were deftly woven into the device architecture, delicately balancing charge injection and transport to achieve an optimal fusion of emissions. In this grand dance of scientific ingenuity, Figure 7.2a elegantly showcases a typical OLED schematic, resplendent with its alluring tapestry of hole and electron transport layers [3]. Then, in 2008, a pivotal moment dawned upon the world – the birth of the very first OLED TV: the Sony XEL-1.

Behold the flexible OLED, a breathtaking symphony of technological ingenuity that unfolds upon a stage adorned with anode, cathode, and an exquisite organic functional layer meticulously interwoven between two flexible substrates. This sandwiched marvel, resplendent in its transparent electrode gracing a single side, emerges as a gateway to a luminous realm, as depicted in the captivating tableau of Figure 7.2c [4]. The realm of flexible OLED stands tall as one of the most promising bastions in the domains of display and lighting technologies, igniting a fervent blaze of research and development within the hallowed halls of display technology. Within the realm of flexible displays, two dominant contenders emerge, each boasting its own unique prowess. The first, akin to a digital rendition of electronic paper, finds solace in the maturity of electrophoretic display technology – a refined artistry that finds its true calling within the embrace of e-reader products. On the other hand, the illustrious OLED/LCD display technology takes center stage, captivating the hearts of electronic aficionados with its kaleidoscopic array of full-color animations. Televisions and smartphones find their voices amplified through this vibrant canvas of technological splendor. As the dynamic landscape of OLED/LCD active display technology continues to evolve, the spotlight falls upon the resplendent OLED display technology – a beacon of innovation that emerges as

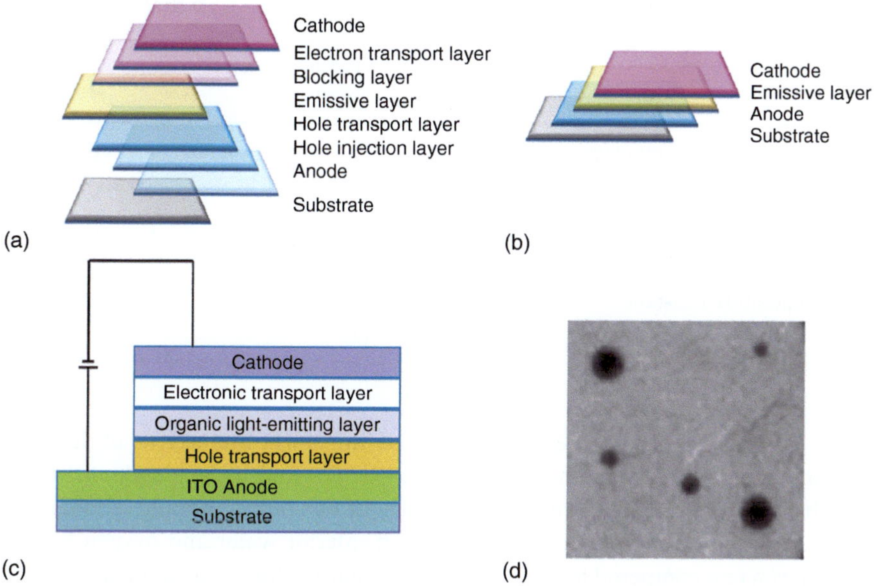

Figure 7.2 (a) Schematic diagram of OLED with typical hole and electron transport layer; (b) schematic diagram of LEC with typical sandwich structure; (c) schematic diagram of OLED device structure; (d) Alq3-based OLED black dots. Source: Adapted from Kalyani et al. [2].

the focal point of research and development, spearheading the mesmerizing realm of flexible active displays. The selection of flexible display substrates bestows upon these luminous marvels an unparalleled ability to transcend traditional boundaries. Metal sheets, delicate thin glass, polymer substrates, and the exquisite union of ultrathin glass and polymers all join hands to offer a kaleidoscope of possibilities. These three noble materials, when harnessed in the creation of flexible OLED displays, breathe life into a realm where bending and rolling into myriad shapes becomes a tangible reality. Yet, it is crucial to recognize the intrinsic relationship between the preparation process, the device structure, and the noble choice of substrate. This delicate trinity reverberates throughout the annals of display technology, dictating the course of action, the cost considerations, and the clarity that graces the resplendent display screens. Within the hallowed pages of Table 7.1, a captivating tapestry of characteristics unfolds, each substrate revealing its unique allure and its potential to shape the destiny of the luminous journey ahead.

At the heart of flexible OLED devices resides an ultrathin organic electroluminescent layer, a coveted core material that possesses a sensitivity so profound that it reverberates with every whisper of water, oxygen, and heat. Alas, this very sensitivity also bears the burden of a lamentable flaw – poor stability. Oxygen, in its audacious presence, wreaks havoc upon luminescence quantum efficiency, device luminescence efficiency, and electron–hole transport capabilities. As if that were not enough, the insidious interplay of moisture finds a treacherous ally in OLEDs, unfurling a tapestry of non-emitting black dots that multiply with the relentless

Table 7.1 Comparison of three types of substrate pattern flexible OLED characteristics.

Characteristics	Polymer substrate	Ultrathin glass-polymer system	Metal flake
Permeability of water and oxygen	×	√	√
Thermal and chemical stability	×	√	√
Mechanical stability	√	√	√
Curlability	√	√	√
High temperature resistance	×	√	√
Weight	√	√	×

passage of time. Gradually, these ominous dots transform into agents of degradation, sabotaging the very essence of the OLED device and its resplendent display. The substrate materials that weave the tapestry of flexibility, although enchanting in their own right, render flexible devices with inferior water and oxygen barrier capabilities when compared to their rigid counterparts. Thus, a formidable shield of hermetic encapsulation emerges as the ultimate safeguard – a meticulous process that unites metal sheets and substrate glass through the alchemical embrace of a UV-curable epoxy resin. Within the ethereal confines of an oxygen- and water-free glove box, the encapsulation ritual commences, shrouding the delicate OLED device in a cocoon of protection. The coatings applied at temperatures below 100 °C bear the sacred mantle of ultralow water and oxygen permeability, an armor forged to repel the insidious intrusion of the elements. The luminaries of the scientific realm, led by the venerable Burrows et al., regale us with tales of hermetic encapsulation techniques involving glass tubes, epoxy resin, and the steadfast presence of nitrogen [5]. In a realm where bottom-emitting OLEDs find solace beneath the seal of epoxy resin and overlay glass, their top-emitting brethren bask in the radiant embrace of the transparent glass cover, allowing their radiant luminescence to cascade upon the world. Desiccant-free, transparent thin-film encapsulation (TFE) emerges as an imperative, bestowing the flexible OLED device with the unobstructed glory it so ardently deserves.

A somber truth looms over the realm of flexible OLED devices – a truth that necessitates the imposition of stringent encapsulation measures. Only through this resolute fortification can the lifespan of the device be extended, its stability fortified, and the coveted dream of mass production be transformed into tangible reality.

In the realm of OLED encapsulation, the scholarly endeavors of Lim et al. shone with remarkable ingenuity, employing silica microparticles to craft meticulously controlled pinholes on the cathode layer [6, 7]. Through their diligent observations, they bore witness to the initial emergence of enigmatic black spots, witnessing their gradual growth over time. A fascinating revelation unfolded before their eyes, as the increase in spot diameter manifested itself in a proportional relationship to the square root of time, a tantalizing dance of growth and temporal progression. Lifka, a visionary in the field, embarked upon a quest to forge a novel TFE that would defy the intrusion of water. With unwavering determination, they unveiled the multilayer marvel known as silicon nitride-silicon oxide-silicon oxide-silicon nitride (NONON) [8]. The water permeability of this miraculous creation stooped to a remarkable level of $<1 \times 10^{-6}$ g/m^2/day for individual NON stacks, a feat accomplished through meticulous depositions carried out at an optimal temperature of 85 °C. The scientific tapestry of OLED encapsulation continued to unfold under the watchful gaze of Jung et al., who synthesized UV-curable cycloaliphatic epoxy-functionalized oligomeric siloxane resins [9]. Their arsenal of resins, birthed through a non-hydrolytic sol–gel reaction method, became the very foundation upon which OLED encapsulation would find its resplendent form. Gi Heon Kim and their diligent team delved into the domain of glass substrate encapsulation, weaving a mesmerizing tale of photopolymerizable blends and crosslinked passivation layers [10]. With pentaerythritol triacrylate (PETIA) and HSP188 (photoinitiator) as their guiding forces, they embarked upon a spin-coating ritual that culminated in the birth of a crosslinked passivation layer. Their unyielding pursuit of knowledge led them to compare the electrical and emission characteristics of OLEDs before and after the formation of this ethereal layer [11]. Stepping further into the labyrinth of encapsulation possibilities, they drew upon the power of poly-*p*-xylene (PPX) and/or poly-2-chloro-*p*-xylene (PCPX) films, casting them as the heroes of thermochemical vapor deposition polymer films (TCVDPF) encapsulation for OLEDs [12]. As these resplendent films draped themselves upon the OLEDs, a remarkable transformation occurred – the encapsulated OLEDs unfurled a lifespan four times that of their unencapsulated counterparts, safeguarded within a realm untouched by the perils of oxygen and moisture. The secret to this longevity lay in the very thickness of the TCVDPF layer, a majestic shield that stood resolute in its quest to bar the intrusion of unwanted elements. Within this mesmerizing tapestry of encapsulation marvels, Giovanni Nisato et al. etched their name in the annals of scientific history as the pioneering heralds of a quantitative measurement of OLED-compatible thin-film encapsulants [5]. With their groundbreaking revelations, they unlocked the door to a realm where encapsulants would be assessed and judged with meticulous precision, paving the way for a future where the wonders of OLED encapsulation could be quantitatively explored and appreciated.

7.2.2 Flexible OLED Packaging Technology

In the intricate realm of OLED devices, encapsulation assumes an awe-inspiring role, where its demands reach extraordinary heights. Practical OLED devices, driven by an insatiable quest for excellence, impose exacting requirements that push the boundaries of possibility. Within this captivating landscape, encapsulation becomes a sacred art, a delicate dance between science and ingenuity. To withstand the rigors of the OLED world, encapsulation must embrace a sublime impermeability. The water vapor transmission rate (WVTR) stands as a formidable criterion, compelling encapsulation materials to exhibit a level of permeability that is less than a mere whisper, less than 10^{-6} g/m^2/day. In this sacred quest, the very essence of moisture becomes a foe to be vanquished, as the encasement steadfastly guards against its treacherous infiltration. Yet, the OLED device's hunger for perfection does not stop there. It beckons encapsulation to shield against another relentless adversary – oxygen. The oxygen vapor transmission rate (OVTR) imposes a demand that transcends the ordinary, calling for a performance that lingers beneath the threshold of 10^{-5} $cm^3/m^2/day$. Oxygen, with its intrusive nature, seeks to corrupt the brilliance within, but encapsulation rises to the challenge, constructing an impervious fortress that preserves the purity of the OLED realm.

The current mainstream encapsulation methods include rigid glass cover encapsulation and TFE. TFE technology has become the current development trend because it does not require a rigid glass cover and enables OLEDs to achieve the flexible function of bending and folding. However, the industrialization of flexible OLED displays is progressing slowly for two main reasons, including the following:

7.2.2.1 Lack of Breakthrough in Encapsulating Technology

The pursuit of flexible OLED encapsulation using TFE technology has presented an intricate challenge, one that demands the achievement of exceptional water and oxygen barrier performance without compromising on the crucial aspect of bending resistance. However, the current landscape reveals that the OLED TFE process is still in its nascent stages, lacking the maturity necessary to strike the delicate balance between water and oxygen resistance, bending performance, process simplicity, and cost-effectiveness. Moreover, while OLED devices have made significant strides in terms of lifespan, surpassing the 10 000-hour mark and boasting storage lives exceeding 50 000 hours, their longevity still falls short when compared to their traditional LCD counterparts. This discrepancy has become a stumbling block, hindering the widespread adoption and large-scale application of flexible OLED displays. The conundrum lies in finding a comprehensive solution that not only addresses the encapsulation challenges but also pushes the boundaries of OLED device lifespan. The industry is acutely aware that surpassing the current limitations is vital to unlocking the full potential of flexible OLEDs and ushering in a new era of mass production. In the quest for progress, researchers, engineers, and industry leaders are pooling their expertise and resources to overcome these barriers. Advanced materials are being explored to enhance water and oxygen barrier performance, while novel encapsulation techniques are being developed to ensure flexibility and durability. Simultaneously, the focus remains on streamlining the encapsulation process and optimizing it for simplicity and cost efficiency. Collaborative efforts, driven by the urgency to realize the vast possibilities of flexible OLED technology, are underway to tackle these obstacles head-on. Research institutions, manufacturers, and industry consortia are engaging in relentless innovation, investing in cutting-edge research and development to conquer the challenges and unlock the true potential of flexible OLED displays. The path to mass production of flexible OLEDs may be intricate and filled with hurdles, but the collective determination of the OLED community propels them forward. As barriers are gradually dismantled, and breakthroughs unfold, the dream of a world adorned with flexible OLED displays becomes increasingly tangible. With each stride forward, the convergence of performance, longevity, encapsulation excellence, and affordability draws nearer, paving the way for a future where flexible OLEDs reign supreme, enchanting the world with their seamless integration into our daily lives.

7.2.2.2 Low Yield Rate

The challenge of achieving high yield in the production of OLED panels has presented a significant hurdle, directly impacting their cost and ultimately driving up prices. This issue becomes more pronounced as the size of flexible OLEDs increases, as it becomes increasingly challenging to maintain precise control over the deposition process, including factors such as deposition rate, substrate temperature, and deposition power, as well as ensuring film uniformity throughout the manufacturing process [13].

The complexities associated with large-scale flexible OLED production have yet to be effectively addressed, impeding the widespread application of this technology. The lack of comprehensive solutions has limited the potential applications of flexible OLEDs, preventing their seamless integration into various industries and consumer devices. Researchers and manufacturers recognize the urgency of resolving these challenges to unlock the full potential of flexible OLED technology. Extensive efforts are underway to refine the deposition process, seeking innovative techniques and methodologies to enhance control and ensure consistent film quality. This includes the exploration of advanced deposition equipment, optimization of process parameters, and the development of novel materials to enhance film uniformity. Collaborative endeavors between academia and industry strive to overcome these obstacles. Research institutions, equipment manufacturers, and OLED panel producers are working hand-in-hand to tackle the complexities associated with large-size flexible OLEDs. The collective expertise and resources directed toward resolving these issues promise to yield breakthroughs that will revolutionize the production landscape, enabling cost-effective manufacturing processes and driving down prices. As progress is made in addressing the challenges of deposition control and film uniformity, the application potential of flexible OLEDs will expand exponentially. These advancements will open doors to a wider range of devices and industries, fostering the integration of flexible OLED technology into smartphones, televisions, wearable devices, and beyond.

While the current limitations in yield and manufacturing complexity have presented barriers, the resilience and dedication of the OLED community remain unwavering. As solutions are developed and implemented, the cost and availability of flexible OLED panels will improve, unlocking new possibilities and paving the way for the widespread adoption of this transformative technology.

Currently, two primary methods are employed for encapsulating flexible OLED devices, with in-line encapsulation being the more prevalent approach. In this method, an inorganic–organic alternating barrier layer is directly coated onto the metal electrode surface of flexible OLED devices, ensuring the avoidance of breaking the vacuum within the OLED devices. This approach offers significant advantages by preventing the metal electrode from coming into contact with water, oxygen, dust, and other impurities, thereby safeguarding the longevity and performance requirements of OLED devices [14].

In-line encapsulation has emerged as the mainstream method due to its ability to ensure long device life and preserve the device's performance characteristics. However, it is worth noting that in-line encapsulation is a more complex process overall, posing challenges for achieving roll-to-roll production. Additionally, the high processing temperature (typically ≥90 °C) required during the preparation of the barrier layer can potentially damage the OLED device, affecting its overall lifespan. Despite the complexities associated with in-line encapsulation, it remains the preferred choice for both domestic and international companies in the OLED industry. The robustness of this method in protecting OLED devices from external environmental factors has made it a trusted and widely adopted technique. Continuous research and development efforts are focused on enhancing the roll-to-roll production feasibility of in-line encapsulation, as well as addressing the challenges related to high processing temperatures. By refining the process, optimizing materials, and exploring novel techniques, the industry is working toward streamlining the in-line encapsulation process and ensuring the compatibility of high-performance OLED devices with mass production requirements. As the technology continues to evolve, it is expected that advancements in in-line encapsulation will further enhance the performance, reliability, and cost-effectiveness of flexible OLED devices. This will contribute to the broader adoption of flexible OLED technology in various applications, from consumer electronics to automotive displays, wearable devices, and beyond.

Another method is offline encapsulation, which involves laminating a high-barrier film onto the metal electrode surface of the OLED device using an adhesive/film. This technique offers simplicity and allows for separate optimization of the OLED device and the high-barrier layer preparation process. However, during the encapsulation process, the OLED device is susceptible to exposure to water, oxygen, and dust, and the encapsulating material may potentially damage the metal electrode. The use of a non-barrier adhesive film can result in side leakage issues in the encapsulated device, allowing water and oxygen to penetrate the device through the adhesive layer and cause damage. Additionally, insufficiently reacted adhesive may react with the OLED device, affecting its lifespan.

The Barix encapsulation technology, initially developed by Vitex System, utilizes an inorganic–organic multilayer encapsulation structure, as illustrated in Figure 7.3 [4]. This approach involves vapor depositing an organic material (such as a polyacrylate film) onto the substrate to enhance film flatness, reduce mechanical damage and defects, and improve flexibility. The inclusion of an organic layer helps relieve interlayer stress and enhances the film's resistance to bending. Subsequently, inorganic materials (e.g. silicon carbide, silicon oxide) are deposited using sputtering or PECVD techniques to enhance the film's barrier properties against water and gas permeation.

During the preparation of large-sized flexible OLED devices, it is inevitable that the film layer surface before cathode plating will contain raised particles. Since the cathode metal layer is extremely thin, these particles can penetrate it and cause pinholes. Consequently, water and oxygen can enter the device through these pinholes,

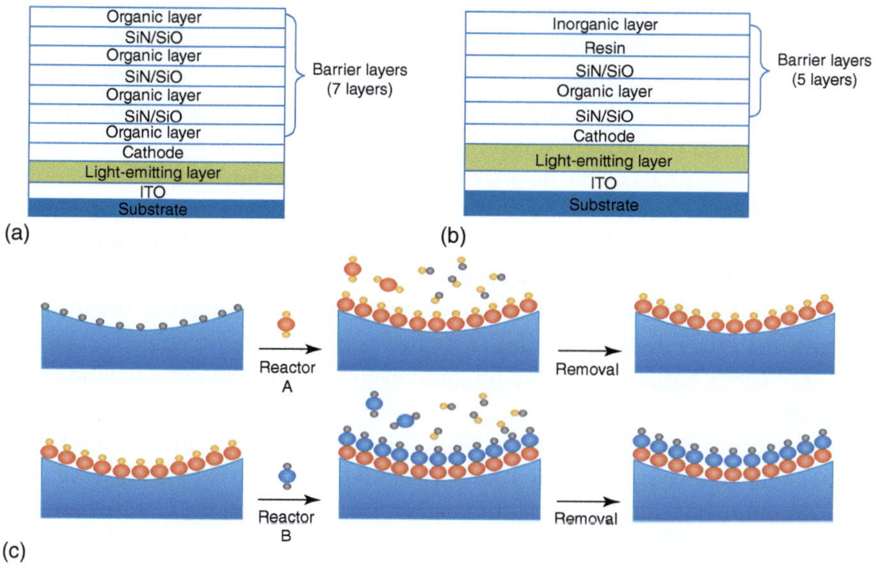

Figure 7.3 (a) Schematic diagram of the barrier layer structure of Barix technology; (b) Samsung improved hybrid encapsulating technology; (c) deposition cycle of an ALD. Source: Reproduced with permission from Liu et al. [4]; © 2017, Nanjing Institute of Electronic Devices.

resulting in damage. The multilayer encapsulation structure improves the device's resistance to bending as well. By employing this inorganic–organic multilayer encapsulation, the water permeability can be reduced to a range of 10^{-4} to 10^{-6} g/m^2/day, while the oxygen permeability can reach 0.005 cc/cm^2/day, significantly extending the device's lifespan.

However, the current process of preparing the inorganic water and oxygen barrier layer poses challenges. The intrinsic cohesive stress of the material makes the film susceptible to hole formation at grain boundaries. If the organic layer is too thin, external water and oxygen can easily penetrate through these holes in the inorganic layer, affecting the performance of the OLED light layer. Therefore, the Barix multilayer stacking technology necessitates dense and uniform inorganic layers while also requiring a sufficiently thick organic layer to mitigate water and gas diffusion. In practical production, the Barix multilayer stacking process involving the alternating deposition of thin films for encapsulation is complex. Additionally, the preparation of multilayer organic–inorganic films leads to high equipment costs. Moreover, the substantial grain boundary stress in the inorganic layer prepared by the PECVD method limits the industrial application of Barix multilayer stacking technology in the flexible OLED encapsulation field.

While the SiN film can serve to cover pinholes in the metal layer, it tends to develop new pinholes itself. To address this issue, an organic polymer layer can be applied on top of the SiN film to create a smoother interface. This polymer layer effectively fills the defects in the SiN layer and extends the pathway for water and oxygen to enter the OLED device. Moreover, the organic polymer layer enhances the device's resistance to bending. Subsequently, another layer of SiN film is coated, which helps mitigate the impact of pinholes on the device by increasing the channel length. This step aims to enhance the barrier properties of the device. By employing a SiN/organic polymer/SiN structure for encapsulation, the device's barrier performance can be significantly improved compared to using SiN alone. This multilayer structure further enhances the device's resistance to water and oxygen.

Samsung Display has enhanced the Barix multilayer TFE structure by implementing a more efficient hybrid encapsulation technology, as depicted in Figure 7.3b [4]. This approach involves a hybrid encapsulation method that combines adhesive resin coating with the application of three overlapping layers of inorganic, organic, and inorganic materials. As a result, the TFE process becomes more efficient, with the number of layers reduced from the original seven layers in the Barix technology to five layers.

The enhanced and optimized Barix encapsulation technology offers several benefits, including reduced process time, increased production capacity, cost savings, and improved product yields. However, the current process still requires more than three barrier layers to meet the requirements of industrial applications. Additionally, the Barix multilayer barrier is prone to peeling off after bending, which poses challenges for manufacturing costs and product yields. Moreover, the preparation process for the multilayer barrier layer involves multiple transfers of samples between different process cavities, and the generation of particle defects during this transfer process significantly impacts the yield. To enable the widespread adoption of Barix multilayer stacked encapsulation technology in mass production lines, further technical innovations are needed. Additionally, there is a need to explore and develop the multilayer barrier layer toward a trend of a single barrier film, which could simplify the manufacturing process and improve overall performance. The ALD (Atomic Layer Deposition) encapsulation technique employs an iterative and alternating sequence of binary reactions. In this process, the precursor material is allowed to react to a certain extent, after which it is abruptly stopped. Subsequently, a large volume of inert gas is rapidly introduced to eliminate the excess precursor material from the chamber, preventing further reactions. This cycle is repeated, as illustrated in Figure 7.3c [4]. By controlling the concentration of the precursor material and the reaction time, the ALD technique achieves precise control over the film's thickness. Most ALD techniques utilize binary reactions, and the aforementioned method enables controlled and quantitative deposition of both reactants. This allows for the precise alignment of atomic layers at the luminescent tip, resulting in the formation of smooth, continuous, and defect-free films.

Vacuum thermal deposition is a conventional vacuum coating technique that requires equipment operating at a pressure of at least 10^{-3} Pa. In this context, Wei Bin proposes the use of vacuum thermal deposition to grow MgF_2 and ZnS films as a hybrid barrier layer. This approach significantly enhances the device's water barrier capability. Furthermore, the incorporation of MgF_2/ZnS as a passivation layer ensures good light transmission. For instance, when an initial luminance of 2000 cd/m^2 is applied without UV resin sealing, a yellow OLED device encapsulated with three pairs of MgF_2/ZnS exhibits a lifetime exceeding 500 hours, with a luminance drop of 56%. By employing UV resin sealing, the device's half-life can be extended to 245 hours. This encapsulation method is straightforward to prepare, cost-effective, and particularly suitable for encapsulating OLED devices with top light-emitting structures. In summary, the ongoing development of in-line encapsulation processes is focused on achieving process simplicity, precise process control, and cost reduction.

Off-line encapsulation technology involves the lamination of a high-barrier film onto the metal electrode surface of flexible OLED devices using an encapsulant/film. Compared to the in-line process, off-line encapsulation is generally less complex to prepare. One advantage of this technology is that it allows for the independent optimization of the encapsulant/film and the high-barrier film. However, during the encapsulation process, there is a risk of exposure to water, oxygen, and dust, which can potentially damage the flexible OLED devices. This exposure is not conducive to achieving high efficiency and long lifespan goals for the devices.

Encapsulation adhesive/film consists of two components: encapsulation adhesive and encapsulation film. Encapsulation adhesive refers to a liquid adhesive used for encapsulation, while encapsulation film refers to a film made of encapsulation adhesive. The encapsulation adhesive is typically solvent-free, UV-curable, and neutral to prevent damage to the metal electrodes, light-emitting layer, and carrier transport materials in flexible OLED devices. Solvents can be detrimental to these components. Encapsulation film can be categorized into hot-melt adhesive film and room-temperature pressure-sensitive adhesive film. When using hot-melt adhesive film for encapsulation, the film needs to be heated to a temperature of 140 °C or higher to reach a molten state for bonding the cell to the high-barrier film. However, the high process temperature of 140 °C can cause thermal deformation of the substrate and damage the OLED light-emitting layer and other functional layer materials, leading to significant degradation of the device's optoelectronic performance. On the other hand, pressure-sensitive adhesive film is equipped with protective films on both sides, which can be bonded to the barrier film and the OLED device after removing the protective films. Compared to hot melt adhesive film, pressure-sensitive adhesive film can be laminated at much lower temperatures, even at room temperature, thus avoiding substrate and functional layer deformation or decomposition caused by heat. In humid environments, high-barrier pressure-sensitive adhesive films can be used to prevent water vapor leakage from the edges of OLED devices. Therefore, the use of pressure-sensitive adhesive film requires the ability to laminate the film at low or normal temperatures, without air bubbles, and preferably with excellent barrier properties to ensure effective encapsulation.

The first-generation flexible barrier film materials, such as polyethylene (PE), polyester (PES), and polyvinyl chloride (PVDC), had limited water barrier capacity. These organic films were used as barrier layers but had drawbacks in terms of barrier performance. The second-generation flexible barrier materials introduced aluminum foil and aluminized film as barrier layers. These materials offered excellent barrier performance and had a relatively simple preparation process [15].

However, they were prone to cracking at creases, which compromised their barrier properties. Additionally, aluminum foil and aluminized film are opaque, which limits their applications.

The development of third-generation high-barrier films began in the mid-1980s with the introduction of inorganic silicon vapor deposition onto polyvinyl alcohol, PET, and polyamide substrates by Mitsubishi Plastics in Japan. In 2001, a technological breakthrough in the silicon oxide barrier layer was achieved with the development of a high-power electron gun in Germany. This breakthrough led to the rapid advancement of oxide barrier film applications. Since 2007, high-barrier film research has been applied to various industries, including solar cell encapsulation, OLED, and quantum dot displays. Barrier film layers have evolved from monolayer to bilayer and multilayer structures. Single-layer barrier films have a simple structure, mature technology, and high light transmission rates. However, they still exhibit poor barrier properties.

Double-layer barrier films have a simple production process and high light transmission rates. While they offer some improvement in barrier performance, they still may not meet the requirements for encapsulating OLED devices. Multilayer barrier films have a more complex structure and manufacturing process. By overlapping polymer and organic layers multiple times on flexible substrates like PET, PEN, or PI, these films achieve high barrier performance. The flattening process applied to the flexible substrate enhances the water and oxygen barrier performance, meeting the requirements for OLED encapsulation. Overall, the development of high-barrier films has progressed through different generations and structural advancements to achieve improved barrier properties for various applications.

7.3 Packaging Applications for Flexible Solar Cells

Currently, silicon-based solar cells dominate the mass production of solar panels, accounting for over 95% of the market. The remaining 5% consists of solar cells made from other materials, including flexible solar cells. Flexible solar cells are part of the emerging technology in the global solar industry and belong to the category of thin-film solar cells. Flexible solar cells offer several advantages compared to traditional solar panel technology. First, they are stable and reliable, providing consistent energy generation over time. They have undergone technological advancements that have improved their efficiency and performance. Moreover, flexible solar cells are relatively low-cost, making them an attractive option for various applications. Their flexibility allows for easy integration into different products and surfaces. This versatility opens up a wide range of possibilities for their use, including solar backpacks, solar convertible vehicles, solar-powered cars, solar sailboats, and even solar aircraft. The development of flexible solar cells has sparked innovation in the solar industry, expanding its reach beyond traditional rigid solar panels. As technology continues to advance, the adoption of flexible solar cells is expected to increase, further contributing to the growth and diversification of the solar energy sector.

7.3.1 Inorganic Flexible Solar Cells

Inorganic solar cells are categorized based on the materials utilized in their production. The main classifications include monocrystalline silicon, polycrystalline silicon, amorphous silicon, and other compound semiconductor materials.

Monocrystalline silicon solar cells are renowned for their high conversion efficiency and well-established technology. In laboratory settings, they have achieved conversion efficiencies of up to 24.7%, while in large-scale production, the efficiency typically reaches around 15%. Monocrystalline silicon cells continue to dominate in large-scale applications and industrial production, but their high cost poses challenges in achieving significant cost reductions. To conserve silicon materials and explore alternatives, there has been a focus on the development of polycrystalline silicon thin-film and amorphous silicon thin-film solar cells. Polycrystalline silicon thin-film solar cells offer a lower-cost alternative to monocrystalline silicon. They exhibit higher efficiencies than amorphous silicon thin-film cells, with laboratory conversion efficiencies reaching up to 18% and industrial-scale production efficiencies around 10%. As a result, polycrystalline silicon thin-film cells are expected to gain prominence in the solar electric market due to their cost-effectiveness and reasonable efficiency. Amorphous silicon thin-film solar cells possess advantages such as low cost, lightweight, and ease of mass production. They also exhibit high conversion efficiencies. However, their practical application has been hindered by issues related to material-induced efficiency decline and stability. Overcoming these challenges and further improving the conversion rates could unlock the great potential of amorphous silicon cells. Nonetheless, the simple production process, low energy consumption, and increasing market share make amorphous silicon thin-film technology an attractive choice. Currently, over half of the thin-film solar cell companies utilize amorphous silicon technology. In summary, monocrystalline silicon solar cells are renowned for their high efficiency and mature technology. Polycrystalline silicon and amorphous silicon thin-film cells offer cost-effective alternatives, with polycrystalline silicon expected to dominate the market and amorphous silicon holding potential for further advancements. The choice of technology depends on factors such as efficiency requirements, cost considerations, and specific application needs.

Silicon, as an indirect energy band semiconductor material, is not considered the most ideal photovoltaic material due to its low light absorption coefficient. Consequently, there has been a growing trend in researching alternative photovoltaic materials. One prominent representative of the second-generation thin-film solar cells is copper-indium-gallium-selenide (CIGS) thin-film solar cells. CIGS cells have gained significant attention due to their numerous advantages, including high efficiency, good stability, adjustable band gap, excellent low-light performance, strong radiation resistance, and compatibility with flexible substrates [16, 17]. After years of research, CIGS technology has gradually entered the industrialization stage. CIGS solar cells have achieved remarkable milestones, with Solar Frontier in Japan setting a record efficiency of 23.35% on a glass substrate [18]. However, the market dominance of crystalline silicon cells, driven by the learning curve effect, has significantly reduced their cost and created a challenging environment for CIGS and other thin-film technologies. Nonetheless, the unique characteristic of CIGS thin-film cells being compatible with flexible substrates has greatly expanded their application possibilities. This feature has opened up broader prospects for CIGS in the field of mobile solar power services, including portable solar backpacks, solar tents, solar wearable devices, and various types of solar-powered aircraft. These applications have become the driving force behind the industrialization efforts of CIGS technology at this stage [19]. In summary, while silicon has limitations as a photovoltaic material, CIGS thin-film solar cells offer a promising alternative with their high efficiency, stability, adjustability, and compatibility with flexible substrates. The industrialization of CIGS technology has been influenced by the cost reductions in crystalline silicon cells but remains driven by its broad application prospects in the mobile solar power sector (Table 7.2).

After years of research and development, flexible CIGS cells have made significant progress. The highest efficiencies achieved so far are 19.4% on stainless steel substrates and 20.8% on PI substrates. Although these efficiencies are still lower than

Table 7.2 Development of flexible CIGS cells.

Substrate	Deposition method	Efficiency (%)	Institution/Company	Year
Mo	H_2Se annealing after electron beam evaporation	8.3	IS ET	1993
	Co-evaporation	14.6	AIST	2010
Ti	Co-evaporation	12	University of Stuttgart, Germany	2000
	NaF Precursor + co-evaporation	15	HMI	2003
	Co-evaporation	17.9	AGU	2009
Stainless steel	Roll-to-roll co-evaporation	11.65	GSE	2002
	SiO_2 + co-evaporation	12.4	Panasonic	2003
	Ti + co-evaporation	17.3	EMPA	2011
	Ti + co-evaporation	17.7	EMPA	2012
	Co-sputtering	19.4	MiaSole	2017
Low-carbon steel	Enamel layer + co-evaporation	17.6	EMPA	2012
	Ni/Cr + co-evaporation	18	EMPA	2018
Polyacyl imine	Co-evaporation	12.8	ETH	1999
	Co-evaporation	13.2	ETH	2004
	Co-evaporation	14.1	ETH	2005
	Co-evaporation	14.7	AIST	2008
	Co-evaporation	18.7	EMPA	2011
	Co-evaporation	20.4	EMPA	2013
	Co-evaporation	20.8	EMPA	2019

the record efficiency of 23.35% on glass substrates, they demonstrate the potential for large-scale applications. Currently, the industrialization of flexible CIGS cells mainly focuses on these two routes. The increasing demand for mobile energy in wearable devices, new energy vehicles, drones, and other portable devices has motivated researchers to invest more in the development of high-efficiency flexible CIGS batteries. However, to enable the widespread application of flexible CIGS cells, several challenges need to be addressed. These include finding suitable flexible substrate materials with high temperature resistance, appropriate expansion coefficients, and lighter weight. Additionally, low-temperature fabrication processes and alkali metal doping technology need to be developed to expand the selection of substrate materials while reducing energy consumption and costs. Moreover, transferring technology from the laboratory to industrial-scale production poses its own set of challenges. This includes the need for high-precision roll-to-roll transmission equipment and high-temperature linear evaporation sources capable of meeting the requirements of large-area deposition.

Once these challenges are effectively addressed, high-performance flexible CIGS cells can be expected to enter mass production, bringing practical changes to people's daily lives. The continuous advancements in flexible CIGS technology hold great promise for the future of portable and mobile energy solutions [19].

7.3.2 Organic Flexible Solar Cells

Organic solar cells have gained significant attention due to their unique properties such as flexibility, lightweight, ultra-thinness, non-toxicity, and color tunability. They can be prepared using high-throughput large-area printing techniques, making them suitable for applications in flexible electronic devices, portable wearables, and photovoltaic building integration. Organic solar cells complement traditional crystalline silicon solar cells and represent a new generation of green energy technology with promising industrial prospects.

During the path to industrialization, the development of organic solar cells should consider not only achieving high efficiency and stability of the active layer materials but also focusing on realizing flexibility and high-throughput printable fabrication methods for practical production and application scenarios. The ability to print organic solar cells over large areas enables cost-effective manufacturing and opens up possibilities for integrating solar cells into various surfaces and products. Efforts are being made to optimize the performance of organic solar cells, improve their power conversion efficiency, enhance their long-term stability, and develop scalable production processes. Research is also focused on exploring new materials, device architectures, and manufacturing techniques to achieve higher efficiency, longer lifespan, and compatibility with flexible substrates. With ongoing advancements in organic solar cell technology, they hold great potential as a renewable energy source, offering a sustainable and versatile solution for various energy applications. Solar cell materials are subjected to certain requirements to optimize their performance. First, they should efficiently utilize solar radiation, meaning that the semiconductor material's bandgap should not be excessively wide, as it would result in low utilization of solar energy. Secondly, high photoelectric conversion efficiency is crucial to maximize the electricity generation from absorbed sunlight. Last, it is essential that the materials used in solar cells are environmentally friendly and do not contribute to pollution. Meeting these requirements enables the development of efficient and sustainable solar cell technologies. Organic solar cells offer several advantages due to the versatility of organic compounds. The chemical structure of organic molecules can be easily modified, allowing for flexible design and synthesis of materials. Additionally, the preparation and purification processes for organic compounds are relatively simple. Organic solar cells can be manufactured into large-area flexible thin-film devices, providing cost advantages and utilizing widely available resources. The preparation of organic solar cells is straightforward and can be achieved through techniques such as vacuum vapor deposition or coating. Furthermore, organic solar cells can be fabricated on flexible substrates, enabling the production of rollable and foldable devices. Compared to silicon solar cells, organic solar cells have the benefits of large-scale manufacturing, affordability, ease of fabrication, and flexibility. With continuous advancements in materials and related technologies, organic solar cells are becoming increasingly promising. In laboratory research settings, organic thin-film solar cells have achieved conversion efficiencies of up to 9.5%. These advancements signify the growing potential of organic solar cells as a viable alternative in the field of photovoltaics.

Researchers at Monash University have made significant progress in developing ultrathin, flexible solar cells that can be mass-produced. This technology also has the potential for applications in printing polymer banknotes, expanding the scope of polymer solar cell development toward low-cost and efficient printing methods. These printable cells offer several advantages over traditional solar panels. They are lightweight and easily transportable, making them suitable for household use. Their flexibility allows them to bend and conform to various surfaces, similar to banknotes. Additionally, their partially invisible nature enables installation in diverse locations, including on roofs, vehicles, windows, or glass panels. Moreover, these cells have the unique property of floating, making them suitable for covering swimming pools or dams. This not only generates energy but also helps reduce evaporation from water bodies. In another notable development, Chen-Yi Yi's group at Tsinghua University's Department of Electrical Engineering has proposed a strategy to minimize defects and lattice micro-stress in chalcogenide films using multifunctional additives. This breakthrough has led to the achievement of the world's highest photovoltaic conversion efficiency (PCE) of 23.6% for single-junction flexible chalcogenide solar cells. By addressing these issues, the research group has made significant advancements in enhancing the performance of flexible chalcogenide solar cells.

Organic–inorganic hybrid chalcogenide materials possess several advantages, including outstanding light absorption capabilities and low processing costs. Flexible chalcogenide solar cells, when prepared on flexible substrates, offer additional benefits such as thinness, flexibility, and a high power-to-mass ratio. These characteristics make them highly promising for various applications, including wearable electronic devices, unmanned aerial vehicles, and photovoltaic

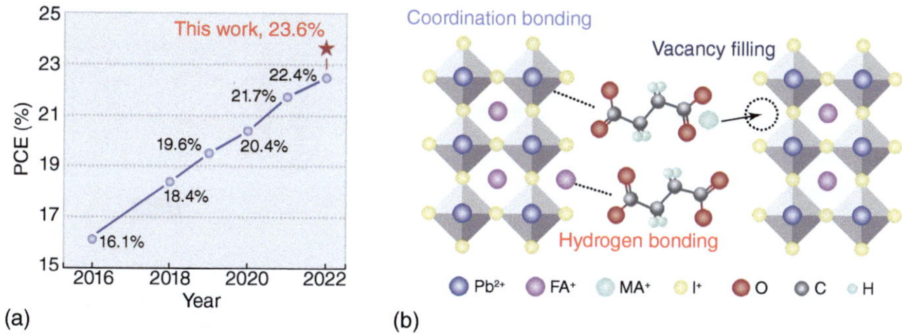

Figure 7.4 (a) Maximum efficiency trend of flexible chalcogenide cell. (b) Schematic diagram of additive action mechanism. Source: Reproduced with permission from Park et al. [15]; © 2014 Elsevier Ltd.

integration in buildings. As a result, these flexible chalcogenide solar cells have garnered significant attention from both academia and industry.

To enable practical applications of these flexible devices, it is crucial to focus on further enhancing their photoelectric conversion efficiency and their ability to withstand bending and mechanical stress. Improving the efficiency of light absorption and charge carrier generation within the device structure can contribute to higher conversion efficiencies. Additionally, optimizing the materials and device architecture to enhance the flexibility and durability of the solar cells will ensure their reliable performance even under bending or mechanical deformation. These research directions are critical for advancing the practical applications of flexible chalcogenide solar cells (Figure 7.4).

7.3.3 Dye-Sensitized Solar Cells

Flexible dye-sensitized solar cells (FDSSCs) distinguish themselves from conventional DSSCs by utilizing bendable substrates for both the photoanodes and counter electrodes, enabling the cells to exhibit flexibility. As a result, the choice of materials for the photoanode and counter electrode has become critical in the development of flexible DSSCs. When flexible substrates are employed, the use of high-temperature sintered titanium dioxide as the photoanode and platinum as the counter electrode may be somewhat limited. Consequently, the selection of appropriate substrates and the development of suitable electrode materials compatible with these substrates have emerged as key areas of focus in the research on FDSSCs [16].

7.3.3.1 Structure of Dye-Sensitized Solar Cells

The dye-sensitized solar cell (DSSC) is composed of three main components: the photoanode, the redox electrolyte, and the counter electrode (refer to Figure 7.5a). The photoanode consists of a flexible conductive substrate, a nanocrystalline semiconductor oxide, and a dye sensitizer. The flexible conductive substrate can be made of materials such as indium tin oxide-coated polyethylene terephthalate (ITO-PET), indium tin oxide-coated polyethylene naphthalate (ITO-PEN), and

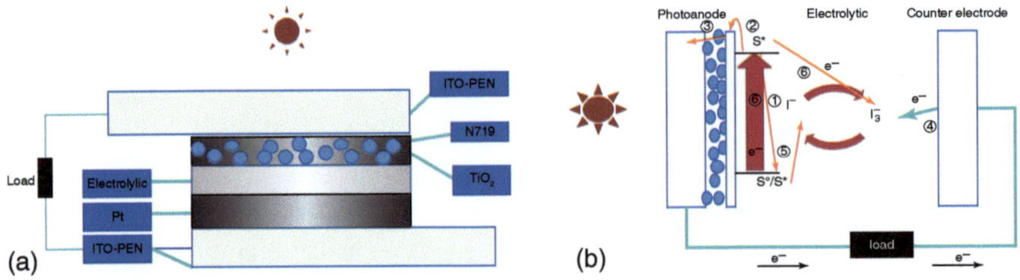

Figure 7.5 (a) Structure of dye-sensitized solar cell; (b) mechanism of operation of dye-sensitized solar cells 2. Operating mechanism of dye-sensitized solar cells. Source: Reproduced with permission from Jia and Zhai [17]; © 2021 Chongqing Southwest Information Co., LTD.

other plastic substrates. It can also include metal substrates like titanium, stainless steel, and titanium foil mesh. The sensitizer commonly used in DSSCs includes N3 and N719. The counter electrode is typically a platinum (Pt) electrode. The redox electrolyte, which often consists of I^- and I_3^- ions, is filled between the photoanode and the counter electrode [17].

The photoanode plays a crucial role in DSSCs by absorbing incident light and initiating the electron transfer process. When the photoanode is exposed to light within a specific visible light range, it absorbs photons and generates electrons. These electrons are then transferred to the external circuit. The counter electrode collects the electrons from the external circuit and returns them to the redox electrolyte, where iodide ions are reduced and transmit the transmitted light. The electrolyte acts as a mediator for electron transfer between the photoanode and the counter electrode, facilitating the regeneration of the dye's excited state. The mechanism of action can be described as follows:

(1) Incident light causes the dye to absorb light energy, resulting in the excitation of electrons from the ground state (S0) to the excited state (S*), generating electrons.
(2) The generated electrons are injected into the conduction band of the nanocrystalline titanium dioxide (nano-TiO_2).
(3) The injected electrons accumulate in the conductive substrate and are transferred through the external circuit to the counter electrode, establishing an electric current.
(4) In the electrolyte, I_3^- ions diffuse to the counter electrode, where they are reduced to I^- as they accept electrons.
(5) The oxidized dye molecules (S^-) react with I^- in the electrolyte, regenerating the dye in its reduced state.
(6) The I_3^- ions, originally reduced to I^-, are re-oxidized to I^- through the electrons generated in the excited state.
(7) The oxidized dye molecules are reduced by the electrons generated in the excited state.

It is important to note that steps (6) and (7) represent compound reactions that can reduce the overall electron utilization rate and decrease the efficiency of the solar cell [2]. The mechanism of action is illustrated in Figure 7.5b.

7.3.3.2 Light Anode

The photoanode in FDSSCs is a crucial component responsible for converting photons into electrons. Its material properties have a significant impact on various aspects such as dye loading, electron transport, and light scattering, ultimately influencing the photovoltaic conversion efficiency of the FDSSC.

In FDSSCs, nanoporous titanium dioxide (TiO_2) films are commonly used as the photoanode material. These films possess a high specific surface area, which enhances the absorption of photons and increases the efficiency of photon-to-electron conversion. The nanoporous structure allows for a larger surface area available for dye sensitization, facilitating a higher dye loading capacity. This, in turn, leads to an increased absorption of incident light and a higher photocurrent generation. Furthermore, the nanoporous structure of the TiO_2 films also facilitates efficient electron injection and transport. The electrons generated upon photon absorption are rapidly injected into the conduction band of the TiO_2, allowing for efficient charge separation and collection. The interconnected network of nanopores ensures effective electron transport pathways, minimizing recombination losses and maximizing the overall electron transmission rate within the photoanode. Overall, the choice and optimization of the photoanode material, particularly nanoporous TiO_2 films, significantly influence the performance of FDSSCs by enhancing photon absorption, improving electron injection and transport, and ultimately increasing the photocurrent and photovoltaic conversion efficiency (Tables 7.3 and 7.4).

In summary, FDSSCs prepared with metal-based photoanode materials exhibit relatively high conversion efficiency [17]. Furthermore, composite oxide semiconductor thin-film photoanodes generally outperform single oxide semiconductor components in terms of their performance. The use of metal-based photoanode materials, such as composite oxide semiconductors, offers several advantages for FDSSCs. These materials possess enhanced electrical conductivity, improved charge transport properties, and optimized energy band alignment, leading to

Table 7.3 Relevant parameters of conductive plastic substrate cells.

Substrate material	Method	Characteristic	Transformation efficiency
ITO/PRN	Ultrasonic spraying technology	Fast	4.00% [20]
ITO/PRN	DC reactive magnetron sputtering method	Temperature controllable	5.30% [21]
ITO/PRT	Hydrothermal method	The particle has high purity, complete crystal form, and no need of sintering	6.54% [22]
ITO/PRT	High-temperature sintering assisted by flat fin heat sink	High flexibility	4.11%

Table 7.4 Relevant parameters of different metal substrate cells.

Substrate material	Method	Characteristic	Transformation efficiency
Stainless steel	Radio frequency magnetron sputtering method	Time and thickness are controllable	6.96% [23]
Titanium foil substrate	Chemical deposition method	The purity and density of the deposited layer are controllable	0.75%
Titanium mesh	Hydrothermal method	The particle purity is high, and sintering is not needed.	5.04%
Titanium wire	Anodizing method and dip coating method	Growth sequence is controllable	3.31%

efficient charge separation and collection. The composite nature of these oxide semiconductor thin films allows for synergistic effects between different components, resulting in improved light absorption and reduced charge recombination losses. Compared to single oxide semiconductor components, composite oxide semiconductor thin-film photoanodes demonstrate superior performance in terms of photon absorption, electron injection, and transmission rates. The incorporation of multiple oxide semiconductors enables a broader absorption spectrum, capturing a wider range of photons and increasing the overall light harvesting efficiency. Additionally, the combination of different materials can enhance electron injection efficiency and facilitate faster charge transport, minimizing losses due to recombination.

By utilizing metal-based photoanode materials, specifically composite oxide semiconductor thin films, FDSSCs can achieve higher conversion efficiency, making them more effective in converting light energy into electricity. These advancements contribute to the overall development and potential commercialization of FDSSCs for various applications.

7.3.3.3 Counter Electrode

Various electrode materials are available besides platinum. Some common alternatives include carbon materials, conductive polymers, and metal compounds. Researchers are also exploring other substrate materials for electrodes, such as PET, PEiN, metal foil, carbon cloth, paper, and cotton textiles. Flexible electrode materials need to meet specific requirements. First, they should exhibit high catalytic performance for I^-/I_3^- and redox reactions. Secondly, they should possess high electrical conductivity to enable rapid collection and transmission of outer circuit electrons. Last, they should have a cheap and straightforward preparation process, enabling widespread application. Consequently, carbon materials, conductive polymers, metal compounds, and their composites have emerged as the most promising flexible counter-electrode materials [16].

7.4 Packaging Applications for Flexible Electronic Devices

The field of flexible electronics is experiencing rapid growth and finding extensive applications in various fields such as medical diagnosis, monitoring, and flexible robotics. Flexible electronics encompass a range of technologies including organic electronics, plastic electronics, bioelectronics, nanoelectronics, and printed electronics. This broad field encompasses diverse applications such as RFID, flexible displays, organic electroluminescent (OLED) displays and lighting, chemical and biological sensors, flexible photovoltaics, flexible logic and storage, flexible batteries, and wearable devices, among others. The continuous advancement of flexible electronics has expanded its scope into new areas, making it a prominent research area in interdisciplinary fields.

In 2000, organic electronics technology was recognized as one of the world's top 10 scientific and technological achievements. The pioneering work of US scientists Alan Hague, Alan Mark Dillmead, and Japanese scientist Eiki Shirakawa in the field of conductive polymers earned them the prestigious 2000 Nobel Prize in Chemistry. Over the past decade, renowned institutions like Cornell University, Princeton University, Harvard University, Northwestern University, and Cambridge University have established dedicated research centers focusing on flexible electronics technology. In China as well, flexible electronics research has garnered significant attention from researchers, leading to a thriving and rapidly evolving field. Consequently, the development of flexible electronics has witnessed substantial progress and continues to be a subject of intense interest.

7.4.1 Basic Structure of Flexible Electronic Devices

Flexible electronic devices consist of four fundamental functional units: components, flexible substrates, crosslinked conductors, and adhesive layers [24]. The diagram in Figure 7.6d illustrates these basic units of flexible electronics.

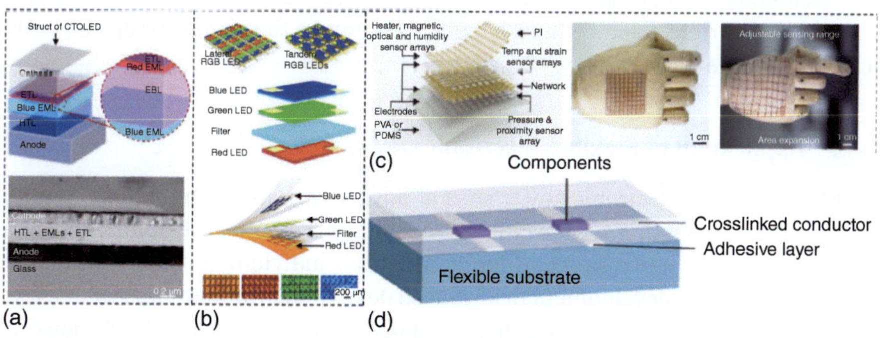

Figure 7.6 (a) CTOLED structure; (b) schematic and optical images of a tandem mLED array using vertical space to increase spatial density; (c) schematic diagram of a multilayer flexible sensor array with seven detection functions (left), optical images of the array before (middle) and after (right) unfolding; (d) basic structure of flexible electronic devices. Source: Adapted from Li et al. [25].

Components in flexible electronics encompass a range of functional units such as resistors, capacitors, inductors, transistors, sensors, batteries, energy storage devices, information storage units, and computing units. These components serve as the core elements responsible for circuit functionality and microsystem operations. The composition of flexible components differs significantly from traditional components. In flexible electronics, organic materials are often employed to achieve the desired functionalities. Alternatively, inorganic materials can be initially positioned on rigid microcellular islands, which are then affixed onto the flexible substrate to prevent damage during bending [24].

Flexible substrates play a crucial role in flexible electronic devices, as they need to possess both flexibility and stable electromagnetic properties akin to those of traditional rigid substrates. These substrates should maintain consistent performance even when subjected to torsional changes in the device structure. This ensures that electronic devices can function properly across various deformation states, allowing for reliable operation in flexible configurations.

Crosslinked conductors in flexible electronics are typically made from metal films or conductive polymers known for their excellent electrical conductivity. These conductors are strategically positioned throughout the flexible substrate to establish connections between different components. The structural design of crosslinked conductors should address the challenge of maintaining stable electromagnetic properties of the components while accommodating the substrate's wide range of expansion and contraction. Overcoming this challenge is one of the key difficulties in the field of flexible electronics technology [24].

To safeguard flexible circuits and systems against various external factors such as dust, gas, electromagnetic interference, radiation, moisture, or chemical exposure, the implementation of protective coverings, also known as encapsulation layers, is necessary. These coverings typically employ composite structures comprising multiple materials to accommodate bending, stretching, or other deformations. They serve to isolate and mitigate the impact of complex environmental factors such as electromagnetic fields, mechanical forces, thermal variations, and temperature/humidity fluctuations. Additionally, they help reduce torsional stress effects at the edges of rigid microcellular islands, preventing their detachment from the flexible substrate [24].

7.4.2 Application of Flexible Electronic Devices

Multilayer flexible electronics offer several advantages, including high-resolution spatio-temporal mapping, a wide range of sensing and actuation capabilities, and the ability to configure and arrange different functions individually. These advantages pave the way for the development of advanced devices such as high-density displays, multifunctional electronic skins, instrumented medical tools, and self-powered systems that integrate multiple functionalities. These applications align with the research domains of optoelectronics, robotics, biomedical devices, and energy devices. In the following section, we will delve into the specific applications of multilayer flexible electronics within these four fields [25–27].

7.4.2.1 Optoelectronics

Research advancements in multilayer flexible electronics have primarily concentrated on enhancing the performance of optoelectronic devices, specifically light-emitting diodes (LEDs). Traditionally, LEDs emit light of a single color, also known as monochromatic light. To overcome this limitation and achieve a broader wavelength range, a common approach employed in commercial displays involves arranging monochromatic LEDs in close proximity in a dense array format [28]. By stacking multiple layers of LEDs, it becomes possible to effectively mitigate the efficiency degradation caused by size reduction while simultaneously expanding the range of luminescence. Furthermore, the implementation of multilayer structures opens up new possibilities for integrating LEDs with other devices, enabling the creation of versatile and flexible systems.

OLEDs are a type of light-emitting device that operates based on current. Traditional OLEDs typically rely on the spatial combination of different types of organic electroluminescent devices (OEEDs) to achieve full-color emission [29, 30]. By incorporating multiple light-emitting layers in a vertical arrangement, tunable color organic light-emitting diodes (CTOLEDs) can be created, as depicted in Figure 7.6a. In this design, the red-emitting layer is much thinner than the blue-emitting layer, resulting in a pronounced contrast in color brightness. An isolation layer is positioned between the two light-emitting layers, serving as an energy barrier. As a result, the CTOLED can transition from emitting red light to emitting blue light at different voltages. When combined with stretchable electrodes and signal amplification circuits, color-tunable OLEDs enable the visualization of electrocardiogram (ECG) signals by representing them with different colors of light [31].

In terms of efficiency, brightness, and stability, inorganic semiconductor-based miniature light emitting diodes (mLEDs) surpass organic-based LEDs [32]. Currently, mLEDs can achieve a resolution of up to 14 000 pixels per inch (ppi) [33]. Enhancements in the manufacturing process and assembly precision of individual LED units can further elevate their performance and spatial density. As illustrated in Figure 7.6b, a tandem mLED array employs vertical space to enhance spatial density. Unlike the conventional single-layer horizontal arrangement, the tandem mLED array can achieve the same resolution even with a unit size three times larger than that of the horizontal mLED array. The multilayer structure, from bottom to top, consists of a red LED (wavelength 628 nm), a long-pass filter (open wavelength 600 nm), a green LED (wavelength 530 nm), and a blue LED (wavelength 475 nm). Each layer has a thickness ranging from 5 to 7 mm, with a 2e5 mm SU-8 epoxy sandwiched between each device, resulting in a multilayer LED with a total thickness of approximately 45 mm. The filters between the red and green LEDs reflect blue and green light while transmitting red light. When observed under fluorescence microscopy, the tandem mLEDs exhibit excellent emission characteristics across different spectral ranges (lower box of Figure 7.6b). Further measurements demonstrate that the chromaticity diagram of the tandem mLEDs overlaps 93% with the International Commission on Illumination standard.

The integration of photonics with wearable sensors in a multilayer format enables the display of information captured by the sensors [34]. Figure 7.6c depicts the structure of a super-adaptive and wearable photonic skin, which consists of multiple layers. The e-skin comprises a photonic sensor layer made of hydroxypropyl cellulose (HPC), an electronic sensor layer composed of carbon nanotubes (CNTs), and an underlying layer consisting of an HPC shape-memory adhesive. When subjected to external mechanical stimulation, the helical pitch of the intermediate-phase HPC undergoes changes, leading to a shift in color. This colorimetric response allows the photonic skin to detect and visually represent various deformations. The color change can be quantitatively assessed by capturing a picture of the photonic skin with a cell phone and subsequently analyzing the image. Moreover, the multilayer system incorporates a piezoresistive multi-pixel strain sensor composed of CNTs. This sensor reflects changes in external pressure through alterations in resistance, resulting in a bimodal mechanical sensing system that provides information through both color and resistance changes.

7.4.2.2 Robot

The human skin, being a remarkable organ, possesses the ability to detect external stimuli like pressure and temperature, enabling safe and effective interaction between humans and their environment [35]. This unique characteristic has served as inspiration for the design of smart prostheses, smart skins, and the advancement of human–machine interfaces (HMIs) and intelligent robotic systems. In human skin, specialized structures such as Pacinian corpuscles, Meissner's corpuscles, Ruffini endings, and Merkel cells are arranged in multiple layers to perceive various types of mechanical stimuli. The concept of electronic skin has emerged to replicate this multilayer structure, enabling large-area, high-resolution, and multimodal measurements [36–38].

The epidermis, dermis, and subcutis in human skin work together to enable its complex sensory functions. In the electronic skin, three functional layers are incorporated to replicate these properties. The design includes two thermistor membranes, namely the hot and cold membranes, positioned both at the top and bottom. The cold membrane not only detects temperature independently but also collaborates with the hot membrane to measure the thermal conductivity of external objects. Moreover, the intermediate porous material's porosity decreases under pressure, resulting in a change in thermal conductivity. The bottom sensor captures this change, reflecting the external pressure applied. A smart hand equipped with this electronic skin achieves an impressive 94% accuracy in identifying seven types of garbage, such as plastic bags, foam, cartons, cans, meal applications, bread, and orange peels. Furthermore, researchers have explored the creation of multilayered electronic skin by simulating a spine-like structure to detect force magnitude and direction [39], as well as collecting temperature and vibration signals simultaneously [40]. These designs offer essential functionalities such as object recognition, precise interaction, and texture perception for robotic arms, prostheses, and anthropomorphic robots [41].

Figure 7.6c showcases a schematic and optical image of a highly stretchable and consistent sensor matrix network capable of scaling the sensing area from 16 to 400 cm^2. The multilayer design employed here enables a high-density integration strategy for sensor arrays in a three-dimensional orientation. This unique feature allows for the integration of seven types of sensor arrays that can simultaneously detect magnetic fields, ultraviolet (UV) light, temperature, relative humidity, in-plane strain, pressure, and proximity. As a result, an artificial skin is created that expands beyond the range of human perception. The modular design of multiple layers also facilitates personalized configuration. Depending on the specific application scenario, users can choose the desired layer to detect different stimuli simultaneously. Furthermore, the stretchable interconnections between each layer enable the multilayer array to unfold within an artificial hand. Similar structural designs have been explored in other research works [42–44], highlighting the development of artificial hands for precise control and feedback. Through the integration of multilayer sensors, electronic skin surpasses the sensory capabilities of human skin, offering unique opportunities for prostheses, humanoid robots, and peripheral nervous systems [36].

7.4.2.3 Biomedical

Flexible electronics have emerged as a valuable technology for biomedical applications as they can establish conformal contact with biological tissues, opening up a wide range of possibilities in the field of biomedicine [45]. These devices find application in continuous monitoring of various physiological signals, intelligent drug delivery, and instrumented surgical tools [27, 46, 47]. While conventional flexible devices offer opportunities for disease prevention, treatment, and recovery, their single-layer structure imposes limitations on spatial resolution and versatility. In contrast, sensors and actuators based on multilayer structures enable simultaneous multimodal measurements or therapies, enhancing the functionality available for biomedical diagnosis and treatment.

Flexible electronics play a crucial role in biomedical applications, particularly in facilitating surgery by integrating sensors and actuators into surgical tools. Minimally invasive surgery offers advantages such as reduced trauma, less pain, and faster recovery compared to traditional surgical approaches [48]. Catheters are commonly used in minimally invasive procedures [49, 50]. Figure 7.7a illustrates the integration of a soft electronic array with a balloon catheter. The flexibility of the electronics enhances the coupling efficiency between the catheter and the tissue, while the multilayer integration enables multimodal functionality, reducing the number of catheters required for the procedure. The multilayer structure consists of three layers: sensors, actuators, and associated electronics. The top layer comprises an electrode array (6.9 mm thick), the middle layer consists of a temperature sensor array (105.3 mm thick with a silicon substrate), and the bottom layer comprises a pressure sensor array (805.5 mm thick with a silicon substrate). Despite the presence of multiple silicone substrates and 3D structures, the total thickness of the multilayer electronics remains below 1 mm. In cardiac surgery, the multilayer flexible electronics are integrated into a balloon catheter (14 Fr during deflation) to provide temporary mapping of temperature and pressure, electrical stimulation, and radiofrequency ablation. The multilayer structure optimizes the vertical space,

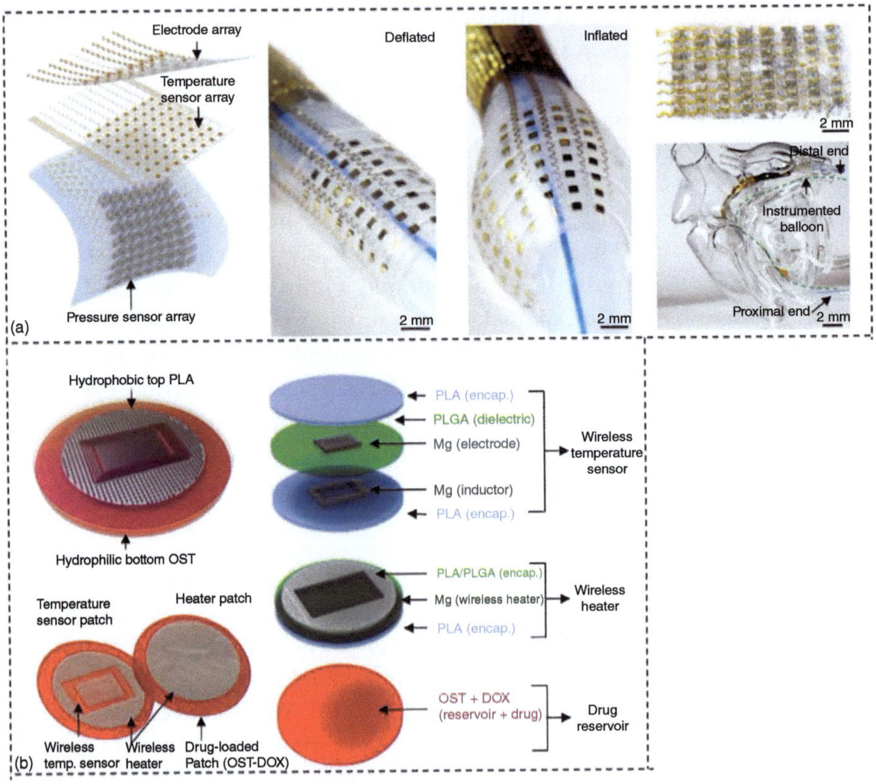

Figure 7.7 (a) Balloon catheter integrated with soft electronic array; (b) a flexible and biodegradable drug delivery device in multilayer format. Source: Han et al. [27]; Reproduced with permission from Springer Nature.

fulfilling the diverse requirements of surgical procedures and enabling surgeons to perform different diagnostic and therapeutic procedures simultaneously using a single surgical tool.

Targeted and on-demand drug delivery is a critical aspect of treating chronic or localized diseases [51]. Smart controlled release technologies are essential in achieving precise drug delivery, and multilayer structures offer opportunities for incorporating endogenous/exogenous stimulus-responsive components and drug release devices. Figure 7.7b illustrates a flexible and biodegradable drug delivery device designed in a multilayer format [52]. This device features distinct layers for the wireless sensing unit, heating unit, and drug delivery unit. Once surgically implanted, an externally applied RF magnetic field triggers the operation of the heater, generating eddy currents and joule heating. The gentle increase in temperature facilitated by the heater promotes controlled drug release and diffusion within the surrounding tissues. This wireless device can also be employed in the treatment of other conditions, like brain tumors. The use of external stimuli to regulate drug release minimizes tissue damage and allows for programmable drug delivery. The multilayer structure employed in this device provides additional temperature control by adjusting the distance and size of the heater, enhancing the precision and versatility of drug release mechanisms.

Literature [52] showcases a multilayer flexible electronic device designed for drug release, specifically in the form of a smart bandage. This multilayer structure integrates touch and temperature sensor arrays, a wireless module, and a drug delivery device. The touch and temperature sensors are positioned at the bottom layer, while the drug delivery layer is located on top. When pressure is applied to the drug delivery pump, the bandage can release medication onto the skin surface through flow channels. This smart bandage is particularly useful during post-trauma recovery, as it detects physiological signals from the wound site and delivers medication as needed to aid in the healing process. Similar designs can be adapted for the treatment of various conditions by incorporating different drugs, devices, or structures, such as microneedles. Automated, controlled drug delivery can be achieved through a closed-loop system, further enhancing the precision and effectiveness of the treatment [53, 54].

In addition to therapeutic applications, continuous monitoring of various physiological signals plays a crucial role in improving human health [55]. Current flexible electronic devices have already demonstrated the capability to measure physiological parameters like heart rate, blood oxygen saturation, blood pressure, and skin temperature. By leveraging multilayer stacking, these flexible systems can simultaneously detect multiple signals, enabling personalized and customizable diagnostics.

7.4.2.4 Energy Equipment

In recent years, there has been a growing interest in converting various forms of environmental energy into electrical energy, leading to advancements in energy devices [56]. Energy harvesters based on multilayer structures have emerged as a promising approach to efficiently convert mechanical energy into electrical energy [57]. By leveraging multilayer designs, energy harvesters can be integrated with energy storage devices and flexible sensors, creating self-powered systems that encompass all the necessary components in a compact and efficient manner [58]. These “all-in-one” systems have the potential to provide sustainable and self-sufficient power for a wide range of applications.

Figure 7.8a depicts the structure and operating principle of a multilayer triboelectric nanogenerator (TENG). TENGs are energy-harvesting devices that utilize contact electrification and electrostatic induction between two materials to generate electrical power [59]. Conventional TENGs operate in four modes: vertical contact-separation mode, contact-slip mode, single electrode mode, and freestanding triboelectric layer mode. While single TENG devices have achieved high power output, stacking multiple TENGs vertically in a multilayer configuration can further enhance power output without increasing lateral dimensions. In the multilayer TENG shown in the left and middle frames of Figure 7.8a, each layer consists of a pair of materials, and the total output power is proportional to the number of layers. This specific multilayer TENG operates in contact separation mode and exhibits a peak power output of 1.044 mW, as shown in the right frame

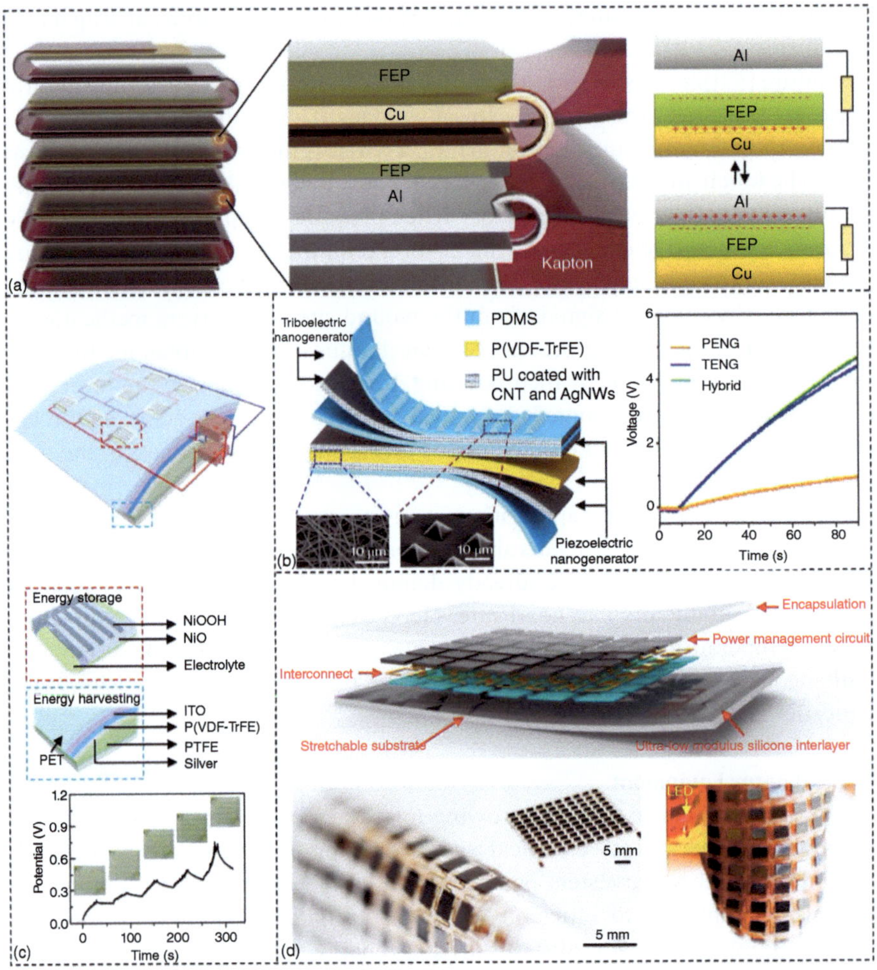

Figure 7.8 (a) Schematic diagram (left), partial enlargement of one layer (middle), and working mechanism of multilayer TENG (right); (b) schematic diagram (left) and charging capability comparison of hybrid nano-generator (right); (c) schematic diagram of multilayer integration of energy harvesting and storage device (top), enlargement of energy storage and storage device (middle), charging curve of device (bottom); (d) multilayer energy harvesting/storage/management system schematic (top) and optical image (bottom). Source: Adapted from Ma et al. [59].

of Figure 7.8a. The flexibility of the materials used in TENGs enables excellent flexibility of the multilayer devices, allowing them to be placed outside the body [60] or even inside the body [61] for wearable mechanical energy harvesting. This flexibility ensures that the multilayer devices do not interfere with the natural activities of the human body.

Figure 7.8b illustrates the structure of a flexible hybrid nanogenerator that combines a triboelectric nanogenerator (TENG) and a piezoelectric nanogenerator (PENG) to improve energy conversion efficiency [62]. PENGs generate a piezoelectric potential when an external force breaks the centrosymmetry of the crystal structure of the piezoelectric material. In the hybrid nanogenerator, the top layer consists of a TENG with a pyramidal microstructured polydimethylsiloxane (PDMS) layer as the starting material and a conducting electrode made of polyurethane (PU) embedded with CNTs and silver nanowires (AgNW). The bottom layer is the PENG, which utilizes a composite of PU, CNT, and AgNW as the electrode and electrospun polyvinylidene fluoride-*p*-trifluoroethylene (P(VDF-TrFE)) nanofibers as the piezoelectric layer. The PDMS layer serves as a separator and electrically isolates the TENG and PENG. Experimental results demonstrate that the hybrid nanogenerator exhibits a higher charging capacity compared to a single TENG or PENG. The right frame of Figure 7.15b shows that the hybrid nanogenerator can charge a 1 mF capacitor to 4.66 V within 80 seconds, indicating its enhanced energy conversion performance.

Figure 7.8c showcases the integration of a nanogenerator and a supercapacitor, enabling the development of a self-powered system [63]. The energy device consists of two parts: a hybrid triboelectric/piezoelectric nanogenerator (TENG/PENG) at the bottom and an electrochromic micro-supercapacitor array at the top. In the bottom energy conversion section, the TENG and PENG generate high output voltage (peak of 150 V) and current (peak of 20 mA), which can be used to charge the supercapacitor through the electrodes. The top layer of the device comprises an array of electrochromic micro-supercapacitors that exhibit a color change when charged. When the charging is completed, the NiO in the energy storage section transforms into NiOOH, indicating the charging process is finished. The lower frame of Figure 7.8c illustrates the charging curve of the supercapacitor. By applying hand pressure to the hybrid nanogenerator, electricity is generated, increasing the voltage of the supercapacitor. Simultaneously, the color of the top layer changes, indicating the charging process. The array structure of the supercapacitor enhances the performance of the system and expands its potential applications.

The application of multilayer flexible electronics in energy devices is depicted in Figure 7.8, showcasing various aspects. In Figure 7.8a, the schematic diagram on the left, the partial enlargement of one layer in the middle, and the working mechanism of a multilayer TENG on the right are illustrated. This TENG operates based on contact energization and electrostatic induction, utilizing a pair of materials to generate power. By stacking multiple layers vertically, the power output can be significantly increased without increasing lateral dimensions.

Figure 7.8b presents the schematic diagram on the left and a comparison of the charging capacity on the right for a hybrid nanogenerator. The hybrid nanogenerator combines TENG and PENG or other types of generators, achieving improved energy conversion efficiency. The structure includes a pyramidal microstructured PDMS layer and a polyurethane (PU) layer embedded with CNTs and silver nanowires (AgNW) as the conducting electrode. The bottom PENG layer consists of a PU-CNTAgNW composite electrode and piezoelectric nanofibers made of P(VDF-TrFE). This hybrid nanogenerator demonstrates higher charging capacity than single TENG or PENG systems, as indicated by the charging of a 1 mF capacitor to 4.66 V within 80 seconds.

In Figure 7.8c, the top schematic diagram illustrates the multilayer integration of an energy harvesting and storage device. The middle part provides an enlarged view of the energy storage and storage device, while the bottom shows the device charging curve. This integration includes a hybrid TENG/PENG at the bottom and an electrochromic micro-supercapacitor array at the top. The TENG and PENG generate high output voltage and current, which charge the supercapacitor through the electrodes. The color change of the NiO in the energy storage section indicates the completion of charging.

Furthermore, Figure 7.8d showcases a schematic view on the top and an optical image on the bottom of a multilayer energy harvesting/storage/management system. This device enhances functionality and increases spatial density. Figure 7.8d represents a typical system that incorporates energy harvesting/storage/management components and a wireless flexible sensor for skin thermal imaging. The system includes semiconductor solar cells (~32 mm total thickness) and chip-scale cells (200 mm total thickness). The devices are made flexible and stretchable using serpentine interconnects and ultralow-modulus silicone interlayers, allowing the system to bend around surfaces such as a cylindrical tube. The stretchable nature of the device enables it to be wrapped around non-expandable surfaces like the index finger without compromising functionality. The folded multilayer structure effectively combines the energy device with electronics at high density, maintaining the stretchability of the system.

7.5 Packaging Applications for Flexible Electronics Sensors

A sensor is a detection device that is capable of sensing and detecting specific information or parameters. It converts the measured information into electrical signals or other required forms of information output, following a specific logic transformation. These signals are then utilized for various purposes, such as information transmission, processing, storage, display, recording, and control.

To protect the internal components and circuitry of a sensor, it is necessary to encapsulate it. The sensor encapsulation structure typically connects the sensing element to the circuit board or other components using metal wires. However, in high-frequency applications, the use of metal wires can lead to electromagnetic interference, which can adversely affect the sensor's performance. Therefore, it is important to explore alternative methods for sensor encapsulation that minimize or eliminate these interference issues, thereby improving overall sensor performance.

Flexible electronics is an emerging electronic technology that involves the fabrication of electronic devices using organic or inorganic materials on flexible and ductile substrates. It represents a departure from traditional rigid electronics by offering greater flexibility and the ability to conform to different working environments, allowing for device deformation to some extent. However, the unique characteristics and requirements of flexible electronics also present challenges, particularly in the development of flexible sensor technology.

One key challenge in flexible sensor technology is achieving stretchability and bendability while maintaining the electronic properties of the circuit. The sensors need to be able to undergo mechanical deformations without compromising their functionality or causing damage to the sensitive components. This requires innovative approaches in terms of material selection, design, and manufacturing techniques.

Another challenge lies in the encapsulation and protection of flexible sensors. As the sensors are subjected to bending, stretching, and other deformations, the encapsulation materials and methods must be able to withstand these mechanical stresses and provide reliable protection to the sensitive electronics inside. Ensuring long-term durability and maintaining the sensor's performance under different environmental conditions are important considerations.

Overall, the development of flexible sensor technology within the field of flexible electronics is a complex and challenging problem that requires addressing issues related to stretchability, bendability, encapsulation, and manufacturing. Overcoming these challenges will pave the way for thc advancement and widespread application of flexible electronics in various industries and sectors.

With advancements in material science, micro, and nanotechnology, the field of flexible sensors is experiencing rapid development. Inspired by the functioning of biological skin, researchers are exploring various conduction methods to convert external stimuli into electrical signals, forming the fundamental basis for flexible sensor research. These sensors are composed of flexible materials, offering excellent flexibility, ductility, and the ability to bend or even fold. Their flexible structure allows for easy adaptation to measurement conditions, enabling convenient and accurate detection in complex scenarios. Consequently, the application of novel flexible sensors has gained significant traction across various fields, including electronic skin, healthcare, electronics, electrical engineering, sports equipment, textiles, aerospace, and environmental monitoring.

Flexible sensors offer diverse and flexible structures that can be tailored to meet the specific requirements of measurement conditions. This flexibility allows for convenient and precise measurements in specialized environments, enabling rapid and accurate detection of unique signals. These advancements address the challenges associated with sensor miniaturization, integration, and intelligent development. As a result, these new flexible sensors play a crucial role in electronic skin, biomedicine, wearable electronics, and aerospace applications.

However, it is important to note that the current technology level for preparing materials such as CNTs and graphene for flexible sensors is still in its early stages. Challenges remain in terms of cost, applicability, and service life. Additionally, commonly used flexible substrates are not resistant to high temperatures, resulting in high stress and weak adhesion between the substrates and thin-film materials. Furthermore, there is room for improvement in the assembly, alignment, integration, and encapsulation technologies of flexible sensors.

In summary, while flexible sensors offer significant potential, there are ongoing challenges that need to be addressed, including material preparation, cost-effectiveness, durability, high temperature resistance, and improving manufacturing processes.

In response to the growing demand for specialized signals and environments, there has been a surge in the development of new flexible sensor technologies. These advancements have led to a wider variety of flexible sensor types and diverse classification systems. When considering their applications, flexible sensors can be categorized into different types, including flexible pressure sensors, flexible gas sensors, flexible temperature sensors, flexible strain sensors, flexible magnetic impedance sensors, and flexible thermal flow sensors. On the other hand, based on their sensing mechanisms, flexible sensors can be further classified into flexible resistive sensors, flexible capacitive sensors, flexible piezomagnetic sensors, and flexible inductive sensors [64]. The focus of this chapter specifically centers around the exploration of new flexible gas sensors, flexible pressure sensors, and flexible humidity sensors [65].

7.5.1 Common Materials of Flexible Sensors

7.5.1.1 Flexible Substrate

To meet the requirements of flexible electronic devices, certain key indicators have emerged for flexible substrates, namely thinness, transparency, flexibility, stretchability, insulation, and corrosion resistance. Among the commonly used flexible sensors, polyimide (PI), polyethylene terephthalate (PET), and polyethylene naphthalate (PEN) are frequently employed as flexible substrate materials.

PI is considered the top-performing organic polymer material, offering exceptional mechanical properties, radiation resistance, and chemical resistance. PET, a saturated thermoplastic polymer, possesses favorable optical properties, and resistance to oils, fats, dilute acids, dilute alkalis, and most solvents. It exhibits excellent abrasion and friction resistance, dimensional stability, and electrical insulation, and is characterized by a low price, high yield, and smooth, glossy surface.

PEN, on the other hand, differs from PET in that the benzene ring is replaced by a more rigid naphthalene ring in its molecular chain. This naphthalene ring structure grants PEN higher physical and mechanical properties, gas barrier properties, and heat resistance surpassing that of PET. Additionally, PEN exhibits excellent characteristics such as UV resistance and radiation resistance.

In summary, PI, PET, and PEN are commonly utilized as flexible substrate materials in flexible sensors due to their desirable properties, including mechanical strength, optical transparency, resistance to various environmental factors, and thermal stability.

7.5.1.2 Metal Materials

Metallic materials, such as gold, silver, and copper, are widely employed as conductive materials in flexible sensors. They are primarily utilized for electrodes and wires within these devices. In modern printing processes, conductive nanoinks are commonly used, which consist of conductive materials such as nanoparticles and nanowires. These conductive nanoparticles offer not only excellent electrical conductivity but can also be sintered together to form thin films or wires. This sintering process allows for the creation of conductive pathways within the sensor structure, facilitating the flow of electrical signals.

7.5.1.3 Inorganic Semiconductor Materials

In the realm of wearable flexible electronic sensors, inorganic semiconductor materials, particularly ZnO (zinc oxide) and ZnS (zinc sulfide), have demonstrated significant potential due to their impressive piezoelectric properties. These materials have been leveraged in the development of flexible pressure sensors that directly convert mechanical energy into optical signals. Through a top-down photolithography process, these sensors achieve a fast response time of less than 10 ms and a spatial resolution of up to 100 μm. This conversion process enables the sensor to accurately record the dynamic pressure at a specific point, allowing for applications such as handwriting recognition and real-time scanning of two-dimensional planar pressure distribution by analyzing emission intensity profiles. The exceptional features offered by inorganic semiconductor materials position them as highly promising candidates for future pressure sensor materials, particularly in terms of their fast response and high-resolution capabilities.

7.5.1.4 Organic Materials

The development of large-scale pressure sensor arrays is crucial for advancing wearable sensor technology. Existing pressure sensors that utilize piezoresistive and capacitive signal mechanisms often encounter the challenge of signal crosstalk, which can result in inaccurate measurements. Overcoming this issue is one of the primary challenges in the development of wearable sensors. Transistors present a promising solution, as they offer excellent signal conversion and amplification properties, which can help minimize signal crosstalk.

Consequently, a significant portion of research in the wearable sensor and artificial intelligence fields is focused on the development of large-scale flexible pressure-sensitive transistors. These transistors hold the potential to address the issue of signal crosstalk and enable the creation of more accurate and reliable pressure sensor arrays on a larger scale. By leveraging the precise signal conversion and amplification capabilities of transistors, advancements can be made in improving the performance and functionality of wearable sensors for various applications.

A typical field-effect transistor (FET) consists of five essential components: source, drain, gate, dielectric layer, and semiconductor layer. These transistors can be classified based on the majority of carriers as either p-type (hole) FETs or n-type (electron) FETs. In the realm of p-type polymer materials used for FET research, thiophene-based polymers have traditionally been prominent, with poly(3-hexylthiophene) (P3HT) being a successful example.

For n-type FETs, naphthalenetetraimide (NDI) and perylenetetraimide (PDI) have demonstrated favorable properties and are extensively studied as n-type semiconductor materials. These materials are commonly employed in small-molecule n-type FETs. Key transistor parameters typically include carrier mobility, operating voltage, and on/off current ratio.

When comparing organic field-effect transistors (OFETs) to inorganic semiconductor structures, OFETs offer advantages such as high flexibility and low preparation costs. However, they also suffer from limitations such as low carrier mobility and high operating voltage. Despite these drawbacks, the unique properties of OFETs make them valuable for various applications, particularly in flexible electronic devices where their flexibility and cost-effectiveness outweigh their limitations in certain scenarios.

7.5.1.5 Carbon Materials

Flexible wearable electronic sensors commonly utilize CNTs and graphene as carbon materials. CNTs possess high crystallinity, excellent electrical conductivity, a large specific surface area, and the ability to control microporous size through synthesis processes. They also exhibit a specific surface utilization rate of up to 100%. Graphene, on the other hand, is known for its thinness, transparency, and good thermal conductivity. Its applications in sensing technology, mobile communication, information technology, and electric vehicles hold significant potential. In the case of CNTs, a conductive polymer sensor was developed by combining multiarmed CNTs with silver and printing techniques. This sensor maintained conductivity of up to 20 S/cm even when stretched by 140%. Additionally, the combined application of CNTs and graphene resulted in the creation of a highly stretchable, transparent FET. This transistor incorporated graphene/single-walled CNT electrodes, single-walled CNT lattice channels, and pleated inorganic dielectric layers. The presence of the folded alumina dielectric layer enabled the transistor to withstand multiple 20% amplitude stretching-diffusion cycles without any change in drain current, demonstrating its excellent sustainability.

7.5.2 Flexible Gas Sensors

The arrangement of gas-sensitive thin-film material on flexible gas sensor electrodes offers several advantages. The substrate used is flexible, lightweight, and easily bendable, allowing for the production of large-area sensors. The thin-film material itself has garnered significant attention due to its higher sensitivity and relatively simple production process. These qualities make it well-suited for portable and low-power gas sensors in specialized environments. This development overcomes the limitations of previous gas sensors, such as difficulties in portability, limited measurement range, small sensing capabilities, and high costs. As a result, it enables simple and accurate detection of gases like NH, NO, and ethanol, attracting widespread interest. Figure 7.9 illustrates the preparation process of the gas sensor, which involves several steps, including fabric substrate pretreatment, inkjet printing of silver nitrate catalyst, chemical deposition of copper electrodes, and extrusion printing of the gas-sensitive material [66].

7.5.3 Flexible Pressure Sensors

Flexible pressure sensors find extensive applications in various fields, including smart clothing, smart sports equipment, and robotic "skin." These sensors utilize

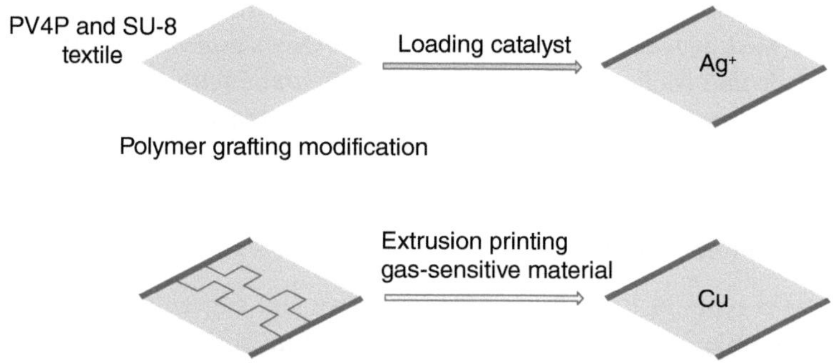

Figure 7.9 Preparation process of gas sensor. Source: Reproduced with permission from Guo et al. [66]; © 2022, Forty-nine Institute of China Electronic Technology Corporation.

substrate materials such as polyvinylidene fluoride, silicone rubber, and PI. Unlike metal strain gauge force sensors or conventional n-type semiconductor chip diffusion-based pressure sensors, flexible pressure sensors offer superior flexibility, electrical conductivity, and piezoresistive properties. These characteristics make them ideal for applications where flexibility and adaptability are crucial.

The detection principle of flexible pressure sensors involves various mechanisms, including piezoresistive, capacitive, piezoelectric, and frictional electric sensing [67]. Piezoresistive pressure sensors operate based on the change in the conductive network of the sensor under the influence of external forces, resulting in a mapping relationship between resistance and pressure. These sensors are widely used due to their simple structure, wide measurement range, and cost-effectiveness, making them the most commonly employed type of pressure sensor.

Capacitive pressure sensors, on the other hand, rely on the compression of the dielectric layer between two electrodes when subjected to external forces. This compression leads to variations in sensor capacitance. Capacitive pressure sensors are extensively utilized in displays and touchscreens, as they only require a slight touch of the screen with a finger to generate electrostatic induction, eliminating the need for applied pressure.

Both piezoresistive and capacitive sensors undergo changes in their geometries when exposed to external forces, resulting in corresponding alterations in electrical signals. These sensors are not limited to specific types of functional conductive materials, providing flexibility in their design and application.

The principle of a piezoelectric pressure sensor is based on the internal polarization of the piezoelectric material when subjected to an external force. This polarization generates opposite charges on the two surfaces of the crystal, resulting in the output of a voltage or current signal to the external circuit. This mechanism enables the conversion of mechanical energy into electrical energy and vice versa, facilitating the measurement of pressure. Piezoelectric pressure sensors are capable of self-powering as they utilize the output signal, but they are primarily suitable for dynamic force measurements.

Frictional sensors, on the other hand, consist of two materials with opposite polarities in contact with each other. When an external force is applied, the surfaces of the materials produce opposite charges. Upon removing the external force, a potential difference is generated between the separated surfaces. One significant advantage of both piezoelectric and frictional sensors is their ability to be self-powered using the output signal. However, these sensors are primarily employed for measuring dynamic forces rather than static pressures.

7.5.4 Flexible Humidity Sensor

There are two main types of humidity sensors: resistive and capacitive. Resistive humidity sensors employ a moisture-sensitive material film that covers the substrate. When water vapor in the air is adsorbed onto the moisture-sensitive film, it causes changes in the resistivity and resistance values of the sensor element. These changes can be measured to determine humidity levels. On the other hand, capacitive humidity sensors typically utilize polymer films as the moisture-sensitive material. Commonly used polymer materials include polystyrene, PI, and caseinate acetate. The humidity measurement is based on changes in capacitance resulting from the adsorption of moisture by the polymer film.

Humidity sensors have evolved from simple moisture-sensitive components to integrated, intelligent devices capable of multiparameter detection. Traditional methods like wet and dry bulb hygrometers or hair hygrometers are no longer sufficient to meet the demands of modern scientific development. Flexible humidity sensors have gained significant attention due to their low cost, low energy consumption, ease of manufacturing, and seamless integration into intelligent system manufacturing. The substrate materials used for flexible humidity sensors are similar to those employed in other types of flexible sensors. Various methods are available for manufacturing humidity-sensitive films, including dip coating [68], rotational coating [69], screen printing, and inkjet printing [70]. These techniques enable the production of humidity sensors with desirable characteristics and properties.

7.5.5 Normal Sensors Compare with Flexible Sensors

Flexible pressure sensors have wide-ranging applications in various industries, including smart clothing, smart sports equipment, and robotic "skin." These sensors are designed using flexible substrate materials such as polyvinylidene fluoride, silicone rubber, and PI. Unlike traditional metal strain gauge force sensors or conventional n-type semiconductor chip diffusion-based pressure sensors, flexible pressure sensors possess exceptional flexibility, electrical conductivity, and piezoresistive properties. These unique characteristics make them highly suitable for applications where flexibility and adaptability are essential.

Flexible sensors offer a versatile and diverse structure that can be customized according to specific measurement conditions. They enable convenient and accurate measurement of special signals in challenging environments, making them valuable in electronic skin, biomedicine, wearable electronic products, and aerospace applications. However, the current technology level for preparing materials such as CNTs

Table 7.5 Comparison of common sensors and flexible sensors.

Category	Common materials	Achievable flexibility	Advantage	Disadvantage
Ordinary sensor	Metal material Ceramic material	5–10% 0–1%	Good electrical conductivity, wear resistance, high-temperature mechanical properties, high hardness, high strength, low price, etc.	Easy to oxidize, rigid materials are not easy to bend, high elastic modulus, no plasticity, and easy to break
Flexible sensor	Semiconductor material	20–80%	The measurement range of special environment and signal is expanded, and it can maintain good performance, portability, and adaptability under deformation such as tension, pressure, bending, and torsion	The relative dielectric constant and piezoelectric coefficient are small, which leads to relatively poor performance, low service life, and high cost

and graphene for flexible sensors is still in its early stages, and challenges remain in terms of cost, application range, and service life.

One drawback of commonly used flexible substrates is their limited resistance to high temperatures, which can result in high stress and weak adhesion between the substrates and thin-film materials. The assembly, alignment, integration, and encapsulation technologies of flexible sensors also require further improvement.

Table 7.5 provides a comparison between ordinary sensors and flexible sensors across various aspects. It demonstrates that flexible sensors have an expanded measurement range in special environments and can maintain good performance even under conditions of pulling, pressing, bending, and twisting. They possess excellent portability and adaptability. However, these advantages are accompanied by certain drawbacks, including lower relative dielectric constant and piezoelectric constant, which can result in relatively poorer performance, shorter lifespan, and higher costs.

References

1 Huajing Intelligence Industry Research Institute (2021). The global flexible electronics market has broad prospects. The development of China's entire industrial chain is uneven. *High-Technology & Commercialization* 27 (7): 36–39.

2 Thejo Kalyani, N. and Dhoble, S.J. (2015). Novel materials for fabrication and encapsulation of OLEDs. *Renewable and Sustainable Energy Reviews* 44: 319–347.

3 Schlingman, K., Chen, Y., Carmichael, R.S. et al. (2021). 25 Years of light-emitting electrochemical cells: a flexible and str etchable perspective. *Advanced Materials* 33 (21): e2006863.

4 Liu, Y., Zhang, B., and Huang, C. (2017). Analysis of the film encapsulation technology of flexible OLED. *Optoelectronic Technology* 37 (2): 139–145.

5 Burrows, P.E., Bulovic, V., Forrest, S.R. et al. (1994). Reliability and degradation of organic light emitting devices. *Applied Physics Letters* 65 (23): 2922–2924.

6 Lim, S.F., Wang, W., and Chua, S.J. (2001). Degradation of organic light-emitting devices due to formation and growth of dark spots. *Materials Science and Engineering B* 85 (2–3): 154–159.

7 Lim, S.F., Ke, L., Wang, W. et al. (2001). Correlation between dark spot growth and pinhole size in organic light-emitting diodes. *Applied Physics Letters* 78 (15): 2116–2118.

8 Lifka, H., Van Esch, H.A., and Rosink, J.J.W.M. (2004). 50.3: Thin film encapsulation of OLED displays with a NONON stack. *Society for Information Display* 35 (1): 1384–1387.

9 Jung, K., Bae, J.-Y., Park, S.J. et al. (2011). High performance organic-inorganic hybrid barrier coating for encapsulation of OLEDs. *Journal of Materials Chemistry* 21 (6): 1977–1983.

10 Kim, G.H., Oh, J., Yang, Y.S. et al. (2004). Encapsulation of organic light-emitting devices by means of photopolymerized polyacrylate films. *Polymer* 45 (6): 1879–1883.

11 Ghosh, A.P., Gerenser, L.J., Jarman, C.M. et al. (2005). Thin-film encapsulation of organic light-emitting devices. *Applied Physics Letters* 86 (22).

12 Yamashita, K., Mori, T., and Mizutani, T. (2001). Encapsulation of organic light-emitting diode using thermal chemical-vapour-deposition polymer film. *Journal of Physics D: Applied Physics* 34 (5): 740–743.

13 Van De Weijer, P., Pcp, B., Unnikrishnan, S. et al. (2017). High-performance thin-film encapsulation for organic light-emitting diodes. *Organic Electronics* 44: 94–98.

14 Park, S.-H.K., Jiyoung, O., Hwang, C.-S. et al. (2005). Ultra thin film encapsulation of organic light emitting diode on a plastic substrate. *Electronics and Telecommunications Research Institute Journal* 27 (5): 545–550.

15 Park, J.-S., Chae, H., Chung, H.K. et al. (2011). Thin film encapsulation for flexible AM-OLED: a review. *Semiconductor Science and Technology* 26 (3): 034001.

16 He, Y., Shen, H., Wang, W. et al. (2018). A review on the electrode materials of flexible dye-sensitized solar cells and flexible perovskite solar cells. *Materials Reports* 32 (21): 3677–3688.

17 Jia, H. and Zhai, X. (2021). Research of photoanodes of flexible dye-sensitized solar cells. *Infrared* 42 (02): 43–48.

18 Zhang F. (2008). Low-temperature preparation and properties of TiO_2 electrodes for flexible DSSC. Master's thesis. Nanjing University of Aeronautics and Astronautics.

19 Liu, X., Zhao, Y., Gong, J. et al. (2020). Technical development and application of flexible Gu (In, GA) Se_2 solar cells. *Vacuum and Cryogenics* 26 (05): 377–384.

20 Han, H.G., Weerasinghe, H.C., Min Kim, K. et al. (2015). Ultrafast fabrication of flexible dye-sensitized solar cells by ultrasonic spray-coating technology. *Scientific Reports* 5: 14645.

21 Lu, J., Xu, S., Du, Y. et al. (2017). Well-aligned TiO_2 nanorod arrays prepared by DC reactive magnetron sputtering for flexible dye-sensitized solar cells. *Materials Letters* 188: 323–326.

22 Wang, Y., Cheng, P., Feng, C. et al. (2019). High performance flexible dye-sensitized solar cells base on multiple functional optimizations. *Solar Energy* 180: 423–428.

23 Lee, K., Lin, L., Suryanarayanan, V. et al. (2014). Titanium dioxide coated on titanium/stainless steel foil as photoanode for high efficiency flexible dye-sensitized solar cells. *Journal of Power Sources* 269 (c): 789–794.

24 Yu, C., Liu, Y., Li, Y. et al. (2019). Application of flexible electronic materials and devices. *Chinese Journal on Internet of Things* 3 (03): 102–110.

25 Li, L., Tang, G., Shi, Z. et al. (2021). Transfer-printed, tandem microscale light-emitting diodes for full-color displays. *Proceedings of the National Academy of Sciences of the United States of America* 118 (18): e2023436118.

26 Jeong, H., Wang, L., Ha, T. et al. (2019). Modular and reconfigurable wireless E-tattoos for personalized sensing. *Advanced Materials Technologies* 4 (8): 1900117.

27 Han, M., Chen, L., Aras, K. et al. (2020). Catheter-integrated soft multilayer electronic arrays for multiplexed sensing and actuation during cardiac surgery. *Nature Biomedical Engineering* 4: 997–1009.

28 Kim, T.H., Cho, K.S., Lee, E. et al. (2011). Full-colour quantum dot displays fabricated by transfer printing. *Nature Photonics* 5 (3): 176–182.

29 Fröbel, M., Schwab, T., Kliem, M. et al. (2015). G et it white: color-tunable AC/DC OLEDs. *Light: Science & Applications* 4 (2): e247.

30 Parthasarathy, G., Gu, G., and Forrest, S.R. (1999). A full-color transparent metal-free stacked organic light emitting device with simplified pixel biasing. *Advanced Materials* 11 (11): 907–910.

31 Koo, J.H., Jeong, S., Shim, H.J. et al. (2017). Wearable electrocardiogram monitor using carbon nanotube electronics and color-tunable organic light-emitting diodes. *ACS Nano* 11 (10): 10032–10041.

32 Huang, Y., Hsiang, E.L., Deng, M.Y. et al. (2020). Mini-LED, micro-LED and OLED displays: present status and future perspectives. *Light: Science & Applications* 105 (9): 2095–5545.

33 Zhou, X., Tian, P., Sher, C. et al. (2020). Growth, transfer printing and colour conversion techniques towards full-colour micro-LED display. *Progress in Quantum Electronics* 71: 100263.

34 Yi, H., Lee, S.H., Ko, H. et al. (2019). Ultra-adaptable and wearable photonic skin based on a shape- memory, responsive cellulose derivative. *Advanced Functional Materials* 29: 1902720.

35 Hammock, M.L., Chortos, A., Tee, B.C. et al. (2013). 25th anniversary article: the evolution of electronic skin (E-skin): a brief history, design considerations, and recent progress. *Advanced Materials* 25 (42): 5997–6037.

36 Hua, Q., Sun, J., Liu, H. et al. (2018). Skin-inspired highly stretchable and conformable matrix networks for multifunctional sensing. *Nature Communications* 9: 244.

37 Li, G., Liu, S., Wang, L. et al. (2020). Skin-inspired quadruple tactile sensors integrated on a robot hand enable object recognition. *Science Robotics* 5 (49): eabc8134.

38 Lipomi, D.J., Vosgueritchian, M., Tee, B.C.K. et al. (2011). Skin-like pressure and strain sensors based on transparent elastic films of carbon nanotubes. *Nature Nanotechnology* 6: 788–792.

39 Boutry, C.M., Negre, M., Jorda, M. et al. (2018). A hierarchically patterned, bioinspired e-skin able to detect the direction of applied pressure for robotics. *Science Robotics* 3 (24): eaau6914.

40 Park, J., Kim, M., Lee, Y. et al. (2015). Fingertip skin-inspired microstructured ferroelectric skins discriminate static/dynamic pressure and temperature stimuli. *Science Advances* 1 (9): e1500661.

41 Li, P., Ali, H.P.A., Cheng, W. et al. (2020). Bioinspired prosthetic interfaces. *Advanced Materials Technologies* 5 (3): 1900856.

42 Kim, J., Lee, M., Shim, H.J. et al. (2014). Stretchable silicon nanoribbon electronics for skin prosthesis. *Nature Communications* 5: 5747.

43 Ho, D.H., Sun, Q., Kim, S.Y. et al. (2016). Stretchable and multimodal all graphene electronic skin. *Advanced Materials* 28 (13): 2601–2608.

44 Lee, Y., Park, J., Cho, S. et al. (2018). Flexible ferroelectric sensors with ultrahigh pressure sensitivity and linear response over exceptionally broad pressure range. *ACS Nano* 12 (4): 4045–4054.

45 Yetisen, A.K., Martinez-Hurtado, J.L., Uenal, B. et al. (2018). Wearables in medicine. *Advanced Materials* 30 (33): 1706910.

46 Nassar, J.M., Mishra, K., Lau, K. et al. (2017). Recyclable nonfunctionalized paper-based ultralow-cost wearable health monitoring system. *Advanced Materials Technologies* 2 (4): 1600228.

47 Honda, W., Harada, S., Arie, T. et al. (2014). Wearable, human-interactive, health-monitoring, wireless devices fabricated by macroscale printing techniques. *Advanced Functional Materials* 24 (22): 3299–3304.

48 Bacha, E. and Kalfa, D. (2014). Minimally invasive paediatric cardiac surgery. *Nature Reviews Cardiology* 11 (1): 24–34.

49 Liu, Z., Ma, Y., Ouyang, H. et al. (2019). Transcatheter self-powered ultrasensitive endocardial pressure sensor. *Advanced Functional Materials* 29 (3): 1807560.

50 Kim, D.H., Lu, N., Ghaffari, R. et al. (2011). Materials for multifunctional balloon catheters with capabilities in cardiac electrophysiological mapping and ablation therapy. *Nature Materials* 10 (4): 316–323.

51 Kim, J., Campbell, A.S., de Avila, B. et al. (2019). Wearable biosensors for healthcare monitoring. *Nature Biotechnology* 37 (4): 389–406.

52 Lee, J., Cho, H.R., Cha, G.D. et al. (2019). Flexible, sticky, and biodegradable wireless device for drug delivery to brain tumors. *Nature Communications* 10 (1): 5205.

53 Lee, H., Song, C., Hong, Y.S. et al. (2017). Wearable/disposable sweat-based glucose monitoring device with multistage transdermal drug delivery module. *Science Advances* 3 (3): e1601314.

54 Gao, W., Ota, H., Kiriya, D. et al. (2019). Flexible electronics toward wearable sensing. *Accounts of Chemical Research* 52 (3): 523–533.

55 Zhao, Z., Zhou, L., Li, S. et al. (2021). Selection rules of triboelectric materials for direct-current triboelectric nanogenerator. *Nature Communications* 12 (1): 4686.

56 Bai, P., Zhu, G., Lin, Z.H. et al. (2013). Integrated multi layered triboelectric nanogenerator for harvesting biomechanical energy from human motions. *ACS Nano* 7 (4): 3713–3719.

57 Chun, S., Son, W., Kim, H. et al. (2019). Self-powered pressure- and vibration-sensitive tactile sensors for learning technique-based neural finger skin. *Nano Letters* 19 (5): 3305–3312.

58 Zhu, G., Pan, C., Guo, W. et al. (2012). Triboelectric-generator-driven pulse electrodeposition for micropatterning. *Nano Letters* 12 (9): 4960–4965.

59 Qin, S., Zhang, Q., Yang, X. et al. (2018). Hybrid piezo/triboelectric-driven self-charging electrochromic supercapacitor power encapsulation. *Advanced Energy Materials* 8 (23): 1800069.

60 Ryu, H., Park, H., Kim, M.K. et al. (2021). Self-rechargeable cardiac pacemaker system with triboelectric nanogenerators. *Nature Communications* 12 (1): 4374.

61 Lee, J.W., Xu, R., Lee, S. et al. (2016). Soft thin skin-mounted power management systems and their use in wireless thermography. *Proceedings of the National Academy of Sciences of the United States of America* 113 (22): 6131–6136.

62 Ma, Y., Zheng, Q., Liu, Y. et al. (2016). Self-powered, one-stop, and multifunctional implantable triboelectric active sensor for real-time biomedical monitoring. *Nano Letters* 16 (10): 6042–6051.

63 Wang, Z.L. and Song, J.H. (2006). Piezoelectric nanogenerators based on zinc oxide nanowire arrays. *Science* 312 (5771): 242–246.

64 Li, L. (2010). Research of information fiber-based flexible sensor theory and applications. PhD's thesis. Donghua University, Shanghai.

65 Duan, J., Li, B., and Li, S. (2015). Research progress on common novel flexible sensor. *Transducer and Microsystem Technologies* 34 (11): 1–4.

66 Guo, H., Song, J., Du, J. et al. (2022). Research on flexible NH_3 gas sensor based on textile. *Transducer and Microsystem Technologies* 41 (03): 39–42.

67 Wang, K., He, Z., and Yu, Y. (2022). Preparation and performance optimization of flexible piezoresistie pressure. *Electronic Components and Materials* 41 (08): 781–793.

68 Yamazo, N. and Shimizu, Y. (1986). Humidity sensors: principles and applications. *Sensors and Actuators* 10 (86): 379–398.

69 Su, P.G., Chen, I.C., and Wu, R.J. (2001). Use of poly(2-acrylamido-2-methylpropane sulfonate) modified with tetraethyl orthosilicate as sensing material for measurement of humidity. *Analytica Chimica Acta* 449 (1/2): 103–109.

70 Lee, C.W., Nam, D.H., Han, Y.S. et al. (2005). Humidity sensors fabricated with polyelectrolyte membrane using an ink-jet printing technique and their electrical properties. *Sensors and Actuators B: Chemical* 109 (2): 334.

8

Testing Standards

Junjie Yuan

Peking University Shenzhen Graduate School, Lishui Road, Shenzhen, 518055, China

The International Electrotechnical Commission (IEC) is a worldwide standardization organization that includes all National Electrotechnical Commissions (IEC National Committees). The goal of the IEC is to promote international cooperation in all standardization issues in the electrical and electronic fields. To this end, among other activities, the IEC publishes international standards, technical specifications, technical reports, publicly available specifications (PAS), and guidelines (hereinafter referred to as "IEC publications"). Their preparation is entrusted to Technical Committees; any IEC National Committee interested in the subject dealt with may participate in this preparatory work. International, governmental, and nongovernmental organizations liaising with the IEC are also involved in this preparatory work, and the IEC cooperates closely with the International Organization for Standardization (ISO) under the conditions established by the agreement between the two organizations.

The international standard IEC 62715-1-1 was developed by Technical Committee IEC 110: Electronic Display Devices. The text of this standard is based on the following documents:

CDV	Report on voting
110/441/CDV	110/480A/RVC

Full information on the vote to adopt this standard can be found in the voting report shown in the table above. This publication has been drafted in accordance with Part 2 of the 1S0/IEC Directive. A list of all components in the IEC 62715 series, under the general heading Flexible Display Devices, can be found on the IEC website.

Flexible Electronic Packaging and Encapsulation Technology, First Edition.
Edited by Hong Meng and Wei Huang.

8.1 Terminology and Alphabetic Symbols

8.1.1 Scope

This part of IEC 62715 gives the preferred terminology, definitions, and symbols for flexible display devices, with the aim of using the same terminology when developing standards in different countries.

8.1.2 Terms and Definitions

For the purposes of this chapter, the following terms and definitions apply.

8.1.2.1 Terminology Classification

The terminology of flexible display devices is classified as follows:

- General terminology
- Terminology related to physical properties
- Terms related to construction features
- Performance and specification-related terms
- Terms related to the production process

8.1.2.2 General Terms

Flexible:

Mechanical compliance to external mechanical stress.

Flexible display devices:

Flexible display panels and flexible modules that can be mechanically bent in one or more steps of substrate handling, fabrication, storage, use, handling, transportation, and repositioning. Bendable display devices are curved, conformable to curved surfaces, bendable, and foldable without creasing.

Flexible display module:

Flexible display devices with external actuators, optical films, and other input/output interaction films (e.g. touch screens or sensing films).

Flexible display panel:

Flexible display devices without external drivers.

Flexible display: Products that contain flexible display modules that are bendable for the end user.

8.1.2.3 Physical Characteristics Related Terms

Ductility:

The flexible display device deforms under compressive stress and maintains the shape of the deformation after the stress is released.

Flexural strength:

Maximum bending stress can be applied to flexible display devices without any deterioration, permanent distortion, or breakage of image quality.

Flexural modulus:

The ratio of the bending stress to the resulting strain in the flexural deformation of the flexible display device.

Periodic stress:

Over time, external stimuli, such as mechanical force or exposure to temperature, are repeatedly applied to flexible display devices.

8.1.2.4 Terms Related to Construction Elements

Gas barrier layer:

A passivation layer type that prevents contaminants in the form of vapor or gas from entering the active device area. The gas barrier layer is usually mounted on the substrate or directly on the active device to prevent water or oxygen from penetrating into the active device area.

Flexible substrates:

The flexible substrate on which the flexible display device is located, such as plastic film, thin glass, metal foil, or paper.

Buffer layer:

The type of passivation layer inserted between the substrate and the electronic components of the flexible display device mitigates the impact of defects on the display manufacturing process and/or operational performance. The polishing layer is usually used on a thin glass substrate to prevent breakage when bending.

Insulation:

A passivation layer type that electrically separates a metal foil substrate from the conductive portion of a device. It can be applied to the substrate surface to create a smooth surface.

8.1.2.5 Symbols Related to Performances and Specifications

Bend radius:

In the bending test, the flex shows the radius of the arc corresponding to the curvature of the centerline between the innermost and outermost surfaces of the device.

Critical bending radius:

The minimum bend radius within which a flexible display device operates without causing image quality degradation, permanent distortion, or breakage.

Strain:

When an external stimulus acts on a flexible display device, the ratio of the amount of change in the length of the device to its original length.

Calcium test:

Test method for permeability of calcium-containing gases. By monitoring the optical transport and changes in electrical properties of the calcium layer, it is possible to measure the permeation rate of the substance used to encapsulate the cell or substrate gas barrier layer. The calcium layer is degraded due to the penetration of moisture or other gases through the substance. The time dependence of the calcium ion change process is inversely proportional to the permeation rate.

Table 8.1 Symbols relating to physical properties.

Term	Symbol	Unit
Strain	E	%
Stress	Σ	Pa

Table 8.2 Symbols related to performance and specifications.

Term	Symbol	Unit
Bending radius	r	m

8.1.2.6 Terms Related to the Production Process

Carrier:

During the manufacturing process of a display, a substance is temporarily attached to a flexible substrate for processing purposes.

Carrier textures:

The flexible substrate is bonded to the carrier by heat and/or pressure.

Carrier stratification:

Peel off a composite substrate or synthesize a flexible display device from the carrier.

Precision winding alignment technology:

The process of continuously manufacturing electronic components of a display device on a roll of flexible substrates.

Lithographic printing:

Lithography technology for electronic display devices that do not use a development process to make patterns.

8.1.3 Alphabetic Symbols (Quantity Symbols/Unit Symbols)

8.1.3.1 Classification

The classification is shown in Tables 8.1 and 8.2.

8.1.3.2 Symbols

Tables 8.1 and 8.2 summarize the symbols used for flexible displays.

8.2 Mechanical Test Method (Deformation Test)

Flexible displays have a different shape compared to any other non-flexible electronic displays. Therefore, a variety of mechanical stress test methods can be used, such as cyclic bending (folding) or dynamic bending test, static bending test, rolling test, and combined mechanical test. The appropriate test method application range should be selected according to the requirements. For each mechanical stress test, the associated test method specification and the purpose of each unique test should be stated. The allowable critical bend radius of the panel depends on the application of the flexible display. Therefore, the required critical bend radius will vary depending on the application range.

When designing a reliable and repeatable method for mechanical folding or deformation testing, several factors need to be considered, as follows:

(1) The motion profile controlled by a single hinge (or pivot) and wound on the mandrel can stretch the sample and, if not precisely driven or controlled, will produce large variations in the test response.
(2) The motion profile controlled by multiple hinges or pivots prevents additional stretching on the specimen and provides better control of the test response.
(3) Clamping the specimen can prevent shearing between layers or multi-layer specimens containing adhesive and greatly affect the test response.
(4) Additional considerations, such as connections too close to vertices or bends, misalignment of the folding shaft, or slight deformation of the shape of the mounting plate, can also affect the reliability and repeatability of the test response.
(5) Samples should be pretreated in a controlled environment for at least 24 hours before testing, as the material properties of binders and polymers may be affected by temperature and water absorption.

8.2.1 Cyclic Bending Test

This step is done by repeatedly bending the specimen to adjust under mechanical stress.

8.2.1.1 Purpose

The purpose of this test is to provide a standard procedure to assess the robustness of flexible displays to cyclic bending stresses, which can typically occur in applications. Bending characteristics may include several typical parameters of the image quality characteristics of the display panel. Typical parameters for display panel image quality may include the presence or absence of luminance, chromaticity, uniformity, line defects, point defects, pixel shrinkage, or cracks.

8.2.1.2 Testing Device

The cyclic bending test device includes a fixture for fixing the bending specimen, moving parts for shuttling, and a control system to adjust the number of cyclic bends, the distance traveled, and the speed of movement during the test. During the test, the specimen should be firmly clamped by a clamped section. There are several cyclic bending test equipment, as shown in Figure 8.1 [8, 11, 14]. It is not necessary to prefer some type of bending test equipment, but a constant bending radius (r) equal

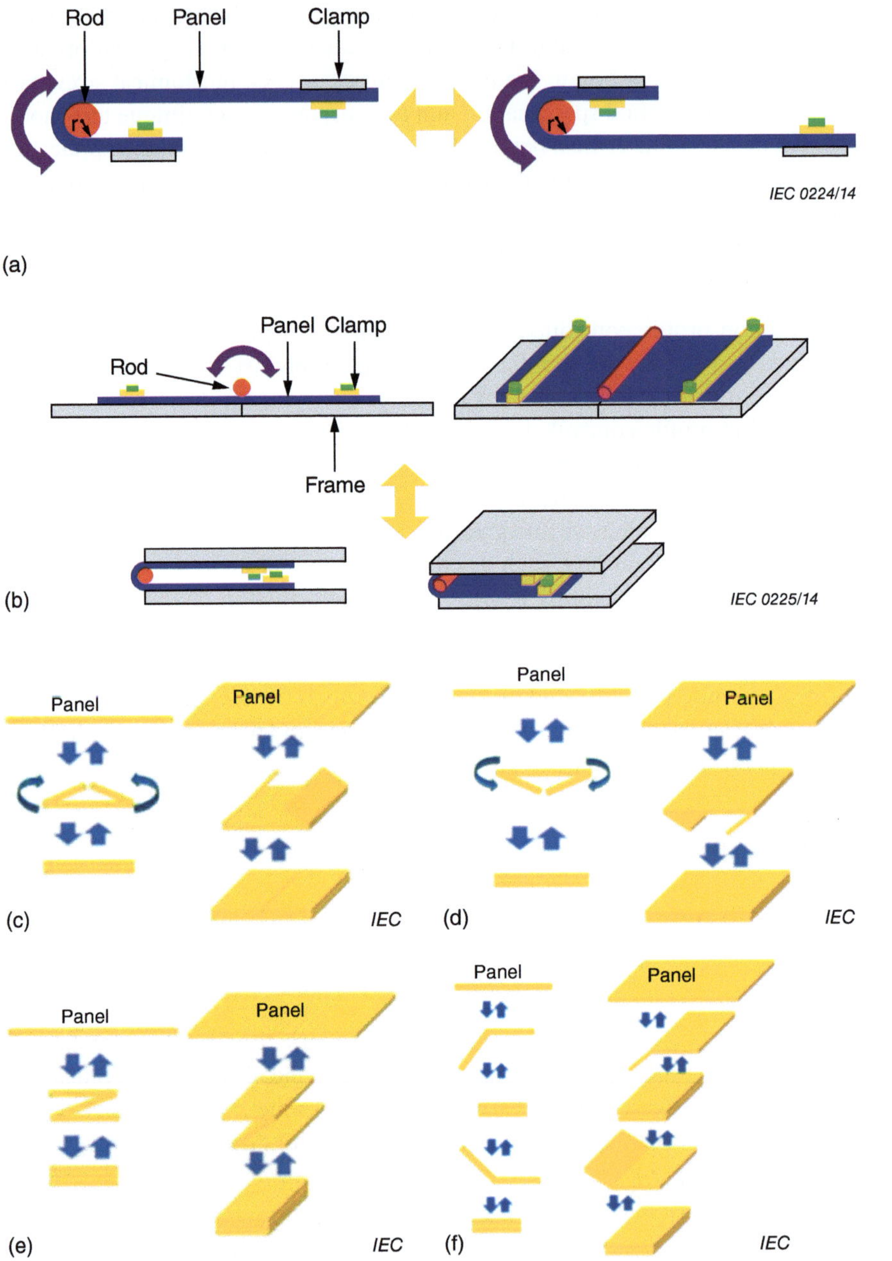

Figure 8.1 Different cyclic bending test devices. (a) cyclic bending (sliding) equipment; (b) cyclic bending equipment. Source: Reproduced with permission from Ref. [26]; © 2018, International Electrotechnical Commission.

to the radius of the rod should be maintained during the bend test. For cyclic bending tests, consider the following devices:

(1) When the specimen shuttles back and forth (Figure 8.1a) or when the specimen is folded and unfolded (Figure 8.1b), the specimen is subjected to bending stress.
(2) When the specimen is bent several times in the inward or outward direction, a device that bends inward multiple times in a cycle (Figure 8.1c) or a device that bends outward multiple times can be used (see Figure 8.1d).
(3) When the specimen is repeatedly bent multiple times in a Z-shape, cyclic multi-bending equipment can be used.
(4) Cyclic inward and outward bending equipment can be used when the specimen is repeatedly bent in both the inward and outward directions (Figure 8.1e and f).

8.2.1.3 Test Procedure

The cyclic bending test uses repetitive motion to move regularly between two points or two states (folded and unfolded) as follows:

(a) prepare the required number of specimens;
(b) initial performance, visual characteristics, testing, and recording of results on the prepared specimen;
(c) one side of the specimen is fixed with a fixture or fixed with adhesive, and the other side is properly supported;
(d) bend the specimen appropriately according to the specified clamping roll angle, bending angular speed, and other conditions;
(e) return the specimen to its initial state before bending;
(f) if necessary, bend the specimen in another direction under specified conditions, such as the rotation angle and bending angular velocity of the clamp roller;
(g) return the specimen to its initial state before bending with the same angular velocity and reverse;
(h) repeat cycles (d) to (g) for the specified number of times;
(i) after the test, the specimen is removed from the instrument;
(j) repeat the test with other specimens after (c) to (j);
(k) measure the stress properties and visual properties of mechanical stress specimens and the number of tests required, and record the results. Compare initial and stress properties and determine the degree of change caused by bending;
(l) individual specimens are tested and performance measurements are recorded in the test report.

8.2.1.4 Test Conditions and Reports

The test conditions are specified as follows:

(a) Bending radius

r (Bend radius): 20, 10, 5, 3, 2, 1, 0.5, 0.2 mm

(b) Bending angle and angular velocity, program a cycle

t (One bend time and interval): 0.5, 1, 2, 3, 5, 10 s

(c) Load
(d) Number of repeat cycles
(e) Acceptance criteria
(f) Number of specimens
(g) The curved inner liner is either the top liner (lace up) or the back liner (face down).
(h) Holding method or holding method and detailed method, such as pressure
(i) Liner conditions (e.g. handling) or specimen and instrument at the same time
(j) The environmental conditions of the test, such as temperature and humidity

8.2.2 Static Bending Test

This test is particularly useful for evaluating the static bending performance of flexible display devices by measuring their performance after maintaining a certain bending time. Each specimen is bent with a fixed bending radius for any length of time.

8.2.2.1 Purpose

This test is to provide a standard procedure for evaluating the static bending performance of flexible display devices when subjected to constant stress for a certain period of time. Each specimen is bent with a fixed bend radius for a controlled length of time.

8.2.2.2 Testing Device

During the test, the body of the display panel should be firmly attached to the surface of the test equipment, where the test equipment should have a circular shape with a certain radius as shown in Figure 8.2. The specimen should be bent at a fixed bending radius for a period of time.

If the bending radius of the specimen is kept constant during the bending test, the bending rod may be removed because the bending rod may scratch the surface of the specimen during the static bending test. During the test, the flexible printed circuit board (FPC) and the driver should be handled carefully so that they are not damaged by bending. In order to avoid bending damage to the driver integrated

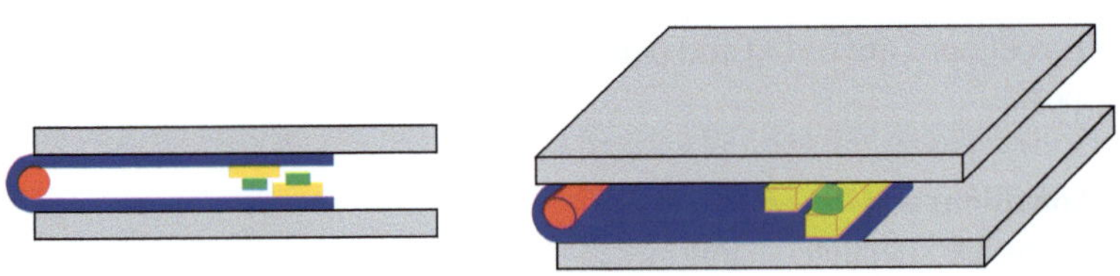

Figure 8.2 Static bending test device. Source: Reproduced with permission from Ref. [26]; © 2018, International Electrotechnical Commission.

circuit (IC) during the bending test, it is best to use the chip-on-film (COF) method.

8.2.2.3 Test Steps

The static bending test should be carried out with a fixed bending radius for a certain period of time, as follows:

(a) prepare the required number of specimens;
(b) initial performance, visual characteristics, testing, and recording of the prepared specimens;
(c) one side of the specimen is fixed by a fixture and the other side is properly supported;
(d) bend the specimen appropriately according to the specified clamping roll angle, bending angular speed, and other conditions;
(e) keep the specimen for a certain bending time;
(f) return the specimen to its initial state before bending;
(g) repeat (d) to (f) perform the prescribed number of tests;
(h) after bending, the specimen is removed from the instrument;
(i) after (c) to (g), repeat the test with other specimens;
(j) Measure the force performance and visual characteristics of mechanical stress specimens and record the test results. The initial properties and stress properties of the materials were compared, and the degree of bending change of the materials was determined.
(k) Test and perform measurements on individual specimens and record them in the test report.

8.2.2.4 Test Conditions and Reports

The test conditions are specified as follows:

(a) Bend radius

r (Bend radius): 20, 10, 5, 3, 2, 1, 0.5, 0.2 mm

(b) Bending angle
(c) Hold duration
(d) Load
(e) Acceptance criteria
(f) Number of specimens
(g) The curved inner surface is either top (face up) or back (face down)
(h) The way to hold the specimen and the detailed holding method, such as pressure
(i) Surface condition of specimens and instruments (e.g. handling)
(j) Environmental condition testing, such as temperature and humidity

8.2.3 Combined Bending Test

This test is especially suitable for evaluating the two bending properties (cyclic bending performance and static bending performance (see Figure 8.3)) of the

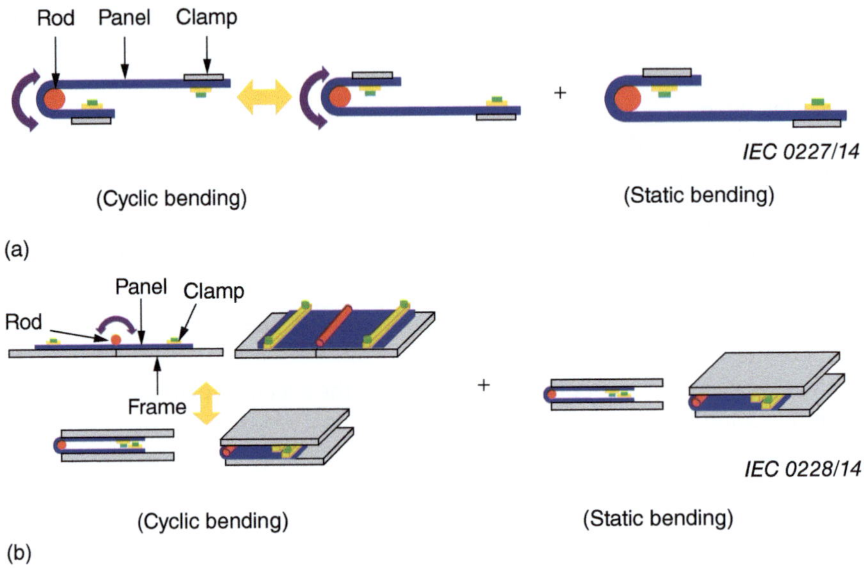

Figure 8.3 Combined bending test device, including cyclic bending test and static bending test. Source: Reproduced with permission from Ref. [26]; © 2018, International Electrotechnical Commission.

flexible display device after maintaining the established shape for a certain period of time and after a certain period of cyclic bending state in the actual use environment.

8.2.3.1 Purpose

The purpose of this test is to provide a standard procedure for evaluating the combined bending properties of cyclic bending and duration bending. The purpose of this test is to take into account the use environment of the real product, and the panel remains bent, folded, or rolled for a long time before the real product is used.

8.2.3.2 Testing Device

The combined bending test uses the same device as the cyclic bending test, as shown in Figure 8.3. The instrument controller should be able to stop the device during each cycle iteration when the specimen is in a bent state.

8.2.3.3 Test Procedure

The cyclic bending test procedure has been modified to include an adjustable duration "hold" between each cycle iteration, where the specimen remains fully bent [8] as follows:

(a) prepare the required number of specimens;
(b) initial performance, visual characteristics, testing, and recording of results on prepared samples;

(c) one edge of the specimen is fixed by a fixture and the other edge is properly supported;
(d) appropriate bending of the specimen under specified conditions, such as the rotation angle and bending angular velocity of the clamping roller;
(e) return the specimen to its initial state before bending;
(f) after a certain number of flexure tests in (d) and (e) to keep the specimen in a fully bent state and then hold;
(g) change a certain holding time to restore the specimen to its initial state;
(h) requires, under specified conditions, such as the rotation angle and bending angular velocity of the clamping roller, to bend the specimen in another direction;
(i) the bending module returns to its initial state before bending with the same angular velocity and reverse direction;
(j) change a certain number of bending tests performed in (h) and (i) so that the specimen is fully bent and then maintained;
(k) change the test to return the specimen to its initial state;
(l) repeat (d) to (k) for a specified number of cycles;
(m) return the specimen to its initial state;
(n) repeat the test with other specimens after (c) to (m);
(o) measure the stress properties and visual properties of mechanical stress specimens with the required number of tests and record the results. The initial properties and stress properties were compared, and the degree of change caused by bending was determined;
(p) test and perform measurements on individual specimens and record them in the test report.

8.2.3.4 Test Conditions and Reports

The test conditions are as follows:

(a) Bend radius

r (Bend radius): 20, 10, 5, 3, 2, 1, 0.5, 0.2 mm

(b) Bending angle and angular velocity, a cyclic program

t (Time between bends and intervals): 0.5, 1, 2, 3, 5, 10 s

(c) The duration of the state of holding
(d) Load
(e) Number of repetitions
(f) Acceptance criteria
(g) Number of specimens
(h) The curved inner surface is either the top (tie up) or the back (face down).
(i) The way to hold the specimen and the detailed holding method, such as pressure
(j) The surface condition of the specimen and instrument (e.g. handling).
(k) Testing of environmental conditions, such as temperature and humidity

8.2.4 Rolling Test

This test is particularly useful for evaluating the rolling performance of a flexible display module after it has been unfolded, rolled in, or held in a coiled display.

8.2.4.1 Purpose

The purpose of this test is to provide a standard procedure for evaluating the robustness of the rolling characteristics of flexible display panels.

8.2.4.2 Testing Device

The specimen should be firmly clamped with clamping components on the drum side and fixed side. The drum should alternately move forward and backward according to the fixed distance, speed, and number of rolls shown in Figure 8.4. There is a slot on the drum side for inserting and clamping the edges of the specimen. The drum is repeatedly shuttled along the drum axis, and the specimen does not touch the equipment plate during the refueling test, as described in Figure 8.4.

During testing, FPCs and drives "should be handled with care so that they are not damaged by distortion." In order to prevent torsional damage to the drive circuit during the bending test, the COF method is adopted.

8.2.4.3 Test Procedure

The rolling test should be performed as follows:

(a) prepare the required number of specimens;
(b) initial performance, visual characteristics, testing of prepared samples, and recording of results;
(c) measure and record each sample size;
(d) One edge of the specimen is fixed to the roller as shown in Figure 8.4, and the other edge fixture is supported by a separately defined load. Align the specimen with the top surface on the inside or outside of the drum;
(e) rolling specimens under specified conditions, such as rolling radius, rolling distance, rolling angular velocity, and load;

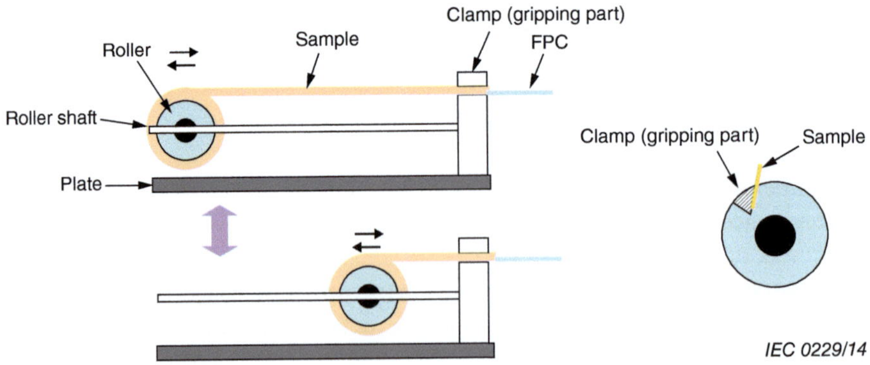

Figure 8.4 Rolling test setup. Source: Reproduced with permission from Ref. [26]; © 2018, International Electrotechnical Commission.

(f) return the specimen to its initial state before rolling with the same angular velocity and opposite direction of rotation;
(g) Repeat the number of cycles specified by (d) to (e). Optionally, another surface side can be tested under the same conditions, which is done by negotiation between the supplier and the customer. In this case, test on the other side of (d) to (g);
(h) change the test, remove the sample from the instrument;
(i) repeat the test with other specimens after (c) to (h);
(j) measure the stress, electrical and optical properties of a mechanical stress module with the required number of tests and record the results. Compare the initial performance and stress performance to determine the degree of change caused by the rolling test;
(k) test and perform measurements on individual specimens and record them in the test report.

8.2.4.4 Test Conditions and Reports

The test conditions are specified as follows:

(a) Rolling radius
(b) Angular velocity, a periodic procedure
(c) Hold duration
(d) Load
(e) Number of repetitions
(f) Acceptance criteria
(g) Number of specimens
(h) Rolling inner surface (top or back)
(i) The way to hold the specimen and the detailed holding method, such as pressure
(j) Surface condition of specimens and instruments (e.g. handling)
(k) Environmental condition tests, such as temperature and humidity

8.2.5 Static Rolling Test

This test is particularly suitable for evaluating the static rolling performance of flexible display modules after the flexible display rolls out, rolls in, or maintains the shape of the roll.

8.2.5.1 Purpose

The purpose of this test is to provide a standard procedure for evaluating the robustness of the static rolling characteristics of flexible display panels.

8.2.5.2 Testing Device

The specimen should be firmly clamped with clamping components on the drum side and fixed side. As shown in Figure 8.4, the roller should alternately move forward and backward with a fixed distance, speed, and number of rolls. There is a slot on the roller side for inserting and clamping the edges of the specimen. The rollers are repeatedly shuttled along the roller shaft, and the specimen does not touch the equipment plate during the rolling test.

During testing, FPCs and drives should be handled with care to avoid distortion damage. In order to prevent twisting damage to the driver IC during the bend test, it is best to use the COF method.

8.2.5.3 Test Procedure

The rolling test should be performed as follows:

(a) prepare the required number of specimens;
(b) Initial performance, visual characteristics, testing, and recording of results on prepared samples;
(c) measure and record the dimensions of each specimen;
(d) As shown in Figure 8.4, one edge of the specimen is fixed to the roller, and the other edge fixture is supported by a separately defined load. Align the specimen with the top surface on the inside or outside of the roll;
(e) rolling specimens under specified conditions, such as rolling radius, rolling distance, rolling angular velocity, and load;
(f) keep the specimen in rolls for a period of time;
(g) restore the specimen to its initial state before rolling with the same angular velocity and opposite direction of rotation;
(h) If necessary, keep the specimen in a roll for a certain period of time;
(i) repeat cycles (e) to (h) for the specified number of times;
(j) Option to test the other side under the same conditions, as agreed between the supplier and the customer. In this case, another opposite surface is tested as required by (d) to (i);
(k) After the end of the test, remove the sample from the instrument;
(l) from (c) to (k), repeat the test with other specimens;
(m) measure the stress performance and electrical and optical properties of a mechanical stress module with the required number of tests and record the results. The initial performance and stress performance were compared, and the degree of change caused by the rolling test was determined;
(n) test and perform measurements on individual specimens and record them in the test report.

8.2.5.4 Test Conditions and Reports

The test conditions are specified as follows:

(a) Rolling radius
(b) Angular velocity, a one-cycle procedure
(c) Duration of Retention
(d) Load
(e) Number of repetitions
(f) Acceptance criteria
(g) Number of specimens
(h) Rolling inner surface (top or bottom)

(i) The manner in which the specimen is held and the detailed holding method, such as pressure
(j) Surface condition of specimens and instruments (e.g. handling)
(k) Testing of environmental conditions, such as temperature and humidity

8.2.6 Torsion Test

This test is suitable for evaluating the damage performance of flexible display modules after suffering damage for a certain period of time. For flexible displays, deformation is easy to occur during use.

8.2.6.1 Purpose

The purpose of this test is to provide a standard procedure for evaluating the robustness of flexible display devices to cyclic torsional stresses that may occur in applications.

8.2.6.2 Testing Device

During the test, the specimen should be firmly clamped and twisted at a certain torsion angle, as shown in Figure 8.5. The specimen should be firmly clamped using appropriate clamping components. During testing, the FPC and drive should be handled carefully to avoid distortion damage. The edge of the specimen where the FPC and driver IC are located on the fixed side should be fixed to prevent torsional damage during the flexure test. When the specimen is torsional, an additional tension is applied to the side of the specimen in addition to the torsional tension. Therefore, during the test, the moving parts on the stationary side should be moved back and forth, as the excess and unnecessary tensile stresses applied to the specimen are eliminated, as shown in Figure 8.5a and b.

8.2.6.3 Test Procedure

The torsion test should be performed as follows:

(a) prepare the required number of display modules;
(b) Initial performance, visual characteristics, testing, and recording of the results of the prepared module;
(c) measure, visually inspect, and record the dimensions of each specimen;
(d) As shown in Figure 8.5, the two edges of the test module are fixed to the fixture assembly;
(e) applying torque under limited conditions such as clockwise angle and oscillation speed;
(f) return the specimen to its initial state prior to testing with the same angular velocity and opposite direction of rotation;
(g) the application of torque under specified conditions, such as counterclockwise angle and oscillation speed;
(h) return the specimen to its initial state before the test with the same angular velocity and opposite direction of rotation;

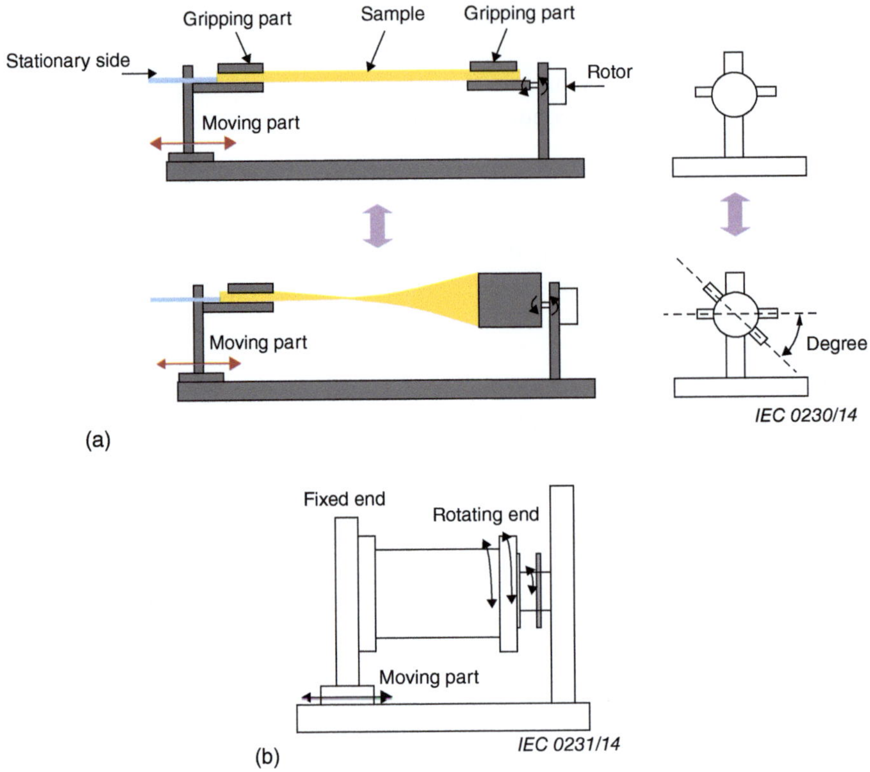

Figure 8.5 Instruments for various torsion tests. Source: Reproduced with permission from Ref. [26]; © 2018, International Electrotechnical Commission.

(i) repeat (e) to (h) or the prescribed number of cycles;
(j) There is an option to test the other side under the same conditions, as agreed between the supplier and the customer. In this case, on the other side of the surface from (d) to (i);
(k) complete the test and remove the sample from the instrument;
(l) from (c) to (k), repeat the test with other specimens;
(m) measure the stress, electrical and optical properties of a mechanical stress module with the required number of tests and record the results. The initial performance and stress performance were compared, and the degree of change caused by the torsion test was determined;
(n) Test and perform measurements on individual specimens and record them in the test report.

8.2.6.4 Test Conditions and Reporting

The test conditions are specified as follows:

(a) Number of specimens
(b) Performance range of initial and stressed electrical and optical properties
(c) Torsional angle and rotational angular velocity

(d) Tensile stress applied
(e) Number of cycles
(f) Acceptance criteria
(g) The way to hold the specimen and the detailed holding method, such as pressure
(h) Surface condition of specimens and instruments (e.g. handling)
(i) Testing of environmental conditions, such as temperature and humidity

8.2.7 Tensile Test

This test is suitable for evaluating the tension characteristics of flexible display modules after being subjected to constant tension at the moment of folding and unfolding, winding, and roll-out changes (see Figure 8.6).

8.2.7.1 Purpose

The purpose of this test is to provide a standard procedure to evaluate the robustness and cyclic tensile stress of flexible display devices, which can often occur in applications.

8.2.7.2 Testing Device

The device consists of a fixed and movable fixture in which the specimen will be fixed for testing. The movable fixture shuttles forward and backward linearly at a set speed, as shown in Figure 8.6. The instrument repeatedly applies and releases tension in the specimen over a set number of cycles. During testing, FPCs and drives should be handled with care to avoid tension damage. The edge of the specimen with FPC and driver IC on the fixed side should be fixed to avoid torsional damage during the bending test.

8.2.7.3 Test Procedure

The tensile test should be carried out as follows:

(a) prepare the required number of display modules;

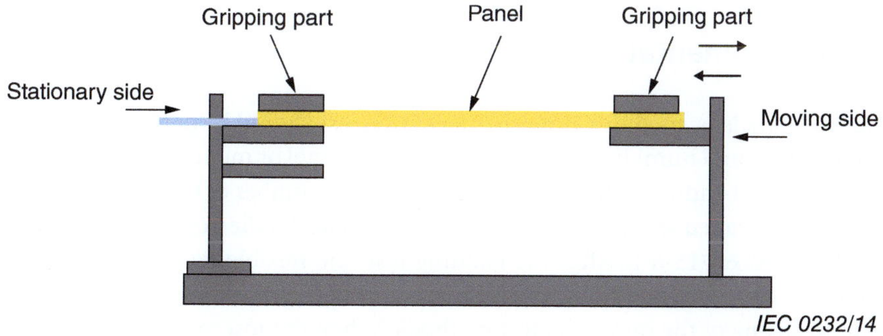

Figure 8.6 Tensile test device. Source: Reproduced with permission from Ref. [26]; © 2018, International Electrotechnical Commission.

(b) measure, visually inspect, and record the dimensions of each specimen;
(c) Initial performance, visual characteristics, testing, and recording of the results of the prepared module;
(d) As shown in Figure 8.6, the two edges of the test module are fixed by the clamping parts;
(e) the application of tension under specified conditions, such as tension or strain distance;
(f) return the specimen to its initial state before tension is applied;
(g) the tension is applied with the same force, the speed decreases, in the opposite direction;
(h) repeat the number of cycles specified by (e) to (g);
(i) after testing, remove the test module from the device;
(j) Measure the stress performance, and electrical and optical properties of the mechanical stress module, and the number of tests required, and record the results. Compare initial and stress properties and determine the degree of change caused by tension testing;
(k) Test and perform measurements on individual specimens and record them in the test report.

8.2.7.4 Test Conditions and Reports

The test conditions are specified as follows:

(a) The maximum tensile force or maximum deformation distance exerted on the test piece
(b) Tension application procedures, such as time-dependent tension and duration between tension application
(c) Number of cycles
(d) Acceptance criteria
(e) Number of specimens
(f) The way to hold the specimen and the detailed holding method, such as pressure
(g) Surface condition of specimens and instruments (e.g. handling)
(h) Environmental condition testing, such as temperature and humidity

8.3 Environmental Test Methods

For flexible panels or modules, a mechanical cycle test should be performed before the temperature and humidity test. First, the flexible panel or module should be subjected to a cyclic bending test. The bending radius and number of bending times are subject to the detailed specifications of the product. The detailed cyclic bending test is based on IEC 62715-6-1. After the bending test, the flexible panel or module is tested for temperature and humidity. Temperature and humidity test methods are often used as one of the durability test methods. When the test methods defined in this document are used as one of these durability test methods, the test cycle or conditions shall be reviewed in accordance with the appropriate requirements specified in the relevant specifications. If the product is curved, the temperature and humidity tests can be carried out directly. If necessary, the effect of simultaneous stress combinations between mechanical cycling tests and other tests should be evaluated according to the specification.

8.3.1 Storage at High Temperature

8.3.1.1 Purpose

The purpose of this test is to check the performance of flexible display panels and modules after high-temperature storage. The test procedure of IEC 60068-2-2 should be applied.

8.3.1.2 Test Conditions

The Test Bb (dry heat of non-heat-dissipating specimens with gradual temperature changes) of IEC 60068-2-2 shall be suitable for the following specific conditions:

(a) temperature

(100, 95, 90, 85, 80, 75, 65, 60, 55, 50, 45, 40, 35, 30) °C ± 3 °C

(b) duration

2, 16, 24, 48, 72, 96, 120, 168, 192, 240, 300, 500, and 1000 h

The temperature and duration of use should be indicated in the report.

8.3.2 Storage at Low Temperature

8.3.2.1 Purpose

The purpose of this test is to check the performance of flexible display panels and modules after low-temperature storage. The test procedure of IEC 60068-2-1 should be applied.

8.3.2.2 Test Conditions

Test Ab (Cooling of Non-Heat Dissipated Specimens with Dual Temperature Variation) of IEC 60068-2-1 shall be applied to the following specific conditions:

(a) temperature

(−50, −45, −40, −35, −30, −25, −20, −15, −10, −5, 0) °C ± 3 °C

(b) duration

2, 16, 24, 48, 72, 96, 120, 168, 192, 240, 300, 500, and 1000 h

The temperature and duration of use should be indicated in the report.

8.3.3 Temperature Change and Storage

8.3.3.1 Purpose

The purpose of this test is to examine the ability of flexible display panels and modules to withstand rapid temperature changes or a specific temperature change rate when the flexible display module is turned off. Test Na of IEC 60068-2-14 shall be suitable for rapid temperature changes, and test Nb shall be suitable for specified temperature change rates. The temperature and duration should be selected according to the specifications.

8.3.3.2 Rapid Temperature Change

(a) high temperature

(100, 95, 90, 85, 80, 75, 65, 60, 55, 50, 45, 40, 35, 30) °C ± 3 °C

(b) microtherm

(−50, −45, −40, −35, −30, −25, −20, −15, −10, −5, 0) °C ± 3 °C

(c) duration

2 h, 1 h, 30 min, 10 min

(d) The transition time is less than 3 minutes
(e) Cycle 5, 10, 20, 30 times

8.3.3.3 Specified Rate of Temperature Change

(a) high temperature

(100, 95, 90, 85, 80, 75, 65, 60, 55, 50, 45, 40, 35, 30) °C ± 3 °C

(b) microtherm

(−50, −45, −40, −35, −30, −25, −20, −15, −10, −5, 0) °C ± 3 °C

(c) duration

2 h, 1 h, 30 min, 10 min

(d) Rate of temperature change

(1 ± 0.2) °C/minute, (3 ± 0.6) °C/minutes or (5 ± 1)°C/minute

(e) Circulate 5, 10, 20, 30 times

8.3.4 Humidity and Heat, Steady State, and Storage

8.3.4.1 Purpose
The purpose of this test is to check the performance of flexible display panels or modules after storage at different temperatures and humidity levels. The test procedure of IEC 60068-2-78 should be applied.

8.3.4.2 Test Conditions
IEC 60068-2-78 applies to the following specific conditions:

(a) temperature

Table 8.3 Examples of humid heat and steady-state test conditions.

	Temperature (°C)	Humidity (% RH)	Duration (h)	Note
H-L	55	30	168	Hot and dry
H-H	60	90	120	Hot and humid
L-H	10	80	120	Cold and humid
L-L	20	10	240	Cold and dry

The temperature should be selected from the values given below, depending on the application.

(60, 55, 50, 45, 40, 35, 30, 25, 20, 15, 10) °C ± 3 °C

(b) Humidity

Humidity should be selected from the values given below, depending on the application.

(10, 15, 20, 30, 80, 85, 90, 93)% RH ± 3% RH

(c) Duration

The duration should be selected from the values given below, depending on the application.

2, 16, 24, 48, 72, 96, 120, 168, 192, 240, 300, 500, and 1000 h

8.3.5 Moist Heat, Circulation, and Storage

8.3.5.1 Purpose

The purpose of this test is to check the performance of the flexible display module under high humidity conditions, combined with cyclic temperature variations depending on the application. The test procedure of IEC 60068-2-30 should be applied.

8.3.5.2 Test Conditions

IEC 60068-2-30 applies to the following specific conditions:

(a) high temperature

The temperature should be selected from the values given below, depending on the application.

(60, 55, 50, 45, 40, 35) °C ± 3 °C

(b) Low temperature

(25 ± 3) °C

(c) Humidity

(93 ± 3) % RH

(d) Circulation

Depending on the application, the number of loops should be selected from the values given below:

1, 2, 5, 10, 20, 30, or 50

8.3.6 High-Temperature Operation

8.3.6.1 Purpose

The purpose of this test is to test the performance of flexible display panels and modules after working under high-temperature operating conditions. The test procedure of IEC 60068-2-2 should be applied.

8.3.6.2 Test Conditions

Test Bd or Be of IEC 60068-2-2 shall be carried out under the following specific conditions. The relevant specification should define the test to be used (Bd or Be), and depending on the application, a combination of temperature and operating time should be considered. Test Bd: Dry heat of a heat-dissipating specimen with a gradual change in temperature and not energized during regulation. Test Be: The dry heat of a heat dissipation specimen with a gradual change in temperature that requires electricity throughout the test.

(a) Temperature

$(80, 75, 70, 65, 60, 55, 50, 45, 40, 35, 30)\,°C \pm 3\,°C$

(e) Working hours

2 a.m., 4 a.m., 8 a.m., 12 p.m., 24 p.m., 48 a.m., 72 p.m., 96 p.m.

8.3.7 Low-Temperature Operation

8.3.7.1 Purpose

The purpose of this test is to examine the performance of flexible display panels and modules after operation under low-temperature operating conditions. The test procedure of IEC 60068-2-1 should be applied.

8.3.7.2 Test Conditions

Test Ad or Ae of IEC 60068-2-1 shall apply to the following specific conditions. The relevant specification should define the test (Ad or Ae) used. Depending on the application, a combination of temperature and operating time should be considered. Test Ad: A heat-dissipating specimen with a gradual change in temperature, energized after the initial temperature is stabilized. Test Ae: Cold heat dissipation sample, temperature gradually changes, need to supply power throughout the test.

(a) Temperature

$(-30, -25, -20, -15, -10, -5, 0)\,°C \pm 3\,°C$

(b) Operating time

2, 4, 8, 12, 24, 48, 72, 96 h.

8.3.8 Humidity and Heat, Steady State, Operation

8.3.8.1 Purpose

The purpose of this test is to check the performance of flexible display panels or modules after operation under high temperature and humidity operating conditions. The test procedure of IEC 60068-2-78 should be applied.

8.3.8.2 Test Conditions

IEC 60068-2-78 applies to the following specific conditions. There must be no condensation on the flexible display panel or module. Depending on the application, a combination of temperature and operating time should be considered.

(a) Temperature

The temperature should be selected from the values given below, depending on the application.

(60, 55, 50, 45, 40, 35, 30) °C ± 3 °C

(b) Humidity

Depending on the application, humidity should be selected from the values given below:

(10, 15, 20, 30, 80, 85, 90, 93)% RH ± 3% RH

(c) Duration

The duration should be selected from the values given below based on the app:
2, 4, 8, 12, 24, 48, 72, 96 h.

8.4 Mechanical Test Methods (Impact and Hardness Tests)

8.4.1 Scope

The purpose of this part of IEC 62715 is to define standard test methods to assess the mechanical stability of flexible display modules, particularly with respect to impact and stiffness. It includes displays such as liquid crystal displays (LCD), electronic paper and organic light-emitting diode (OLED) displays, and resistance to external forces applied to the panel.

8.4.2 Sample Preparation

The sample should be a display module because the final evaluation is based on the quality of the panel image, such as luminescence, color chroma, uniformity, line defects, and point defects. Mechanical robustness testing can lead to degradation of the image quality of the panel [17–25]. During impact and stiffness tests, the display module under test should be flat to obtain correct test results. The module should be fixed to the panel using other instruments or materials (e.g. adhesives or brackets).

Neither the adhesive nor the holder should interfere with the measurement. The flatness and size of the sample should be determined by the supplier and the customer. Temperature, humidity, and storage time (e.g. 24 hours) prior to sample preparation tests should be controlled and reported to obtain reproducible test results.

8.4.3 Ball Drop Test

This test is particularly suitable for assessing the mechanical robustness of flexible display devices by measuring the change in performance after the ball is dropped.

8.4.3.1 Purpose

The purpose of this test is to evaluate the mechanical robustness of the sample under the point impact that occurs when the flexible display product is actually used. The ball drop test can cover several typical characteristic parameters of display panel image quality. Typical parameters of display panel image quality may include luminescence, chromaticity, uniformity, line defects, and dot defects.

8.4.3.2 Testing Device

The ball drop test equipment consists of a metal (or rubber or plastic) ball and a hard plate, as shown in Figure 8.7. The specimen body should be placed on the hard panel. The test conditions for the ball drop test, such as the height of the ball from the panel (e.g. 5–10 cm), which can differ from the panel size (e.g. moving size, flat size, or larger), the type (mass) of the ball (e.g. metal, rubber, or plastic), the ball diameter, the weight of the ball (e.g. 10–50 g), the position of the panel at which the ball is struck during the ball drop test, as shown The characteristics of the panel image quality may depend on the above test conditions In order to obtain a consistent ball drop height during the test, special equipment can be used to drop the ball at a reproducible height, such as mechanical grips and magnetic plates. The different designs and materials of the hard panels below the specimen have a significant impact on

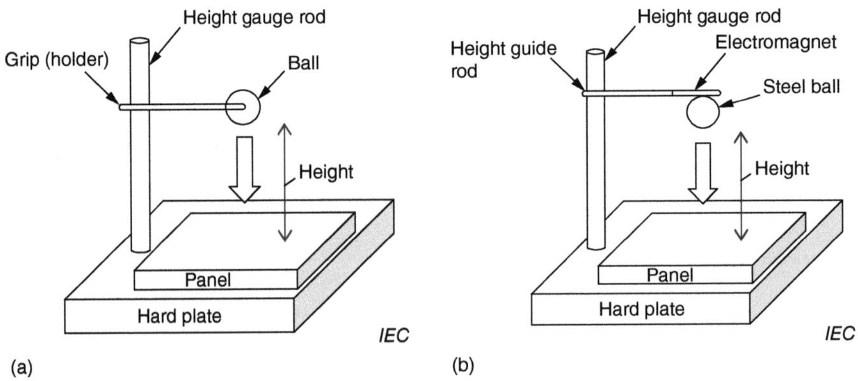

Figure 8.7 Ball drop test equipment setup. (a) Mechanical ball drop test and (b) electromagnetic assisted ball drop test. Source: Reproduced with permission from Ref. [26]; © 2018, International Electrotechnical Commission.

the test results. The design and material of the hard plate should be precisely mentioned or defined, as shown in Figure 8.7. An electrically assisted ball drop test can be performed in which the steel ball can be connected by a solenoid. After the first strike, the ball must not continuously hit the panel to prevent additional damage from the second strike, which could lead to inconsistent results.

8.4.3.3 Test Procedure

The ball drop test should be performed by dropping a metal (or rubber or plastic) ball at a constant height and striking the panel vertically. To properly assess the mechanical robustness of the panel, several locations on the panel should be selected equally (e.g. a total of nine points per panel and three points each in the left, center, and right areas) to represent the overall robustness of the panel; if necessary, the same ball drop test can be performed on multiple panels to obtain a more accurate overall performance. After the first hit, the ball must not hit the panel in succession to prevent additional damage from a second hit, which could lead to inconsistent results. The evaluation of the panel image can be performed sometime after the ball drop test (time interval), which should be mentioned.

8.4.4 Impact Test

This test is particularly suitable for evaluating the mechanical robustness of flexible display devices. The test is often referred to as the "impact test" because it simulates the "touch" or "push" of a finger.

8.4.4.1 Purpose

The purpose of this test is to evaluate the mechanical robustness of the specimen under repetitive external loads that represent repeated touches or pushes of a finger or stylus on a flexible display panel, as would occur in the actual use of a flexible display product. The hit test will cover several parameters typical of display panel image quality characteristics. Typical parameters of display panel image quality may include luminescence, chromaticity, uniformity, line defects, and point defects.

8.4.4.2 Test Equipment for Impact Testing

Impact measuring equipment is a device used to test the impact resistance of an object to evaluate the impact resistance of flexible electronic products in daily use by simulating the force and direction of impact in a real environment. In this paper, we will introduce the methods and importance of utilizing impact measurement equipment to test the impact performance of flexible electronic products.

First of all, a series of impact measurement equipment, such as impact table, impact hammer impact sensor, etc., need to be prepared before conducting impact testing. These equipments can simulate different impact conditions and forces, so as to make a comprehensive assessment of the impact performance of flexible electronic products. An impact test device is briefly described in Figure 8.8.

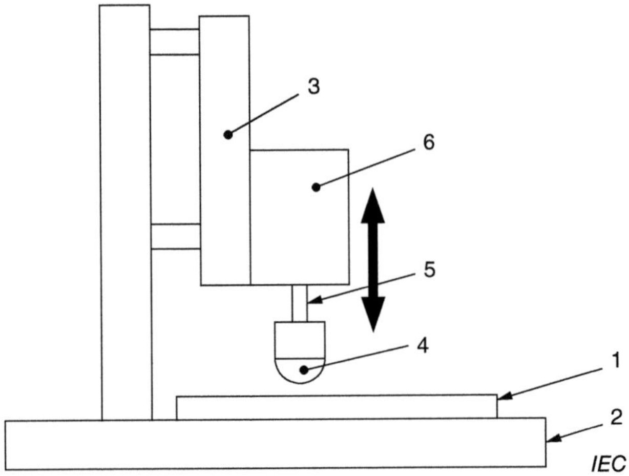

Key

1	Flexible display	4	Hitting tip
2	Hard plate	5	Hitting rod
3	Reciprocation mechanism	6	Force measuring device

Figure 8.8 Example of impact test device. Source: Reproduced with permission from Ref. [26]; © 2018, International Electrotechnical Commission.

8.4.4.3 Test Process

The impact test should be performed as follows:

(a) Selecting a suitable tip for the test, such as the material and shape of the tip.
(b) Selecting the location of the test specimen to be struck.
(c) The tip is set to the appropriate distance from the specimen.
(d) Reciprocating motion with the desired force. Figure 8.9 shows an example of tip movement and applied force:

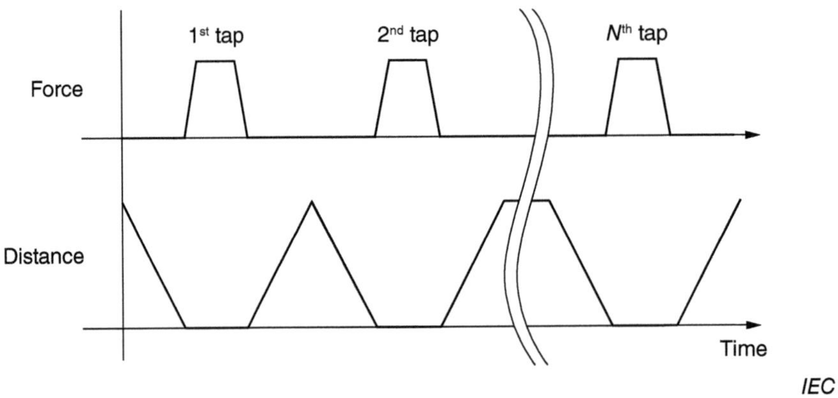

Figure 8.9 Example of tip motion and applied force. Source: Reproduced with permission from Ref. [26]; © 2018, International Electrotechnical Commission.

(e) Striking the specimen at the required period.
(f) Inspect the specimen for defects after impingement.
(g) Repeat (d) to (f), if required.
(h) If required, change the parameters, such as (a), (b), or (c).

8.4.5 Pendulum Side Impact Test

This test is particularly suitable for assessing the mechanical robustness of flexible display devices by measuring the change in performance after a side impact, since flexible display devices include very thin dielectric and metal layers that are very flexible but relatively weak against side impacts.

8.4.5.1 Purpose

The purpose of this test is to evaluate the robustness of the specimen under external side impact to represent the side impact environment, which would occur in the actual use of a flexible display product. The pendulum side impact test may cover several typical characteristic parameters of display panel image quality. Typical parameters of display panel image quality may include luminescence, chromaticity, uniformity, line defects, and point defects.

8.4.5.2 Testing Device

The pendulum side impact test equipment includes metal (or plastic) as the pendulum material, a ball (or cylindrical pendulum) as the shape of the pendulum, and a hard plate, as shown in Figure 8.10. The pendulum arm may be made of a rigid material such as metal. Clamps should be used to secure the specimen to the hard plate. The material of the pendulum arm attached to the ball or striking component should be rigid to reproduce a consistent impact force. Test conditions for the pendulum side impact test should be mentioned, such as angle of the ball (plastic or metal), type (mass), weight of the ball, diameter of the ball, length of the pendulum arm, distance from the ball drop position to the ball impact panel position,

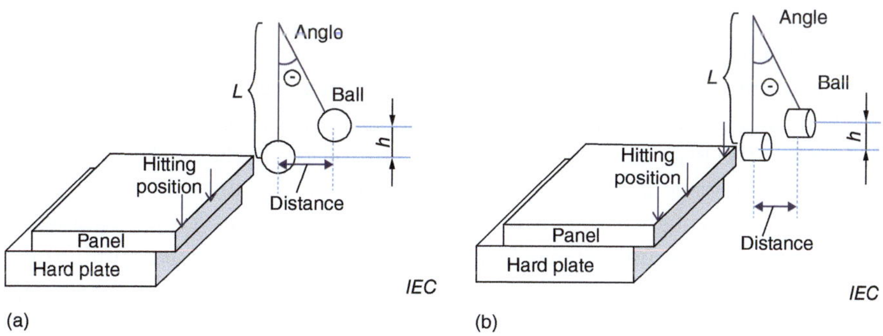

Figure 8.10 Pendulum side impact test device. (a) Ball type pendulum and (b) cylindrical type pendulum. Source: Reproduced with permission from Ref. [26]; © 2018, International Electrotechnical Commission.

strike position of the panel, and number of strikes, as the characteristics that show the image quality of the panel may depend on the above test conditions. To obtain better and more reliable results, several ball-striking positions should be tested. The kinetic energy of impact should be estimated for each test to better ensure repeatability. The kinetic energy can be calculated from the mass of the ball and the length of the swing arm. The calculated energy is the energy applied at impact, not the energy that causes damage, as energy will be lost due to pendulum bounce. Said pendulum shape should allow the use of other shapes, such as the long end of a wedge-shaped cylinder. This is because there are different shapes that can affect the panel during the manufacturing process, such as impact-locating pins.

Assuming that the sample edge is located at an angle of 0 in the pendulum arm, all the energy of the pendulum is kinetic energy at the time of impact. The total energy of the pendulum can be calculated using Eq. (8.1). For the method to be repeated, the pendulum should always deliver the same impact energy to the sample.

$$E = \frac{1}{2}mV^2 = mgh = mgL(1 - \cos\theta) \rightarrow V = \sqrt{2gL((1 - \cos\theta))} \tag{8.1}$$

where m is the mass of the pendulum and V is the velocity of the pendulum at $\theta = 0$.

8.4.5.3 Test Steps

A pendulum side impact test should be performed by dropping a ball from the side and striking the panel sideways. In order to properly assess the robustness of the panel, several locations of the panel should be equally selected to represent the overall robustness of the panel. If necessary, multiple pendulum side impact tests (e.g. $n = 2$–5 times) can be performed at the same location of the panel to obtain a more accurate average performance. Panel images can be evaluated at certain times after the pendulum side impact test.

8.4.6 Stylus Scratch Test

The stylus scratch test is designed to simulate scratches from sharp or hard objects that may occur in actual use. However, in contrast to pencil hardness testing, stylus scratch testing is not sensitive to differences in the hardness of surface materials.

8.4.6.1 Purpose

The purpose of this test is to evaluate the scratch resistance of the sample when the flexible display product is actually used. Push the stylus onto the surface. Gradually increase the load until visible defects appear on the surface. The lowest load at which any damage is observed should be reported. This test may cover several parameters typical of display panel image quality characteristics. Typical parameters of display panel image quality may include luminescence, chromaticity, uniformity, line defects, and point defects.

8.4.6.2 Testing Device

The instrument (Type A) specified in ISO 19252 shall be used (see Figure 8.11a)). The test procedure shall be in accordance with ISO 19252. Another instrument (Type B)

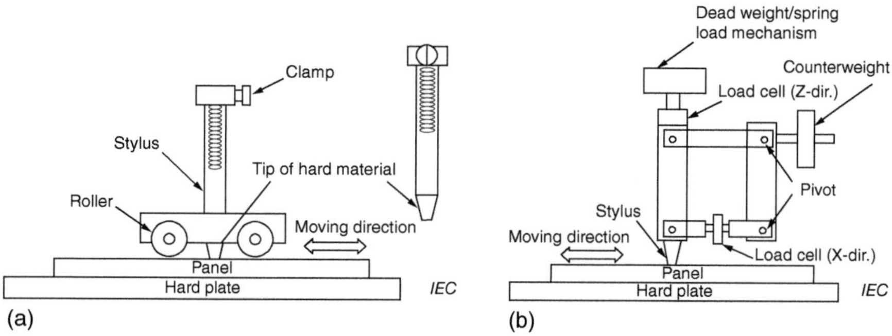

Figure 8.11 Instrument for stylus scratch testing using stylus holders with different pressures by adjusting the spring load. (a) Stylus scratch test (Type A). (b) Stylus scratch test (Type B). Source: Reproduced with permission from Ref. [26]; © 2018, International Electrotechnical Commission.

as specified in ASTM D7207-13 shall also be used (see Figure 8.11b)). The stylus shall be made of a hard material such as diamond, sapphire, or tungsten alloy. Test results vary depending on the material of the stylus; therefore, it is not possible to compare test results for different stylus materials.

8.4.6.3 Test Steps

The radius of the stylus should be recorded (e.g. radius ranges from 0.2 to 1 mm), and the stylus should be moved a fixed number of times in one direction at a constant distance. The experiment should start with a lower load to a higher load (N) (e.g. from 1 to 10 N). The load should be increased until any damage to the surface is observed. The angle of the stylus to the surface of the specimen should be normal ($90° \pm 1°$) and the speed of the scratch should be recorded (e.g. speed ranges from 5 to 10 mm/s). The stylus testing device should be controlled manually or automatically, and the stylus should be cleaned before each test. Depending on the agreement between the supplier and the customer, other test conditions may be used; however, any deviations from the standard test conditions described above should be reported. The minimum load at which any damage occurred shall be reported. Any deviations from the standard test conditions shall be reported, as well as the material of the stylus.

8.4.7 Steel Wool Wear Test

The wire wool wear test is typically used to evaluate surface damage to display panels that occur in actual use.

8.4.7.1 Purpose

The purpose of this test is to evaluate the abrasion resistance of the specimen when the flexible display product is actually used. The surface of the specimen should be abraded using a specified load of steel wool. The surface is abraded a specified number of times, and the robustness of the surface is evaluated after these abrasions. This test will result in visible defects such as scratches and dents and may include several

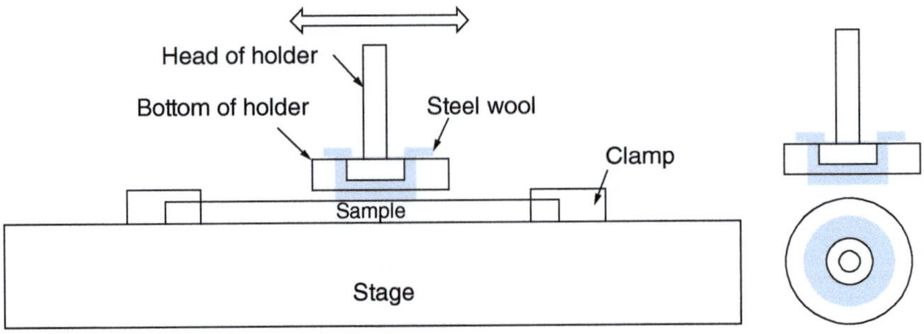

Figure 8.12 Steel wool wear test device. Source: Reproduced with permission from Ref. [26]; © 2018, International Electrotechnical Commission.

parameters typical of the image quality of the display panel. Typical parameters of display panel image quality may include luminescence, chromaticity, uniformity, line defects, and point defects.

8.4.7.2 Testing Device

Wire wool No. 0000 should be used as an abrasive. The steel wool shall be attached to the wear head and tightly interlocked between the top and bottom of the holder (see Figure 8.12). The bottom surface of the wear head should be flat and should be round or square. The area of wear should be recorded (e.g. area ranges from 0.8 to 5.0 cm^2).

8.4.7.3 Test Procedure

On the specimen, a piece of steel wool is placed inside the weight, and the surface is wiped horizontally at a constant speed (e.g. in the speed range of 40–120 mm/s). The steel wool is firmly fixed between the top and bottom of the gripper by inserting it firmly into the narrow gap between the bottom of the gripper and the head of the gripper. The steel wool must not come loose during the test. The scrubbing length should be recorded (e.g. a length range of 60–100 mm). The load per unit area should also be reported (e.g. load range 1–20 N/cm^2), where area (cm^2) is the area of the wear head. The number of wear cycles should be recorded (e.g. number of wear cycles from 1 to 100,000). The test can be stopped when any damage is observed. Depending on the agreement between the supplier and the customer, other test conditions may be used; however, any deviations from the standard test conditions described above should be reported. The applied load (N) and the number of wears that resulted in visible defects shall be reported. All testing processes shall be automated to obtain repeatable and reliable results. The steel wool should be replaced after each test to obtain a correct evaluation.

References

1 IEC 60027(All parts). Alphabetic symbols used in electrical technology.

2 IEC 60050-845. International electrotechnical Terminology, Part 845: Lighting.

3 ISO 80000-1. Quantities and units - Part 1: General provisions.

4 ISO 9241-302:2008. Human-system interaction ergonomics. Part 302: Terminology for electronic visual displays.
5 Suo, Z. et al. (1999). Mechanics of rollable and foldable film-on electronics. *Applied Physics Letters* 74 (8): 1177–1179.
6 Gleskova, H., Wagner, S., and Suo, Z. (1999). Failure resistance of amorphous silicon transistors under extreme in-plane strain. *Applied Physics Letters* 75 (19): 3011–3013.
7 Gleskova, H. et al. (2002). Electrical response of amorphous silicon thin-film transistors under mechanical strain. *Journal of Applied Physics* 92: 6224–6229.
8 Grego, S. et al. (2007). A method to evaluate mechanical performance of thin transparent films for flexible displays. *Thin Solid Films* 515: 4745–4752.
9 Menard, E., Nuzzo, R.G., and Rogers, J.A. (2005). Bendable single crystal silicon thin film transistors formed by printing on plastic substrates. *Applied Physics Letters* 86 (9): 093507.
10 Chen, Z. et al. (2001). A mechanical assessment of flexible opto-electronic devices. *Thin Solid Films* 394: 202–206.
11 Servati, P. and Nathan, A. (2005). Functional pixel circuits for elastic AMOLED displays. *Proceedings of the IEEE* 93 (7): 1257–1264.
12 Rolland, A. et al. (1993). Electrical properties of amorphous silicon transistors and MIS-devices: comparative study of top nitride and bottom nitride configurations. *Journal of the Electrochemical Society* 140 (2): 3679–3683.
13 Shur, M.S. et al. (1997). SPICE models for amorphous silicon and polysilicon thin film transistors. *Journal of the Electrochemical Society* 144 (8): 2833–2839.
14 Cairns, D.R. and Crawford, G.P. (2005). Electromechanical properties of transparent conducting substrates for flexible electronic displays. *Proceedings of the IEEE* 93 (8): 1451–1458.
15 IEC 62341-6-1. Part 6-1: Methods for measuring optical and electro-optical parameters. Organic light-emitting diode (OLEO) displays.
16 Electronic paper displays - Part 4-2: Environmental test methods.
17 Suo, Z., Ma, E.Y., Gleskova, H., and Wagner, S. (1999). *Applied Physics Letters* 74: 1177.
18 Gleskova, H., Wagner, S., and Suo, Z. (1999). *Applied Physics Letters* 75: 3011.
19 Gleskova, H., Wagner, S., Soboyejo, W., and Suo, Z. (2002). *Journal of Applied Physics* 92: 6224.
20 Grego, S., Lewis, J., and Temple, D. (2007). *Thin Solid Films* 315: 4745.
21 Menard, E., Nuzzo, R.G., and Rogers, J.A. (2005). *Applied Physics Letters* 86: 093507.
22 Chen, Z., Cotterell, B., Wang, W. et al. (2001). *Thin Solid Films* 394: 202.
23 Servati, P. and Nathan, A. (2005). *Proceedings of IEEE* 93: 1257.
24 Rolland, A., Richard, J., Kleider, J.P., and Mencaraglia, D. (1993). *Journal of the Electrochemical Society* 140: 3679.
25 Shur, M.S., Slade, H.C., Jacunski, M.D. et al. (1997). *Journal of the Electrochemical Society* 144: 2833.
26 (2018). *IEC 62715-6-1, Flexible Display Devices.* International Electrotechnical Commission.

9

Analysis of Flexible Electronic Packaging Enterprise

Zhenrong Wei

Advanced Materials School, Peking University Shenzhen Graduate School, Lishui Road, 518055 Shenzhen, China

9.1 Flexible Electronic Packaging Enterprise

9.1.1 Samsung SDI-Korea

Samsung SDI is a leading manufacturer of batteries and electronic materials based in Yongin, Gyeonggi Province, South Korea. It operates through two divisions: Energy Solutions and Electronic Materials. The Energy Solutions division produces rechargeable batteries for IT equipment, automobiles, and energy storage systems (ESS), while the Electronic Materials division produces materials for semiconductors and displays. Samsung SDI is actively developing its secondary battery business for electric vehicles, IT, and ESS, as well as its electronic materials business for semiconductors and displays. Its employees strive to achieve the highest quality and profitability advantages through superior technology and qualitative growth. Moreover, Samsung SDI is dedicated to fulfilling its social responsibility for the environment and safety through Environmental, Social, and Governance (ESG) management, as well as strengthening social contribution and symbiotic cooperation. It also provides ESS services [1] (Figure 9.1).

In the organic light-emitting diode (OLED) industry, the production of thin-film encapsulation (TFE) material serves to safeguard OLED materials from external moisture and oxygen, functioning as an encapsulation material.

9.1.1.1 Product Appearance

The product has a transparent appearance like water, and a large amount of the product is transparent yellow in color. Samsung SDI's TFE products are broadly divided into products for vapor deposition and inkjet and are currently widely used in flexible devices.

9.1.1.2 Business History

2010–2011: Development of organic layer for thin-film packaging materials
2012: SDI TFE monomer received customer approval

Flexible Electronic Packaging and Encapsulation Technology, First Edition.
Edited by Hong Meng and Wei Huang.

Figure 9.1 Samsung SDI product structure diagram. Source: Reproduced with permission from https://www.samsungsdi.com.cn.

2014: SDI TFE monomer (for vapor deposition A) approved by customer/mass production/development of SDI TFE monomer (for vapor deposition B)

2015: SDI TFE monomer (B for vapor deposition) approved by customer/volume production inkjet process (C for inkjet) approved/volume production.

9.1.1.3 Product Features

Samsung SDI succeeded in mass production of TFE organic material for flexible screens for the first time in the world and has the only mass-produced product in the world. The performance is excellent in terms of the meticulousness of the evaporation and inkjet process and the reliability of the panel products (Figure 9.2).

Glass encapsulation materials → TFE materials for flexible screens (e.g. smartphones, smartwatches, etc.).

9.1.1.4 Product Specifications

Some Patents

CN108249764 A Composition for Encapsulating an Organic Light-Emitting Diode and a Display Prepared Using the Same [2].

The present invention provides a composition for encapsulating an OLED and a display prepared therefrom. The composition comprises non-silicone

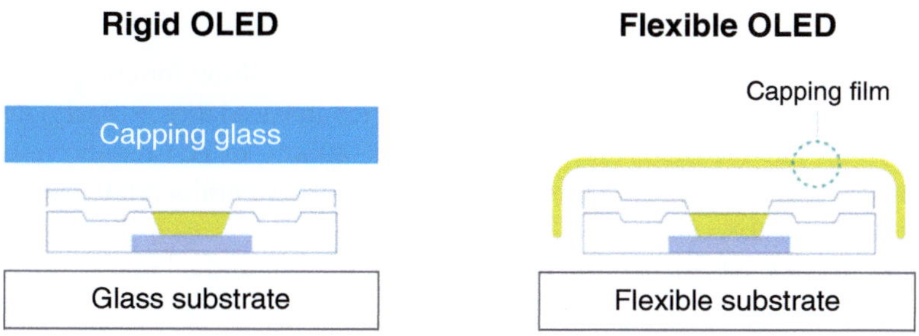

Figure 9.2 Organic and inorganic multilayer structure. Source: Reproduced with permission from https://www.samsungsdi.com.cn.

di(meth)acrylate (A) in a range of 10–70 wt%, silicone di(meth)acrylate (B) in a range of 20–70 wt%, mono(meth)acrylate (C) in a range of 5–40 wt%, and initiator (D) in a range of 1–10 wt%, wherein the (B) silyl di(meth)acrylate is represented by Figure 9.3. The composition exhibits a high light curing rate, leading to improved light transmission through the organic barrier layer (Table 9.1).

Content of the Invention
Technical Problem

This invention presents a composition for encapsulating an OLED, allowing an organic barrier layer with elevated plasma resistance to bolster the reliability of the OLED.

Another aspect of the present invention provides a composition for encapsulating an OLED, said composition enabling an organic barrier layer exhibiting high transmittance as well as high light curing rate.

A further aspect of the present invention provides a composition for encapsulating an OLED, said composition facilitating the formation of an organic barrier layer by using, for example, deposition, ink jet printing, and the like.

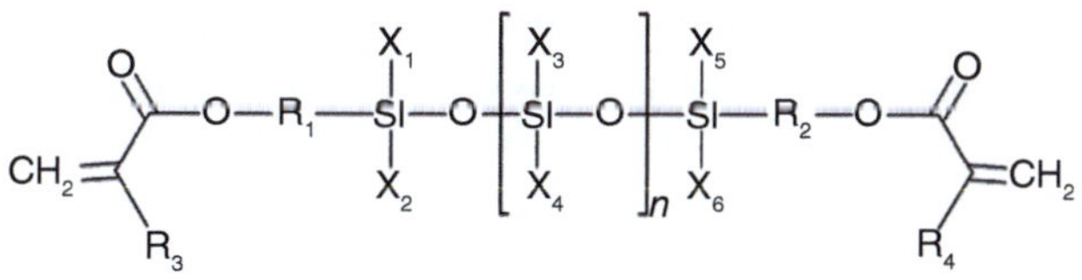

Figure 9.3 Silyl-di(meth)acrylate.

Table 9.1 SDI TFE grades.

Grade	A grade for evaporation B grade for evaporation	C grade for inkjet
Tec	TFE organic layer evaporation process	TFE organic layer inkjet process

Technical Solutions

According to an aspect of the present invention, the composition for encapsulating an OLED may comprise: 10–70 wt% of (A) non-silicone-based di(meth)acrylate; 20–70 wt% of (B) silicone-based di(meth)acrylate; 5–40 wt% of (C) mono(meth) acrylate; and 1–10 wt% of (D) initiator, wherein (B) silyl-di(meth)acrylate can be represented by Figure 9.3:

(wherein R1, R2, R3, R4, X1, X2, X3, X4, X5, X6, and *n* are described in the following detailed description).

Another aspect of the present invention provides that an OLED display may consist of: an OLED; and a barrier stack formed on the OLED, which consists of an organic barrier layer and an inorganic barrier layer, wherein the organic barrier layer may be formed using a composition for encapsulating an OLED.

Beneficial Effects

The present invention provides a composition for encapsulating an OLED, wherein the composition can achieve an organic barrier layer exhibiting plasma resistance. As a result, the OLED exhibits enhanced reliability.

The present invention provides a composition for encapsulating an OLED, wherein the composition exhibits a high light-curing rate. As a result, the organic barrier layer exhibits enhanced light transmission.

Further, the present invention provides a composition for encapsulating an OLED, wherein the composition facilitates the formation of an organic barrier layer by, for example, deposition, ink jet printing, etc.

9.1.2 LG Chem-Korea

LG Chem has three main business and product areas: Basic Materials and Chemicals; Information Technology and Electronic Materials; and Energy Solutions [3].

9.1.2.1 Basic Materials and Chemicals

LG Chem is a supplier of petrochemical products ranging from base distillates to specialty polymers. For example, it is a large producer of common plastics such as acrylonitrile butadiene styrene (ABS), styrene-acrylonitrile resin (SAN), and polyvinyl chloride (PVC). It also produces raw materials and fluids, including plasticizers, specialty additives, alcohols, polyolefins, acrylics, synthetic rubbers, styrenics, high-performance polymers, engineering plastics, elastomers, conductive resins, and other chemicals.

9.1.2.2 Information Technology and Electronic Materials

LG Chem supplies display and optical films, printed circuit materials, and toner. It also provides LCD polarizers, which are multilayer films applied to the top and bottom surfaces of TFT-LCD panels to allow light from the backlight unit to pass through the panel, and 3DFPR (thin-film patterned delayers) films, which allow three-dimensional viewing.

9.1.2.3 Energy Solutions

LG Chem completed development and began mass production of Korea's first lithium-ion battery in 1999. By the end of 2011, LG Chem was the third-largest manufacturer in the world with an annual production capacity of 1 billion cells. It is also a supplier of automotive batteries for electric vehicles, such as the Focus, Chevrolet Volt, and Renault ZOE. The plant can produce enough batteries each year to make 50,000–200,000 packs for electric and hybrid vehicles, such as GM's Chevrolet Volt and Ford Focus Electric, and other automakers' upcoming plug-in electric vehicles. Its research and development arm, LG Chem Power, is located near Troy, Michigan. LG Chem Power and LG Chem Michigan began as a company called Compact Power, Inc. that made batteries in South Korea and later switched to batteries made at LG Chem Michigan's Dutch plant.

In terms of packaging materials, LG Chem's FSPM film is an OLED sealing material that prevents damage and wetting of water and oxygen-intolerant OLED components and protects the components from physical impacts. This has led to epoch-making improvements in component life for OLED displays (Figure 9.4).

Temporary process film (TPF) is a protective film used to protect OLED components and sealing layers in the manufacturing process of plastic OLED (POLED) panels (Figure 9.5; Table 9.2).

Characteristics of TPE.

(1) Low adhesion without damaging OLED panels.
(2) good wettability when the film is attached.
(3) Minimal residue after removal of the film.

Some Patents

CN106463649 B Encapsulation Film and Method of Encapsulating Organic Electronic Devices Using the Encapsulation Film [4].

The present application provides an encapsulation film, a product for encapsulating an organic electronic device (OED) using the encapsulation film, and a method for encapsulating an OED. This encapsulation film prevents moisture or oxygen from entering the OED from the external environment, prevents bonding failure and damage to the organic film due to volume expansion caused by a reaction between moisture and hygroscopic agents, and provides a high level of reliability due to the OED's increased life and durability.

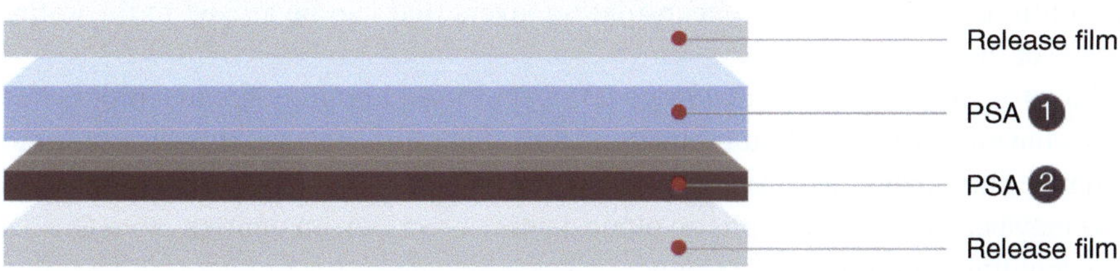

Figure 9.4 OLED encapsulation film (FAS film) structure. Source: Reproduced with permission from https://www.lgchem.com.

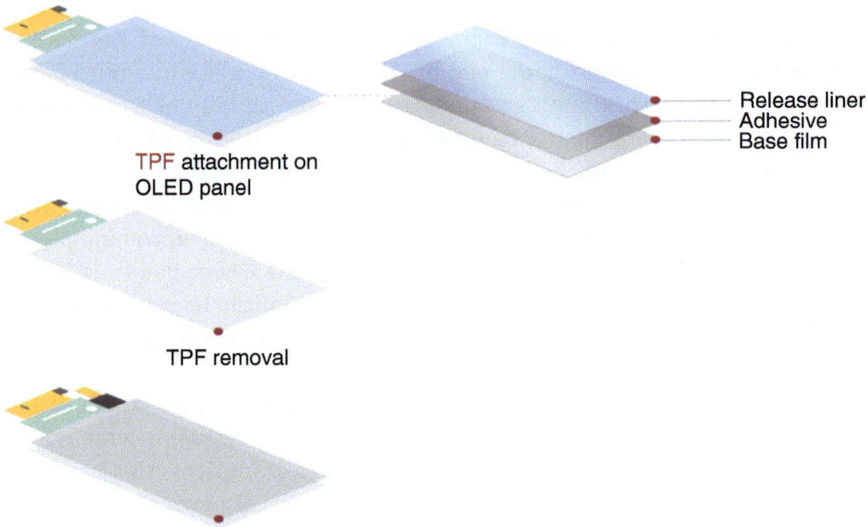

Figure 9.5 The use of TPF in POLED panels. Source: Reproduced with permission from https://www.lgchem.com.

Table 9.2 Product classification.

Grade	Features	Application
LTF-C series	* Top-end OLEDTPF protection film * Be capable of fitting cell's dimensions (MaxHalf6G)	Plastic OLED panel
LTF-B series	* Decrease PI's process film * Be capable of fitting Cell's dimensions (MaxHalf6G)	Plastic OLED panel

Source: Reproduced with permission from https://www.lgchem.com.

Contents of the Invention

One aspect of the present application provides a film for encapsulating OEDs. According to an exemplary embodiment of the present application, said encapsulating film includes multiple encapsulation layers that can be affixed to the entire surface of the OED.

The term "organic electronic device (OED)" as used herein refers to a product or device having a structure that includes a layer of organic material that uses holes to exchange charge with electrons between a pair of electrodes facing each other. Said OED may be, for example, a photovoltaic device, a rectifier, an emitter, or an OLED, but is not limited thereto.

In an exemplary embodiment of the present application, said OED may be an OLED.

According to an exemplary embodiment of the present application, there is provided an encapsulation film comprising an encapsulation layer, said encapsulation layer comprising a moisture barrier layer and a crack barrier layer. Said encapsulating film may comprise at least one moisture barrier layer and at least one crack barrier layer, further comprising a separate layer in addition to said moisture barrier layer and crack barrier layer.

In one embodiment, the encapsulation film according to the present application is an encapsulation film for encapsulating OEDs comprising a crack barrier layer having a tensile modulus of 0.001–500 MPa and a moisture barrier layer having a tensile modulus of 500–1000 MPa, and said crack barrier layer may be formed to contact the OEDs when encapsulating the OEDs. That is, said fracture barrier layer may be formed to encapsulate the entire surface of the OED when encapsulating the OED. The tensile modulus of said fracture barrier layer may be lower than the tensile modulus of the moisture barrier layer.

The materials used for the fracture barrier and the moisture barrier according to the present application are not particularly limited, as long as they meet the range of said tensile modulus. Unless otherwise specifically limited, the tensile modulus used herein is the tensile modulus measured at a temperature of 25 °C. Alternatively, unless otherwise specifically limited, tensile modulus as used in this specification may refer to the tensile modulus measured after curing of the curable component. In one embodiment, said tensile modulus may refer to the tensile modulus measured after curing for about 120 minutes at about 100 °C, the tensile modulus measured after exposure to UV radiation at a radiation dose of about U/cm^2 or more, or the tensile modulus measured after additional heat curing after UV radiation.

0.001–440 MPa, 0.001–430 MPa, 0.001–420 MPa, 0.001–410 MPa, 0.001–400 MPa, 0.05–450 MPa, 0.1–450 MPa, 0.2–450 MPa, 0.3–450 MPa, 0.5–450 MPa, or 0. Alternatively, said moisture barrier layer may have a tensile modulus at room temperature of 500–1000 MPa, 530–1000 MPa, 550–980 MPa, 580–950 MPa, 600–930 MPa, 630–900 MPa, or 650 MPa to As described above, it is preferred for large installations where the tensile modulus of said crack barrier is lower than the tensile modulus of the moisture barrier, and where the ratio between the moisture absorbent layers, as will be described below, can be controlled to provide effective water resistance to said membrane. As used herein, the term "hygroscopic agent" refers to a material that chemically prevents moisture and water vapor from penetrating into the said encapsulated membrane. Generally, when a hygroscopic agent reacts with moisture in a membrane, stress is generated as the hygroscopic agent reacts with the moisture to expand in volume. Thus, when said hygroscopic agent does not have an elasticity sufficient to reduce the swelling stress of removing moisture, said membrane can be peeled off from the adhesion. Poor interfacial adhesion of any of said multilayer encapsulation layers to the organic diode facestock may result in failure or rupture of the interlayer. For example, peeling due to stress can be avoided when the elastic modulus of said encapsulation layer is controlled at a low level. That is, when two layers with different elastic moduli are stacked and the crack barrier layer with a low elastic modulus is configured to be in contact with the OED, the crack barrier layer can be stress-reduced despite the reaction of the moisture-absorbing agent of the crack barrier layer with moisture, thereby generating stress. In addition, other physical properties of said film, such as durability, can be met.

In an exemplary embodiment of the present application, the water vapor transmission rate (WVTR) of said encapsulation layer measured at 100 °F and 100% relative humidity along the thickness direction of the membrane (prepared as a 100 μm thick membrane) is below 50, below 40, below 30 g/m^2/day or less, 20 g/m^2/day or less, or 10 g/m^2/day or less. Said encapsulation layer is formed to have such a WVTR that, when applied to the encapsulation structure of an electronic device can effectively prevent the penetration of moisture or oxygen from the external environment, thereby obtaining an encapsulation structure that can stably protect the diode. The lower limit may be, but is not limited to, for example, 0 g/m^2/day, because of the low WVTR, which can exhibit excellent water resistance.

The term "encapsulation layer" as used herein may refer to the various layers that make up the encapsulation film. That is, the term "encapsulation layer" as used herein may refer to various moisture barrier layers, crack barrier layers, and laminating layers, as will be described below, as well as multilayer structures comprising all of the said layers.

Said encapsulating films may have various structures, including structures in which the fracture barrier layer is disposed on one surface of the moisture barrier layer and structures in which the fracture barrier layer is disposed on both surfaces of the moisture barrier layer. Here, among the structures in which the fracture barrier is provided on one or both surfaces of the moisture barrier layer, there may be structures in which the fracture barrier is directly attached to the moisture barrier layer and structures in which the fracture barrier is indirectly attached to the moisture barrier layer through a barrier layer. As used herein, the term "encapsulating resin" refers to the ferrous base resin that forms the moisture barrier layer, the crack barrier layer, or the laminate. For example, said fracture barrier may comprise an encapsulating resin with a glass transition temperature of 0 °C or less. Alternatively, said moisture barrier layer may contain an encapsulating resin with a glass transition temperature of 85 °C or more. However, this application is not limited to the above scope, but as long as the aforementioned modulus of elasticity is met, said fracture barrier layer can contain an encapsulating resin with a glass transition temperature of 85 °C or more, or said moisture barrier layer can contain an encapsulating resin with a glass transition temperature of 0 °C or less. The OED can be protected from chemical or physical damage. The said encapsulation layer can have a multilayer structure by separating the crack barrier layer from the moisture barrier layer by any of the methods known in the art for forming multilayer encapsulation layers without limitation. In addition, the types and contents of the components of the crack-blocking layer, the moisture-blocking layer, and the laminated layer constituting said encapsulating layer, such as the encapsulating resin, the moisture-absorbing agent, the tackifier, the curing agent, other additives, or the filler, may be the same or different from each other.

That is, any well-known material that provides an encapsulation layer satisfying the abovementioned tensile modulus as a component of the encapsulation layer can be used without limitation. Alternatively, materials that do not meet said tensile modulus when alone but do meet said tensile modulus when combined may be used as components of the encapsulation layer.

9.1.3 3M-United States

Founded in 1902, 3M's global headquarter is in St. Paul, Minnesota, United States of America. As a world-renowned diversified technology innovator, 3M products and technologies have been deeply integrated into people's lives for more than 100 years, developing more than 60,000 products in a wide range of fields, from housewares to medical products, from transportation and construction to business, education, electronics, and communications. Globally available products.

The first exclusive product, Three-Mite abrasive cloth, was introduced in 1914.

In 1921 the world's first – waterproof abrasive sandpaper, Wetordry, was born and registered a patent, creating a whole new era for industrial grinding.

In 1924, 3M began formal product development.

In 1939, 3M invented the world's first reflective traffic signs and world's first audio cassette tape, especially the synthetic rubber soles used by the American astronaut Armstrong to step on the moon also came from 3M.

3M also invented Scotch masking tape, Scotch cellophane tape, Scotch vinyl electrical insulation tape, replicable diaper tape, and other innovative products.

Especially the Post-it notes, which were born in 1980, have brought about a revolutionary change in information communication. The emergence of an endless stream of innovative products testifies to each of 3M's advances. For example, Post-it, Scotchgard, Scotch, Thinsulate, Filtrate, Scotch-Brite, Nomad, Dyneon, and Nexcare are continuously, one after another, introducing a variety of products under one famous brand name to the world, influencing and changing the way of life and habits of human beings [5].

3M has developed more than 60,000 products, with an average of 500 useful products each year. As a world-renowned diversified technology company, 3M products abound from handy newspaper stickers to traffic signs and colorful advertising films, from household products to medical supplies, and from cell phones to laptops to cars. 3M can be found in almost every field (Figures 9.6 and 9.7; Tables 9.3–9.7).

Some Patents

CN103079816 B Blocking Assembly with Encapsulant and Photovoltaic Cell [6].

The present invention relates to an assembly, said assembly comprising an encapsulant, a first polymeric film substrate disposed on said encapsulant, a barrier film disposed on said first polymeric film substrate, a pressure-sensitive adhesive layer disposed on said barrier film, and a second polymeric film substrate disposed on said pressure-sensitive adhesive. Said first polymeric film substrate has a first coefficient of thermal expansion. Said second polymeric film substrate has a second coefficient of thermal expansion higher than said first coefficient of thermal expansion. Said second polymeric film substrate is generally resistant to UV degradation.

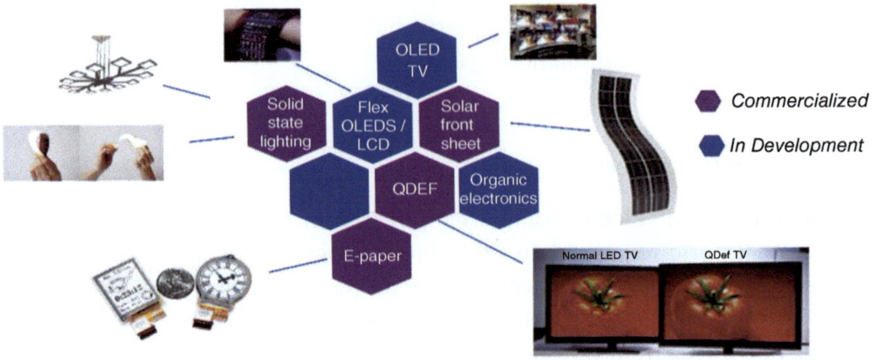

Figure 9.6 3M flexible packaging market. Source: Reproduced with permission from 3M / https://www.3m.com.cn/3M/zh_CN/company-cn/?utm_medium=redirect&utm_source=vanity-url&utm_campaign=www.3m.com.cn / last accessed 02 August 2023.

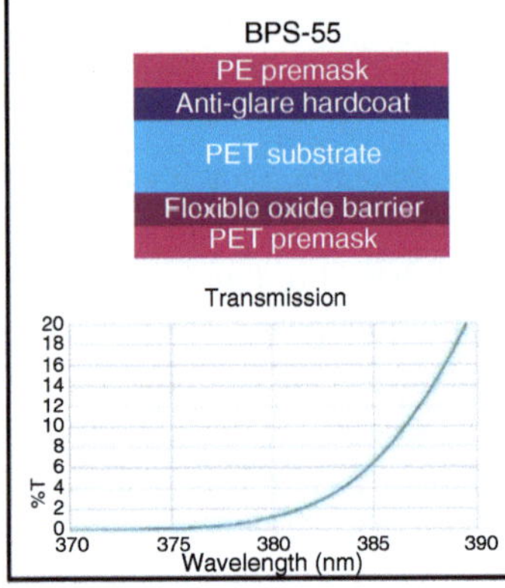

Physical and optical characteristics

Film property	3M BPS-55
Thickness	55 μm
WVTR specification*	$<1 \times 10^{-3}$ g/m²/day, 20°C/100%RH
WVTR, typical values†	$<1.2 \times 10^{-5}$ g/m²/day 20°C/100%RH
Pencil hardness	≥3H
Transmission	>88%
Transmission @ 380 nm	<3%
Haze	6%

Anti-Glare coating
Increased Contrast Ratio over BPS-270

Figure 9.7 Hydrogen barrier protective film (BPS-55). Source: Reproduced with permission from https://www.3M.com.cn.

Table 9.3 Package product line.

Flexible Transparent Barrier (FTB3)	Barrier Protection Sheet (BPS)
• WVTR: $6.2 \times 10^{-6} - 1.2 \times 10^{-5}$ g/m²/day Extrapolated from 50°C/100% RH → 20 C/100% RH • Barrier coated Optical grade PET based films • Typically used within a stack/device • Barrier Elongation: up to 2%, 200k cycles	• FTB3 family features, plus: • Front surface protection: Anti-Glare Hard coat: >3H UV blocking: <3% @ 380 nm

Source: Reproduced with permission from https://www.3M.com.cn.

Table 9.4 List of encapsulated films.

	FTB3-50	FTB3-UV-50[b)]	FTB3-125[a)]	BPS-270	BPS-55[b)]
Thickness	50 μm	50 μm	125 μm	270 μm	55 μm
WVTR ($g/m^2/day$) @20C/100%RH (release criteria)	$<1 \times 10^{-3}$ $g/m^2/day$	$<1 \times 10^{-3}$ $g/m^2/day$	$<1 \times 10^{-3}$ $g/m^2/day$	$<1 \times 10^{-3}$ $g/m^2/day$	$<1 \times 10^{-3}$ $g/m^2/day$
Flexibility	No damage (2 mm mandrel and 100k bends); (2% elongation and 200k cycles)	No damage (2 mm mandrel and 100k bends). (2% elongation and 200k cycles)	Flexible	Flexible, but boardy	Flexible
Haze	<2.5%	<2.5%	<2.5%	9 ± 3%	6 ± 2%
Gloss60	N/A	N/A	N/A	65	~85
Transmission	>88%	>88% (<3% at 380 nm)	>90%	>88% (<2% at 380 nm)	>88% (<3% at 380 nm)
Color (b*)	<1	<1	<1	1–2	<1.5
Other functionality	Heat Stable		Heat Stable	Anti-GlareHard coat (>3H)	Anti-Glare Hard coat (>3H)
Est. Commercial timing	Commercially Available	Q4 2019	Q4 2019	Commercially Available	Q4 2019

- WVTR release tested via Moccn Permatran equipment at 50 °C100%RH and extrapolated to 2 °C/100%RH.

- FTB3-50 typical performance from Aquatran2 (extrapolated from 50 °C/100%RH): $6.2 \times 10^{-6} - 1.2 \times 10^{-5}$ $g/m^2/day$ 20C/100%RH

A4 sheet samples available.
a) In Commercial Product.
b) Development.
Source: Reproduced with permission from https://www.3M.com.cn.

Table 9.5 Hydrogen barrier protective film (BPS-270).

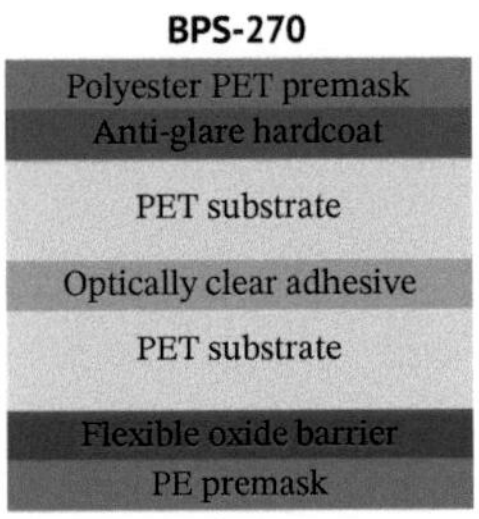

Physical and optical characteristics

Film property	3M BPS-270
Thickness	269 ± 13 pm
WVTR specification	$<1 \times 10^{-3}$ g/m²/day, 20°C/100%RH
WVTR, typical values	$<1.2 \times 10^{-5}$ g/m²/day 0°C/100%RH
Pencil hardness	≥3H
Transmission	>88%
Transmission @ 380nm	<2%
Haze	9 ± 3%

Anti-Glare coating.
Source: Reproduced with permission from https://www.3M.com.cn.

Table 9.6 Flexible transparent encapsulation film (FTB3-UV-50).

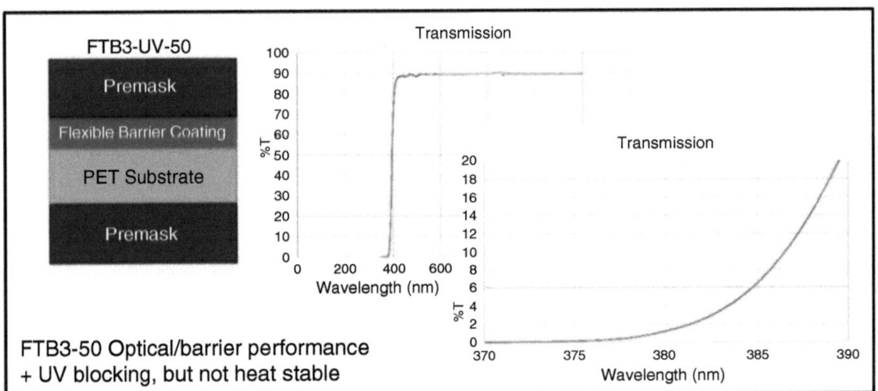

Source: Reproduced with permission from https://www.3M.com.cn.

Said component can transmit visible and infrared light. In some embodiments, said encapsulant layer is on a photovoltaic cell.

Content of the Invention

Despite the advances in encapsulation technology, the prayer for blocking and durability in solar applications remains a challenge, and prays for additional work to bring effective flexible encapsulation solutions to the solar market. The present invention provides components that can be used, for example, to encapsulate solar devices. The components disclosed herein are generally flexible, transmit visible and infrared light, and have excellent blocking properties.

While fluoropolymer films and other weather-resistant films are said to be available components for encapsulation systems for flexible photovoltaic cells (e.g. in U.S. Patent Application Publication No. 2003/002949), fluoropolymer films can have

Table 9.7 Flexible transparent packaging film (FTB3-50).

FTB3-50

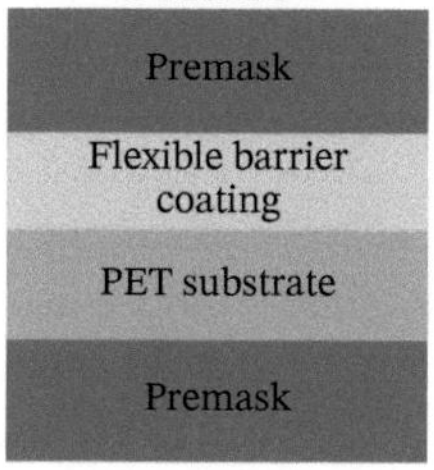

Typical physical and optical characteristics

Film property	3M™ FTB3-50 heat stabilized
Thickness	52 ± 5 μm
WVTR Spec	<1 × 10^{-6} g/m²/day, 20 °C/100%RH
WVTR(20°C/100%RH)	6.2 × 10^{-6}–1.2 × 10^{-5} g/m²/day
Transmission	>88%
Haze	<2.5%
Clarity	>95%
Shrinkage (85 °C)	0.00% MD, −0.01% TD[b)]
Shrinkage (135 °C)	−0.00% MD, −0.02% TD[b)]
Barrier Elongation	2% upto 200 k cycles

a) Based off limited data from Aquatran 2, extrapolated from 50°C/100% RH. Release criteria <0.001 g/m²day.
b) MD is machine direction, TD is transverse direction.
Source: Reproduced with permission from https://www.3M.com.cn.

a coefficient of thermal expansion more than 100 ppm/K (percent/Kelvin), while flexible carriers available for photovoltaic cells (e.g. polyimide films or metal fillings) can have less than 30 ppm/K. This mismatch in the thermal expansion coefficient between the fluoropolymer film and the flexible carrier can cause thermal stress in the encapsulated flexible photovoltaic cell. It has been found that under certain thermal lamination conditions (e.g. 50 °C), the development of encapsulation systems that include multilayer barrier films on fluoropolymers laminated to low-CTE flexible substrates with thermally cured encapsulants (e.g. ethylene vinyl acetate) can lead to delamination thermal stresses.

Flexible photovoltaic cells (e.g. CIGS) have relatively high profiles (e.g. high profiles in proportion to organic electroluminescent (EL) devices). Typically, thin-film CIGS cells have sink and interconnection bands that may be, e.g. 0.15 mm above the surface of the cell. These high-profile assemblies can act as stress collectors that may exacerbate the problems caused by CTE mismatch.

The barrier assemblies according to the present invention may be used, for example, to minimize the effect of CTE mismatch between the climate-resistant top layer and the flexible PV module that may result from high-temperature lamination.

In one aspect, the present invention provides an assembly comprising an encapsulant layer; a first polymeric film substrate, said first polymeric film substrate having a first surface, a second surface opposite said first surface and a first coefficient of thermal expansion, wherein said second surface of said first polymeric film substrate is provided on said encapsulant layer; a barrier film, said barrier film being provided on said first surface of said first polymeric film a pressure-sensitive adhesive layer, said pressure-sensitive adhesive layer having a first surface and a second surface opposite said first surface, wherein said first surface of said pressure-sensitive adhesive is provided on a barrier film opposite said first polymeric film substrate; and a second polymeric film substrate, said second polymeric film substrate being provided on a second surface of the pressure-sensitive adhesive layer, wherein said second polymeric film substrate has a second thermal coefficient of expansion, and wherein said second coefficient of thermal expansion is higher than said first coefficient of thermal expansion. Said component can transmit visible light and infrared light. In some embodiments, the encapsulant layer is on the photovoltaic cell.

In this patent application, terms such as “a,” “a,” and “said” are not intended to refer to a single entity only, but rather include a broad category, specific examples of which may be used as examples. The terms “a,” “one,” and “described” may be used interchangeably with the term “at least one.” The phrases “at least one (one)” and “including at least one (one)” followed by a list refer to any of the items in the list and any combination of two or more of the items in the list. Unless otherwise specified, all ranges of values include their endpoints and the non-integer values between the endpoints.

9.1.4 UDC-United States

Universal Display Corporation (UDC), founded in 1994 and listed on NASDAQ in 1996, is one of the world leaders in research, development, and commercialization in the OLED industry. The company is headquartered in Ewing, New Jersey, with subsidiaries in China, Hong Kong, Ireland, Japan, and Korea, and a wholly owned subsidiary, Adesis. Inc, located in New Castle, Delaware. The company's core business is the production and sale of organic light-emitting semiconductor (universal PHOLED) materials and the transfer of proprietary technology [7].

Sales of OLED phosphorescent materials are the primary and core business, accounting for approximately 98% of the company's revenue (2019Q2), with the remaining portion of revenue coming from subsidiaries engaged in the research, development, and commercialization of organometallic composites. As the phosphorescent luminescent materials sold by the company have advantages over fluorescent materials – higher luminescence efficiency, lower thermal energy dissipation, etc. – they are the mainstream luminescent materials adopted in the market, so the company's main business is highly competitive. According to the company's announcement, as a global leader in the OLED high-tech industry, UDC currently holds over 5000 patents worldwide. The company invested a lot of money in R&D technology and products at the early stage of its establishment, and to maintain its advantage in the field of luminescent materials, the company's R&D expense ratio (R&D expense as a percentage of operating revenue) is maintained at around 20% every year. The company's R&D includes both in-house R&D and external cooperation with university research institutes.

UDC has developed a novel film encapsulation technology for ultra-flexible plastic substrate systems. This product, UNIVERSALBARRIER® TFE, offers the following advantages (Figure 9.8):

- Effective permeation barrier with organic–inorganic hybridization properties
- Transparent and flexible film properties
- Effective perimeter coverage for virtually edgeless or “borderless” packaging

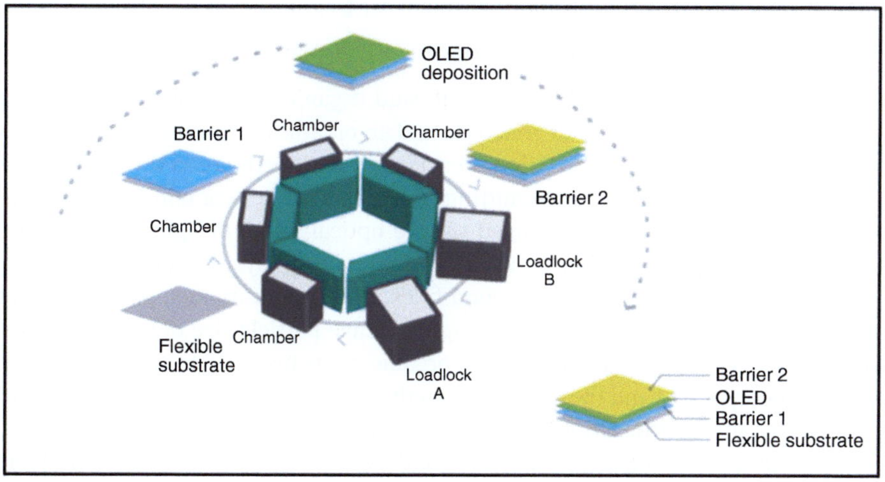

Figure 9.8 UDC thin-film encapsulation technology. Source: Reproduced with permission from https://oled.com.

- Enables ultra-flexible barriers without shredding, with a bending radius <1mm
- Cost-effective process, compatible with roll-to-roll processing
- Ideal for OLED, PV, battery, and other thin-film devices.

Some Patents

CN101901825 B Organic Electroluminescent Display Device [8].

The present invention provides an organic EL display device, said organic EL display device comprising: an organic EL element and a fines-containing layer, said organic EL element comprising transparent electrodes, counter electrodes and an organic compound layer disposed between these electrodes, said organic compound layer comprising a light-emitting layer, said fines-containing layer being arranged on the light path of light emitted from said light-emitting layer, wherein said fines-containing layer comprises fines layer contains an organic resin material, a first fines and a second fines, said second fines having a weight average particle size greater than the weight average particle size of said first fines, wherein the refractive index n_1 of said organic resin material to which said first fines are added and the average refractive index n_2 of said organic compound layer satisfy the relationship $|n_1 - n_2| < 0.25$, and wherein the first fines are added to the center of which said organic resin material and the refractive index n_3 of said second fines satisfy the relationship $n_3 - n_1 > 0.2$.

Contents of the Invention

The present invention is intended to solve the conventional problems described above and is designed to achieve the following objectives. More specifically, the present invention aims to provide an organic EL display device with excellent light extraction efficiency.

The means used to solve the above problem are as follows:

(1) An organic EL display device said organic EL display device comprising.

An organic electroluminescent element, said organic EL element comprising a transparent electrode, a counter electrode and an organic compound layer disposed between said transparent electrode and said counter electrode, said organic compound layer comprising a light-emitting layer, and a fines-containing layer, said fines-containing layer being arranged on the optical range of light emitted from said light-emitting layer, wherein said fines-containing layer contains an organic resin material, a first fines and a second fines, said weight average particle size of said second fines is greater than the weight average particle size of said first fines, wherein the refractive index n_1 of said organic resin material to which said first fines are incorporated and the average refractive index n_2 of said organic compound layer satisfy the relationship$|n_1-n_2| < 0.2$ s, and wherein the refractive index n_1 of said organic resin material to which said first fines are incorporated and the refractive index n_3 of said second fines satisfy the relationship n_3. refractive index n_3 satisfies the relationship $n_3-n_1 > 0.2$s.

(2) The organic EL display device according to (1), wherein said second fine particle is an inorganic fine particle.
(3) The organic EL display device according to (2), wherein said inorganic fines comprise at least one of TiO_2, ZnO, and ZrO_2.
(4) The organic EL display device according to any one of (1)–(3), wherein said fines-containing layer has a thickness of less than 5 μm.
(5) The organic EL display device according to any one of (1)–(4), wherein said fines in said fines-containing layer are primary particles.
(6) The organic EL display device according to (5), wherein said fines in said fines-containing layer are surface treated with organic material.
(7) Organic EL display device according to any one of (1)–(6), wherein said fines-containing layer further contains a fines dispersant.
(8) The organic EL display device according to any one of (1)–(7), wherein said first fines have a weight average particle size of less than 50 nm and said second fines have a weight average particle size of 50 nm to 5 μm.
(9) The organic EL display device according to any one of (1)–(8), wherein said organic resin material comprises at least one selected from among an imide resin, an acrylic resin, an acyl resin, and a silicone base resin.

According to the present invention, conventional problems can be solved, and an organic EL display device with excellent light extraction efficiency can be provided.

9.1.5 Amcor-United States

Amcor, one of the world's top packaging companies, was founded in Australia in 1860 and now has nearly 200 plants in 43 countries worldwide, employing more than 33,000 people, with a global revenue of $9.319 billion in fiscal 2017. Food, beverage, healthcare, personal and home care, tobacco, and other industrial packaging are among the downstream customers [9].

In the field of flexible electronics packaging, Amcor has developed multilayer structured ultrahigh-barrier films. The multilayer ultrahigh-barrier film has two types of export: one is provided directly to the flexible electronic device manufacturer for lamination protection packaging of the manufactured devices, and the other is provided to the flexible conductive film manufacturer, which sputtered TCO conductive layer on the barrier film layer to make high barrier conductive film, and then provided it to the flexible electronic device manufacturer for use as electrodes for flexible electronic devices for device fabrication (Figure 9.9; Table 9.8).

Some Patents

CN107969124 A Flexible Multilayer Packaging Film with Ultrahigh-Barrier Performance [10].

The present invention relates to a flexible multilayer packaging film with high gas barrier properties comprising: one or more support layers (1, 10); one or more barrier layers (6, 60), each of the one or more barrier layers (6, 60) comprising an organic layer (2, 20) and an inorganic layer (3, 30), wherein said multilayer film has a barrier layer of less than 0.1 $cm^3/m^2/24\,h/atm$, preferably less than 0.05 $cm^3/m^2/24\,h/atm$, measured at 23 °C and 50% relative humidity F. $cm^3/m^2/24\,h/atm$, preferably less than 0.05 $cm^3/m^2/24\,h/atm$, and most preferably less than 0.03 $cm^3/m^2/24\,h/atm$,

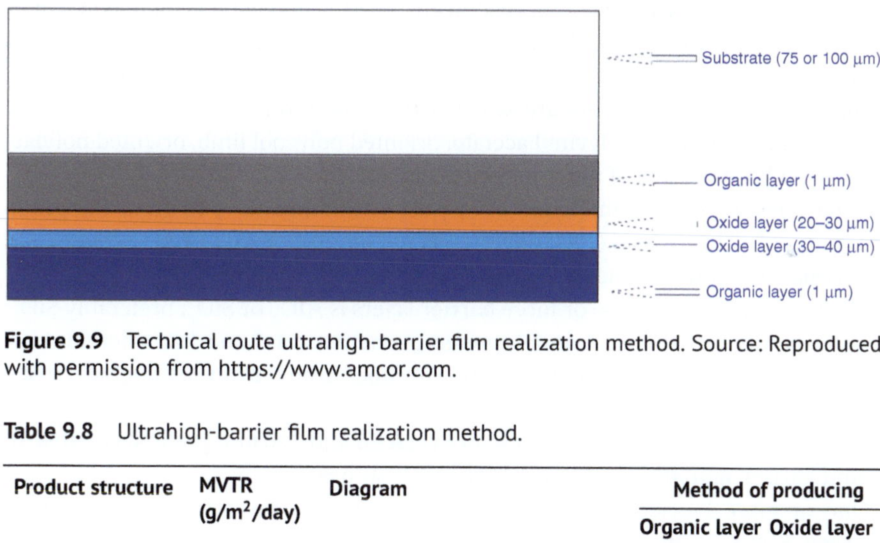

Figure 9.9 Technical route ultrahigh-barrier film realization method. Source: Reproduced with permission from https://www.amcor.com.

Table 9.8 Ultrahigh-barrier film realization method.

Product structure	MVTR ($g/m^2/day$)	Diagram	Method of producing	
			Organic layer	Oxide layer
Dry composite of N barrier films	$<10^{-2}$	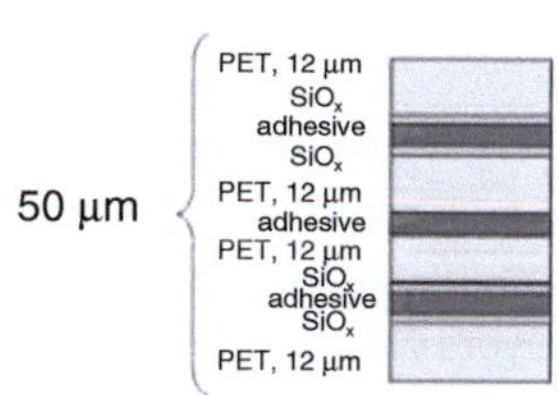	—	CVD

Source: Reproduced with permission from https://www.amcor.com.

and most preferably less than 0.05 $cm^3/m^2/24\,h/atm$. 0.03 $cm^3/m^2/24\,h/atm$ of oxygen transmission rate.

Contents of the invention

The present invention discloses a flexible multilayer packaging film having high gas barrier properties comprising: one or more support layers (1, 10); one or more barrier layers (6, 60), each of the one or more barrier layers (6, 60) comprising an organic layer (2, 20) and an inorganic layer (3, 30) (or consisting thereof), wherein said laminate of the multilayer film has a surface area of less than 0.1 $cm^3/m^2/24\,h$ at 23 °C and 50% relative humidity 0.1 $cm^3/m^2/24\,h/atm$, preferably less than 0.05 $cm^3/m^2/24\,h/atm$, most preferably less than 0.03 $cm^3/m^2/24\,h/atm$. [0020] Preferred embodiments of the present invention disclose at least one or a suitable combination of the following features:

The oxygen transmission rate of the laminate of the multilayer film increases by not more than 30%, preferably not more than 20%, and most preferably not more than 10%, after a Gelbo Flexmechanical constraint test in accordance with ASTM F39274;

One or more barrier layers are sandwiched between an additional sealant layer (5) and one or more support layers (1, 10): the organic layer of the one or more barrier layers is polyvinyl alcohol, specifically an ethylene vinyl alcohol copolymer containing between 25 and 50 mol% alcohol, preferably between 27 and 45 mol% alcohol.

One or more support layers are selected from the group consisting of oriented polypropylene, oriented polyvinyl acetate, oriented polycool limb, oriented polylactic acid, and copolymers of these polymers, paper, and cardboard (carton, carton).

Sealant layer having a thickness of at least 10 μm and selected from the group consisting of polyethylene (PE), polypropylene, amorphous polyvinyl acetate, and copolymers of these polymers.

The inorganic layer of one or more barrier layers is AlO_x or SiO_x, preferably SiO_x;

The x in SiO_x is between 1 and 2, preferably between 1.5 and 1.8, wherein AlO_x has the formula Al_yO_z, with a ratio y/z of a number from 0.2 to 1.5, preferably y of 2 and z of 3;

The inorganic layer has a thickness between 20 and 120 nm, preferably between 30 and 100 nm, and most preferably between 40 and 80 nm.

One or more barrier layers comprise an aluminum layer having a thickness between 5 and 50 μm, preferably between plus 5 and 25 μm, and most preferably between 5 and 15 μm.

The present invention further discloses a method for producing a flexible multilayer packaging film according to the present invention (or a method for producing a flexible multilayer packaging film according to the present invention), comprising the steps of providing one or more support layers (1, 10); coating one or more organic layers (2, 20) or inorganic layers (3, 30) of one or more barrier layers (6, 60) on one or more support layers (1, 10).

Preferred embodiments of the method of the present invention disclose at least one or a suitable combination of the following features:

Coating one (or one) additional sealant layer (5) over one or more coated barrier layers (6, 60); providing at least the first and second support layers (1, 10): an organic layer (2) of the first barrier layer (6) coated on the first support layer (1) and an inorganic layer (30) of the second barrier layer (60) coated on the second support layer (10): or an inorganic layer (3) of the first barrier layer (6) coated on the first support layer (1) and an inorganic layer (3) of the second barrier layer (10) coated on the second support layer (60) of the organic layer (20) on the second support layer (10).

An additional sealant layer (or layers) (5) is applied to the coated second barrier layer (60).

Coating of the inorganic layer by electron beam vacuum deposition.

Coating of inorganic layers by electron beam vacuum deposition at a coating voltage of 5–40 kV at pressures below 103 mbar.

The inorganic layer of one or more barrier layers is coated on the surface of the organic layer of one or more barrier layers, or the organic layer of one or more barrier layers is coated on the surface of the inorganic layer of one or more barrier layers, preferably the inorganic layer of one or more barrier layers is coated on the surface of the organic layer of one or more barrier layers, and wherein the organic layer is a polyvinyl alcohol layer, in particular comprising between 25 and 50 mol% of alcohols, preferably an ethylene vinyl alcohol copolymer comprising between 27 and 45 mol% of alcohols.

The inorganic layer (3) of the first barrier layer (6) is coated on the surface of the organic layer (2) of the first barrier layer (6), and the organic layer (20) of the second barrier layer (60) is coated on the surface of the inorganic layer (30) of the second barrier layer (60): or the organic layer (2) of the first barrier layer (6) is coated on the surface of the inorganic layer (3) of said first barrier layer (6), and the inorganic layer of the second barrier layer (60) of the inorganic layer (30) is coated on the surface of the organic layer (20) of said second barrier layer (6.): and wherein the organic layer is a polyvinyl alcohol layer, in particular an ethylene vinyl alcohol copolymer comprising between 25 and 50 mol% of alcohol, preferably between 27 and 45 mol% of alcohol.

Subjecting the support layer to a capacitating treatment to enhance the compatibility between said organic layer of said one or more support layers and said one or more barrier layers.

The capacitating treatment is selected from plasma treatment, corona treatment, flame treatment, or a combination of one or more of these treatments.

Also disclosed is a package comprising a flexible multilayer film according to the present invention.

9.1.6 Vitriflex-United States

Vitriflex Inc. holds 26 international patents related to high-barrier films and has developed the highest grade encapsulation barrier film with a WVTR of 10^{-6} g/m^2/day to meet the encapsulation requirements in different fields such as flexible photovoltaic cells, flexible display, flexible electronics, flexible lighting, flexible sensing, etc. Vitriflex is also developing a gas barrier solar technology for solar panels. It specializes in the field of renewable energy and solar energy [11].

Some Patents

CN105647324 A New Blocking Layer Stack and Method and Composition Thereof Content of the Invention [12].

In accordance with the foregoing, the present invention provides, on the one hand, a barrier layer stack. The barrier layer stack comprises: (i) a barrier material comprising a defect; (ii) a sealing layer covering the defect in the barrier material; (iii) wherein the presence of the sealing layer reduces or prevents diffusion of vapor and/or gas molecules through the defect in the barrier material when the barrier layer stack is in an operational state. In preferred embodiments of the present invention, the defect comprises a notch located within the barrier material, and the sealing layer does not protrude into the said notch.

By way of example, said sealing layer has a thickness between about 0.1 micron and about 10 microns, and said barrier material has a thickness between about 10 nanometers and about 1 micron. Said sealing layer comprises a polymer and/or an inorganic material. Preferably, said sealing layer may comprise at least one of acrylic vinyl, epoxy resin, and polyurethane vinyl. Said barrier material may be an inorganic layer comprising at least one of a metal, a metal oxide, a metal nitride, a metal oxynitride, a metal carbon nitride, and a metal carbon oxide. In accordance with an aspect of the present device, said inorganic barrier layer comprises at least one of aluminum, silver, silicon, series, tin, building, knob, saw, sawing, graft, uranium, machine, and confinement. Said barrier material may be a transparent material.

In one embodiment of the present invention, said barrier material comprises two barrier layers and a reaction layer sandwiched between the two barrier layers. In this embodiment, each of the barrier layers may be made of one or more amorphous materials.

On the other hand, the present invention provides a combination device. The combination device comprises: (i) a lower layer; (ii) an encapsulant capable of at least partially encapsulating the lower layer; and the encapsulant comprises: (a) a barrier material including defects; (b) a sealing layer covering the defects in the barrier material, when the barrier layer stack is in operation, the presence of the sealing layer can reduce or prevent the diffusion of steam or gas molecules through defects in the barrier material. In this aspect, the combination device may include one of a solar cell module, a light-emitting module, a light-emitting diode display, an electrolyzer, or a reflective display module.

In another aspect, the present invention provides an anticorrosive package. The anticorrosive package comprises: (i) an object that is sensitive to vapor and/or gas molecules; and (ii) an encapsulation body capable of at least partially encapsulating the object: and the encapsulation body comprises (a) a barrier material including defects; (b) a sealing layer covering the defects of the barrier material: wherein the presence of the sealing layer reduces or prevents the diffusion of vapor and/or gas molecules through the barrier material when the barrier layer stack is in operation. (b) a sealing layer covering defects in the barrier material, wherein the presence of the sealing layer reduces or prevents diffusion of vapor and/or gas molecules through defects in the barrier material when the barrier stack is in operation. Preferably, the object includes one of a foodstuff, a medical product, or a biologically active substance.

In another aspect, the present invention provides a method of manufacturing a barrier layer stack. The method comprises: (i) obtaining a barrier material; (ii) applying a protective film adjacent to said barrier material; and (iii) curing the protective film to form a barrier stack.

Preferably, said applying the protective film comprises applying at least one of acrylic vinyl, epoxy resin, or polyurethane vinyl. The curing of the protective film can be done at a temperature range between about 10 and about 100 °C and/or by treating said protective film with UV radiation.

In a preferred embodiment of the present invention, the process of obtaining the barrier material comprises: (i) loading the flexible substrate onto a coater; (ii) placing the flexible substrate or a portion of the coater such that the flexible substrate is in a first position within the coater; and (iii) while the flexible substrate is in the first position, creating one or more barrier layers on the flexible substrate, the barrier layers comprising defects that allow vapor and/or gas molecules to be transported.

In another aspect, the present invention provides a composition for a stack of barrier layers. The composition comprises: (i) a barrier material having defects that allow gas and/or vapor to pass through, the barrier material comprising at least one of a metal, a metal oxide, a metal nitride, a metal oxynitride, a metal carbon nitride, and a metal oxynitride; and (ii) a sealing layer comprising a polymeric material, the polymeric material being at least one of acrylic vinyl, epoxy resin, and polyurethane vinyl.

The concentration of the barrier material may be between about 1% by weight and about 99% by weight, and the concentration of the sealing layer may be between about 1% by weight and about 99% by weight.

However, when read in conjunction with the accompanying drawings, the method of construction and operation of the present invention and the additional objects and advantages may be best understood from the description of the following specific embodiments.

9.1.7 TBF-Singapore

Tera-Barrier Films (TBF) was established in June 2009 as a separate venture from the Institute of Materials Research and Engineering (IMRE), an institute of the Singapore Agency for Science and Technology Development. TBF develops barrier materials and barrier stacking technology for OLEDs and solar cells, and its transparent and flexible films are constructed in layers consisting of inorganic layers and organic layers dispersed with nanoparticles. The nanoparticles are responsible for reacting with water molecules and confining them inside the film, and for filling the cracks created by the inorganic layer. Based on the above structure, the film can achieve a water vapor permeability of only 10^{-6} g/m^2/day at 60 °C and 90% humidity by laminating two micron-thick layers of inorganic and organic layers each. Unlike other companies' ordinary transparent flexible films, moisture easily penetrates inside through the cracks created in the inorganic layer, so the inorganic and organic layers are each laminated with three to four layers. The film has the highest water vapor barrier performance to date and has been tested by the Center for Process Innovation in the UK. The tests show that the new film is 1000 times more moisture-resistant than existing technology [13].

TBF has developed the technology to manufacture the developed product in a roll-to-roll process. Based on this technology, the film can be formed in 300 m in 1 minute and will be commissioned to manufacturers in the United States and Germany to start supplying at a scale of 300 m^2 per month. In the future, TBF plans to set up its own pilot production line in Singapore to produce 50,000 m^2 per month in 2012, and in 2013, TBF will set up a mass production line to produce 2.5 million m^2.

Selected Patents

US20190027414 A1 Encapsulated Barrier Stack Consisting of Nanoparticles Encapsulated By Dendritic Molecules [14].

Overview of the Invention

The present inventors were surprised to find that a sealing layer consisting of nanoparticles encapsulated by a dendritic polymer provides the advantage of one of the following functions or properties, or any combination thereof, when used as a barrier layer: (A) the high filling density of the polymer design of the dendritic polymer-nanoparticle film (sealing layer) reduces the porosity of the nanoparticle film and is able to prevent the diffusion of moisture oxygen through the nanoparticle seal layer encapsulated by the dendritic polymer; (B) crosslinking with other components of the sealing layer (e.g. nanoparticles, oligomers, polymers) provides mechanical stability and increases the bond strength between the nanoparticles; (C) the chemistry of the composite, and thus the chemoselectivity of the barrier layer is expected to take advantage of the chemistry within and/or on the surface of the dendrimer; d) the surface of the dendrimer-encapsulated nanoparticle sealing layer of the present invention has a "spherical blanket" conformation, which has a larger contact surface than the sealing layer with embedded nanoparticles (e.g. those disclosed in WO 2005/0249901 A1 and WO2008/057045). The larger contact surface allows for a better barrier to moisture, making the sealing layer more effective.

The present inventors were also surprised to find that nanoparticles encapsulated by dendritic polymers are able to seal or plug defects.

An encapsulated barrier stack is provided according to the present invention, which is a low-cost device with multifunctional properties including UV blocking and good anti-reflective properties.

Thus, in one aspect, the present invention provides an encapsulated barrier stack capable of encapsulating articles sensitive to moisture and/or oxygen and comprising a multilayer film, wherein the multilayer film comprises: one or more barrier layers having low moisture and/or oxygen permeability, and one or more sealing layers provided to contact a surface of at least one of the barrier layers so as to cover and/or plug defects present in the barrier layer, wherein the one or plurality of sealing layers comprises a plurality of nanoparticles encapsulated in dendritic macromolecules, the nanoparticles being active because they are capable of interacting with moisture and/or oxygen to impede penetration of moisture and/or oxygen, wherein the nanoparticles are wholly or partially encapsulated in the dendritic macromolecules and/or dendrites. Preferably, the dendrimer-encapsulated nanoparticles are crosslinked to each other, i.e. crosslinked dendrimer-encapsulated nanoparticles.

In another aspect, the present invention provides an electronic module comprising an electronic device sensitive to moisture and/or oxygen, wherein the electronic device is arranged within an encapsulated barrier stack according to the invention.

[0015] In another aspect, the present invention provides a method of manufacturing an encapsulation barrier layer having one or more sealing layers comprising nanoparticles encapsulated in a dendritic polymer.

In embodiments of the method of fabricating an encapsulated barrier layer, the method comprises providing one or more barrier layers and forming one or more sealing layers, wherein forming one or more sealing layers comprises

(i) mixing, in the presence of a polymerizable compound and/or a cross-linkable compound, an encapsulating material comprising a dendrimer or precursor thereof (optional) with a plurality of nanoparticles that are reactive because they are capable of interacting with moisture and/or oxygen to form the sealing mixture.
(ii) applying the sealing mixture to the barrier layer under conditions that allow the nanoparticles to be wrapped or encapsulated in the dendritic macromolecules, thereby forming a sealing layer.

The polymerization step of the polymerizable compound or the crosslinking of the cross-linkable compound is performed when the polymerizable compound or the crosslinking of the cross-linkable compound is present in the encapsulating material.

Preferably, said polymerizable compound of the encapsulating material is a monomer. Said encapsulating material may also comprise organic materials such as silanes, surfactants, and other additives. In addition, it may also comprise a suitable solvent.

Preferably, "cross-linked dendritic polymer encapsulated nanoparticles" are formed.

Alternatively, in a second embodiment of a method of manufacturing an encapsulated barrier layer, the method comprises providing one or more barrier layers and forming one or more sealing layers, wherein forming the one or more sealing layers.

(i) providing an encapsulating material comprising or consisting of nanoparticles encapsulated by a dendritic polymer, the nanoparticles being active because they are capable of interacting with moisture and/or oxygen.
(ii) optionally mixing the encapsulating material with a polymerizable compound or a cross-linkable compound so as to form a sealing mixture, and (iii) applying the sealing mixture to the barrier layer under conditions that allow the nanoparticles to form a sealing layer.

Preferably, said encapsulating material comprises a dendritic macromolecule and a polymerizable compound.

Preferably, said polymerizable compound of the encapsulating material is a monomer. Said encapsulating material may also comprise organic materials such as silanes, surfactants, and other additives. In addition, it may also comprise a suitable solvent.

Preferably, the "cross-linked dendritic polymer-encapsulated nanoparticles" are formed.

The encapsulated barrier layer according to the present invention has encapsulated nanoparticles. The dendrimers, dendrimers, and their precursors, optionally combined with polymerizable compounds and crosslinking agents, are used as encapsulating materials or for the functionalization of the nanoparticles. Dendrimers, dendrites, or their precursors, optionally combined with polymerizable compounds and crosslinkers of the encapsulating material, react with the nanoparticles to form the "encapsulating material." Thus, in the present case, the "encapsulating material" is the material that precedes the reaction(s) leading to the formation of the encapsulation and sealing layer. The "encapsulating material" is the material that encapsulates the nanoparticles once the reaction leading to encapsulation has occurred.

In the present case, the "dendrimer-encapsulated nanoparticles" are either encapsulated by dendrimers, surrounded by dendrimers, or are dendrimer nuclei formed after dendrites are attached to their surfaces.

Furthermore, in this context, it is worth noting that the term "encapsulated" does not necessarily mean that the entire surface of the active nanoparticle is encapsulated by the encapsulating material of the present invention. Instead of the surface of the nanoparticles being 100% encapsulated, the present invention also includes that only about 50% or more, or about 60% or more, or about 75% or more, or about 80% or slightly more, or about 85% or more, or about 90% or more, or about 95% or more of the surface of the active nanoparticles is encapsulated, or in other words, passivated by the encapsulating material after the formation of the encapsulation. For example, by interconnecting dendritic molecules or dendrites or curing or crosslinking polymerizable compounds. The present inventors were also surprised to find that dendritic macromolecule-encapsulated nanoparticles can seal or plug defects, and they can also enhance gas barrier properties. In addition, the encapsulated barrier layer according to the present invention is a low-cost device with multifunctional properties such as UV blocking and good anti-reflective properties.

The encapsulation barrier layer of the present invention can have a porous barrier layer, which can be an oxide film or a sealing layer. The sealing layer may contain functionalized nanoparticles that are encapsulated or passivated by dendrimers or dendrimer/polymer blends.

In some embodiments, the sealing layer may be a single layer. In some embodiments, the barrier layer of the said package has a single-sealing layer. In some embodiments, the encapsulated barrier layer comprises a plurality of sealing layers.

The present invention provides a barrier stack with improved flexibility, gas barrier, weathering, optical, mechanical, and reliability properties, and also provides a cost-effective solution.

According to the first aspect, the present invention provides an encapsulated barrier stack. Said barrier layer is capable of encapsulating articles that are sensitive to moisture and/or oxygen. Said encapsulation barrier layer comprises a multilayer film. The multilayer membrane includes one or more barrier layers and one or more sealing layers comprising nanoparticles encapsulated in a dendritic polymer providing low moisture and/or oxygen permeability. The multilayer membrane further includes one or more sealing layers. Said one or more sealing layers are provided to be in contact with the surface of said at least one barrier layer. Thus, one or more sealing layers cover defects present in the barrier layer. One or more sealing layers comprise a plurality of dendritic macromolecules and/or dendritic and organic species, for example, nanoparticles encapsulated by dendritic macromolecules. The nanoparticles are active because they are capable of interacting with moisture and/or oxygen, thereby impeding the passage of moisture and/or oxygen through the defects in the barrier layer. [0028] According to a second aspect, the present invention provides an electronic device. The electronic device includes active components that are sensitive to moisture and/or oxygen. The active components are arranged in an encapsulated barrier stack according to the first aspect.

According to a third aspect, the present invention provides a method of manufacturing an encapsulation barrier layer according to the first aspect. The method includes providing one or more barrier layers. The method also includes forming one or more sealing layers. In a first embodiment of the third aspect, forming one or more sealing layers comprises mixing an encapsulating material according to the present invention, the material comprising or including one of: a dendrimer or precursor thereof, a dendrimer/polymerizable compound mixture, a dendrimer/cross-linkable compound mixture, dendrimer/cross-linkable compound mixtures with multiple nanoparticles or functionalized nanoparticles. The polymerizable or cross-linkable species include monomers, polymers and/or oligomers, or combinations thereof.

Alternatively, in a second embodiment of the third aspect, one or more sealing layers are formed comprising an optional blend of an encapsulating material comprising dendrimer-encapsulated nanoparticles with polymerizable or cross-linkable compounds. The polymerizable or cross-linkable species include monomers, polymers, and/or oligomers or combinations thereof.

According to a fourth aspect, the present invention relates to the use of dendritic polymer-encapsulated active nanoparticles for the preparation of sealing layers for barrier layers. The nanoparticles are active because they are capable of interacting with moisture and/or oxygen, thereby impeding the passage of moisture and/or oxygen through defects in the barrier layer.

According to a fifth aspect, the invention relates to the use of dendritic polymer-encapsulated reactive nanoparticles for encapsulating electronic devices or for food packaging, pharmaceutical packaging, or medical packaging.

9.1.8 Fraunhofer ISC-Germany

More than 20 years ago, Fraunhofer ISC invented a new material containing an organic/inorganic composite barrier material called ORMOCER® (short for Organic

Modified Ceramics), an inorganic–organic hybrid polymer. This is a new material consisting of a submicron layer-thick organic barrier material and an inorganic barrier material simultaneously laminated to a biaxially oriented polyester (BOPET) substrate, with high gas and moisture barrier properties that are claimed to be a substitute for glass. Since the development of such materials, expertise in the synthesis, functionalization, and further processing of ORMOCER®s has been continuously developed and adapted to a wide range of applications [15].

Ormocer®s are first coated with 100–1000 nm-thick organically modified ceramics on the surface of BOPET films by a roll coating process. The organically modified ceramics are reticulated silica cross-linked with organic cross-linkers and then compounded with 20–100 nm-thick inorganic alumina or silica to obtain super barrier films (Figure 9.10).

Selected Patents

EP2424655 A2 Method for Encapsulating Liquid or Paste-Like Substances in Crosslinked Encapsulation Materials [16].

The invention relates to an encapsulation method for encapsulating liquid or paste substances and to capsules obtained with the aid of this method.

(Micro)encapsulation is the principle of encapsulating a particulate solid, liquid, or gas in a small part of an envelope. This technique of packaging (active) ingredients is popular in the fields of pharmaceutical/medicinal, cosmetic, nutritional/food, and other consumer fields and is important for textile, chemical, agricultural, and environmental fields, with a fast-growing trend. With the possibility for. B. conversion of liquid substances into powders, – encapsulation of volatile substances (aromas, flavors), substances before the reaction z. protection with air, light, or liquid, controlled release of active ingredients (e.g. drugs), fixation of toxic substances, separation of active substances from each other, – improvement of tolerance of drugs and physical properties of powders open up not only new product quality but also new product ideas and strategies.

Among the numerous encapsulation techniques, many are drug-specific, e.g. tailored to a specific kind of active ingredient (e.g. coalescence). In other cases, heat is required (thermoplastic melting) or solvents are needed so that sensitive materials can be damaged. One principle (e.g. combination of material and process) combines particularly gentle and clean encapsulation without temperature coloration, solvent input, contamination, or damage of sensitive components by foreign substances with high encapsulation efficiency and maximum possible variability of encapsulable components.

This depends on its physical and chemical properties.

Due to range of applications and the intended function of the active substance, there is a wide variety of encapsulation materials and encapsulation techniques. The most important processes dominated by this equipment are extrusion, spray drying, rotary disc, and liquid bath technologies. In most technologies, the application of heat, solvents, etc. is unavoidable. Processes using chemical or physical phenomena, usually based on self-forming processes (emulsion technology, coalescence), have higher advantages compared to sequential technological processes.

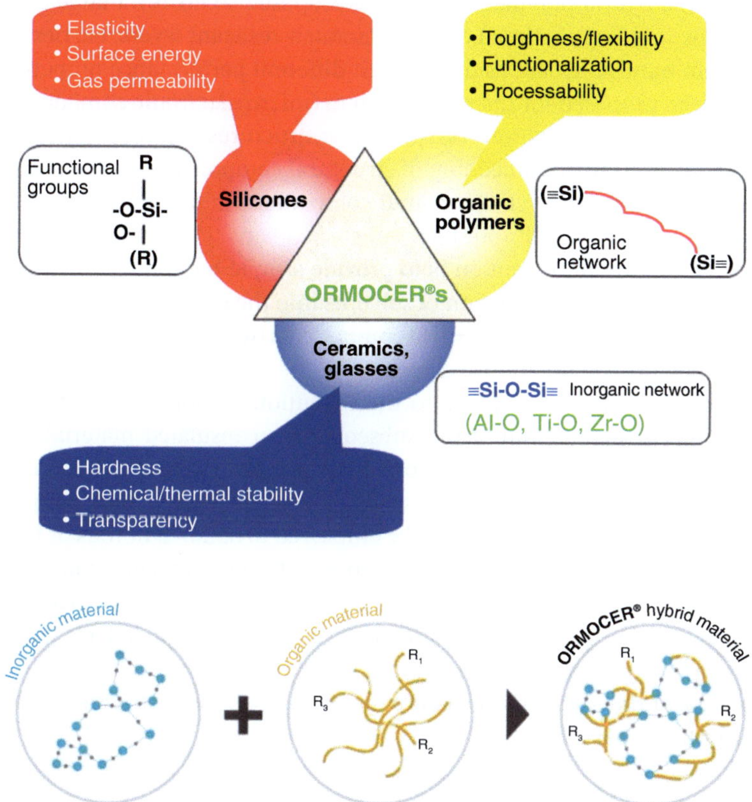

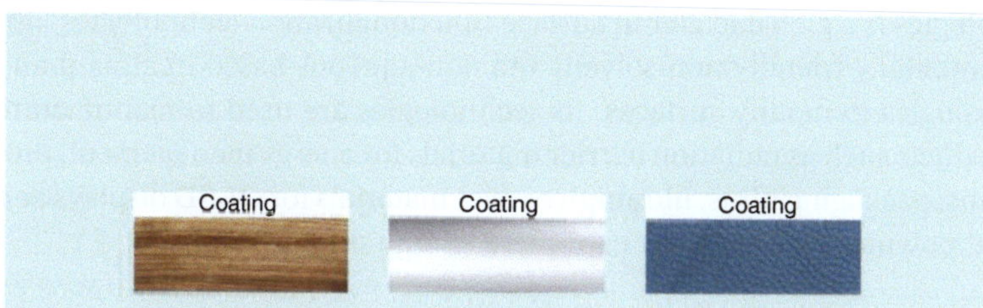

Figure 9.10 ORMOCER® coatings can act as a protective coating on a wide range of material surfaces. Source: ORMOCER.

On the other hand, the economy is very reactive substance-specific and very strictly limited in terms of encapsulation materials, usually with the disadvantage of low encapsulation efficiency and purity. In the case of emulsification processes/coalescence, the active ingredient is first mixed into the later shell material. By curing the shell material from the edges, the active ingredient/constituents are gradually concentrated inside. In turn, silica-based materials require, i.e. very strong precipitation base conditions. This may contaminate or destroy the active ingredient/component.

Inorganic–organic hybrid polymers have been proposed so far for a large number of applications such as functional coatings (scratch-resistant layers, barrier layers, etc.) and membrane applications due to their different performance properties. They have also been proposed for the passivation of (micro)electronic components, especially as final encapsulation layers on multilayer structures. The term "encapsulation" refers to the coating or embedding of structural components, rather than to the formation of self-supporting shells around components in different aggregated states.

The purpose of the present invention is to provide a rapid hardening encapsulation method that avoids the introduction of solvents into the encapsulated material, avoids contamination or damage to sensitive components by foreign substances, and generally does not require temperature changes.

The encapsulation method according to the invention comprises co-extruding the crosslinked precursor material of the subsequent encapsulated material with the respective composition, whereby the crosslinked precursor material is extruded around the composition as an outer jacket (e.g. in a ring). Depending on the requirements, suitable process controls can be employed to obtain a fibrous or granular structure in which the composition is surrounded by the precursor material. Immediately or shortly after the extrusion step and preferably in relative proximity to the extrusion device, the fibers or particles are preferably guided by gravity through an area where the precursor material is crosslinked by external action. The resulting crosslinked product contains components that fill these structures, such as capsules or hollow fibers.

9.1.9 Sigma Technologies-The United States

Sigma Technologies is a global leader in surface functionalization technologies, utilizing environmentally friendly non-solvent and non-aqueous-based treatment and coating technologies to modify surfaces. Its technologies are used to manufacture web-based products such as radiation barrier materials for energy management, protective uniforms, packaging films, ultrahigh barrier materials for OLED displays and solar cells, and polymer dielectrics for capacitive energy storage [17].

9.1.9.1 Monolayer Barrier Films

A single protective coating over any inorganic barrier layer using SigMaterials vapor deposition technology can significantly improve barrier performance. Sigma monolayer barrier films consist of a single metallization layer and a functional protective layer for improved barrier performance and are ideal for most food and electronic packaging applications.

9.1.9.2 Multilayer Barrier Films

Sigma multilayer barrier films consist of multiple inorganic and organic layers that provide truly breakthrough barrier performance for industrial applications such as flexible displays and photovoltaics. Based on Sigma's technology of depositing functional nanolayers in vacuum, Sigma barrier films benefit from a variety of multilayer structures, including polymeric nanolayers applied before, after, or before and after the deposition of active barrier layers, as well as nanoparticle sealing layers that create both a tortuous path for water vapor and the equivalent of a moisture trap into the multilayer structure. The result is extremely superior barrier performance, better durability, and enhanced printability.

Some Patents

CN1960811 A Large-Area Electroluminescent Light-Emitting Device [18].

An EL light-emitting device was manufactured in a semi-continuous process (80, 40, 64) using a vapor-phase deposition technique to reduce the thickness of the dielectric layer (72, 52). The phosphor layer (34), the dielectric layer (52), and the electrode layer (58) are sequentially deposited on a flexible lamellar substrate (30), preferably PET (14) coated with conductive ITO (12), which passes continuously through the deposition zone (74, 32, 54, 82, 84). By depositing the dielectric layer in a vacuum (50), very thin layers are possible, which produce increased transparency and capacitance. The resulting multilayer structure is therefore suitable for the manufacture of large-area EL devices.

Contents of the invention

Considering the preceding, the present invention relates to the development of a semi-continuous process, which is mainly based on the application of polymer multilayer (PML) technology. According to one aspect of the invention, the deposition of (one or more) dielectric layers is performed by depositing and curing pure radiation-curable monomers under vacuum. As a result, the dielectric layer is formed as a very thin film, thereby increasing its transparency, compared to the thicker dielectric layer previously deposited by screen printing or equivalent processes under atmospheric conditions. Furthermore, the capacitance of the resulting EL device is correspondingly increased by the small distance between the two electrodes in the device. In the preferred embodiment, the dielectric layer is deposited on both sides of the phosphorescent layer. Alternatively, a pure single thin-film dielectric layer can be deposited on the front or rear side of the phosphor layer.

In all cases, these layers are deposited on a flexible lamellar substrate, preferably PET coated with conductive ITO, which passes continuously through each deposition zone. Traditionally, the deposition of phosphor layers is performed by screen printing or roll-on coating, or by depositing and curing phosphor powder mixed with a radiation-curable monomer binder under atmospheric conditions. After vacuum deposition of the phosphor layer (or layers) on the dielectric layer, the resulting multilayer structure is coated with a high-conductivity layer to form a back electrode (having a resistivity of less than 0.1 ohm/square, preferably on the order of 0.01 ohm/square). This step is preferably performed by vapor-phase deposition in a vacuum chamber. Alternatively, the metal layer can be precipitated and cured under atmospheric conditions as a mixture of metal powder and a radiation-curable binder. According to another aspect of the invention, all steps of each deposition stage are performed continuously on a flexible sheet, which is wound on a series or located on a sheet continuously fed from the core. Thus, since the basic length of the lamellar material is contained in the clad (or sheet), the size of the final device being fabricated is limited only by the width of the lamella (or sheet), which makes it possible to produce large EL displays in a semi-continuous manner. At each stage of deposition, the pick-up roll (or core of the sheet) is used as a feed roll in the next stage and is rewound on another pick-up roll to produce the final roll of the finished product. As a result of this method, all the steps required to manufacture the EL light-emitting device can be performed inline in two or three consecutive stages of operation, with only discontinuities arising from the need to move the pickups from one stage to the feed roll position in the next stage.

If desired, the final vacuum zone can include cells for depositing protective polymer layers on both sides of the structure. The multilayer compositions thus produced can then be divided as desired to obtain separate devices.

According to another aspect of the invention, a mask or equivalent device can be used to perform successive deposition of the phosphor and dielectric layers on the ITO electrode layer to prevent deposition on a predetermined portion of the ITO layer, preferably the edge swath on one or both sides of the sheet. Metal deposition of the back electrode layer is then performed to cover these exposed portions of the front ITO electrode, resulting in relatively large and continuous conductors along the edges of the ITO layer that can be used to increase the overall conductivity of the front layer. The back metal layer is then split as needed to isolate this edge from the portion intended for the back electrode. Thus, the rear electrode deposit also provides an extended conductor to increase the ability of the front electrode to illuminate a large area of the EL device.

Various other objects and advantages of the present invention will become clearer through the description in the following specification and through the novel features specifically noted in the appended claims. Accordingly, for the purposes described above, the invention comprises the features illustrated in the accompanying drawings, fully described in the detailed description of the preferred embodiments, and specifically identified in the claims. However, the accompanying drawings and description disclose only one of the many ways in which the invention may be practiced.

9.1.10 Toppan Printing-Japan

(Toppan), also known as Toppan, is a global printing company. The company was founded in 1900 and is headquartered in Kanda and Izumi-cho, Udaeda-ku, Tokyo, Japan. Toppan's products include commercial and publishing printing services, product packaging, functional material products, interior materials, photomasks, optical filters, touch sensor substrates, and more [19].

(1) Toppan Printing Co., Ltd.'s business segments and products include

9.1.10.1 Information Network

Securities and cards: investment security certificates, passbooks, product coupons, gift certificates, lottery tickets, data printing, IC cards.

Commercial printing: posters, catalogs, brochures, flyers, direct mail, calendars, yearbooks, corporate communication tools.

Book printing: weekly and monthly magazines, books, art books, dictionaries, textbooks, e-publications.

Business forms: RFID and NFC solutions, digital media-related business.

9.1.10.2 Living Environment

Packaging: flexible packaging materials, paper containers, cups, molded plastic products, complex liquid containers, cardboard containers, marketing planning, product planning, functional packaging materials development and manufacturing.

Functional products: solar cell backsheets, molded plastic products for electronic devices, materials for recording information, secondary battery components, interior decoration materials: decorative paper/sheets, wallpaper, flooring materials, interior lamps, decorative panels.

9.1.10.3 Electronics

Displays: LCD color filters, anti-reflection film, plasma display electromagnetic wave shielding mesh.

Semiconductor: semiconductor photomasks, large-scale integrated circuits, OEM equipment, lead frames, BGA/CSP substrates, color filter arrays, image sensors and small display devices, etched products, design services for printed circuit boards.

(2) GL Series Films

Toppan Printing Co., Ltd. began producing aluminum vapor deposition products in the early 1980s, and in the late 1980s, utilizing vapor deposition technology, we began producing silicon oxide vapor deposition products and began producing GL-type films.

GL-AE is a ceramic vapor deposition-processed PET film with excellent transparency and barrier properties. It is widely used as a standard type of ceramic vapor deposition film to replace VPDC-coated films or to rationalize the structural layer of packaging materials.

GL-AEH is a steaming type of product, which is a transparent steaming packaging material with high barrier property of glass.

GL-AEY is a biaxially oriented nylon film processed by ceramic vapor deposition that has excellent oxygen barrier performance and is an alternative to PVDC-coated biaxially oriented nylon films.

OP-M is a chlorine-free barrier material with OPP as the base material and special treatment to give the function of adding. It is the same price as KOP and has high performance as an oxygen and water vapor barrier.

GL-E is the only glass evaporation product in the GL series, with the same or higher barrier performance as GL-AE.

GL-AEO is an alternative grade to medium-high barrier KOP, and technical development has been completed.

Features of GL series products.

(1) Excellent water vapor barrier, oxygen barrier, and fragrance retention, which can effectively protect the contents of the bags.
(2) High transparency, so that the packaged goods can be seen.
(3) Unlike aluminum foil and aluminized laminated film, microwave heating can be used.
(4) With excellent durability, GL-AEH is suitable for steaming use. Good adaptability to the environment
(5) No toxic gas is produced when burning, and almost no toxic residue remains; good adaptability to processing
(6) It can be printed directly, which helps to rationalize the combination between layers of packaging materials; it is suitable for PL method requirements
(7) Because aluminum foil is different from vacuum-coated aluminum film, it can effectively avoid detection by metal detectors.
(8) New Conductive Film

We have recently developed a new type of conductive film by combining an optical barrier film and an organic conductive film, which achieves high transparency and bendability and can also be produced in large quantities (Table 9.9).

Applications.

- Protection of organic EL lighting, EL quantum dots, and other components.
- Protection of electronic materials (dimming films, electronic paper, etc.).
- Protection of organic semiconductor materials (inductor applications, etc.).
- Electromagnetic wave shield.

Some Patents

CN1791510 B Transparent Gas Shielding Laminated Film, Electroluminescent Element, Electroluminescent Display Device, and Electrophoretic Display Panel [20].

(1) The invention relates, for example, to transparent laminated films having high gas shielding to oxygen and water vapor and usable as packaging materials requiring high transparency for, for example, food, pharmaceuticals, electronic components, and optical parts; and to EL elements, EL display devices, and electrophoretic display panels utilizing said laminated films.
(2) Content of the Invention.
(3) The present invention solves the abovementioned problems of the prior art; it is an object of the present invention to provide gas-shielding laminated films with high gas shielding properties, such as for oxygen and water vapor, and also with high transparency.

Table 9.9 Characteristics of the new conductive film.

Dimension (MAX)	Thickness	Structure	WVTR	Conductivity
1300 mm (width) × 2000 mm (length)	50–135 μm	Heat-insulting film + organic conductive film (substrate:PET)	10^{-3} g/m²/day 40 °C, 90%	150 Ω/□

(4) Another object of the present invention is to provide EL elements, EL display devices, and electrophoretic display panels that are sealed with a gas shielding laminate film having high gas shielding properties such as oxygen and water vapor and also high transparency.

(1) According to a first aspect of the present invention, there is provided a transparent gas shielding laminate film comprising a first transparent gas shielding film, a self-adhesive layer of 1–30 microns formed on the first transparent gas shielding film, and a second transparent gas shielding film formed on the self-adhesive layer.
(2) According to a second aspect of the present invention, there is provided an EL element comprising.
(3) a laminate structure comprising a transparent electrode layer, an EL layer formed on the transparent electrode layer, a dielectric layer and one of the insulating layers formed on the EL layer, and a back electrode layer formed on the dielectric layer and one of the insulating layers, and a transparent gas shielding laminate film sealed to cover said laminate structure, the laminate film comprising a first transparent gas shielding film, a self-adhesive layer of 1–30 micron spleen formed on the first transparent gas shielding film, and a second transparent gas shielding film formed on the self-adhesive layer.
(4) According to a third aspect of the present invention, there is provided an EL display device comprising.
(5) A laminate structure comprising a color filter layer, a transparent electrode layer formed on the color filter layer, an EL layer formed on the transparent electrode layer, one of a dielectric layer and an insulating layer formed on the EL layer, and a back electrode layer formed on the dielectric layer, and the insulating layer and a transparent gas shielding laminate film sealed to cover said laminate structure, the laminate film comprising a first transparent gas shielding film formed 1–30 micron thick self-adhesive layer, and a second transparent gas shielding film formed on the self-adhesive layer.
(6) According to a fourth aspect of the present invention, there is provided an electrophoretic display panel comprising
(7) a display unit comprising a transparent electrode layer, a display area containing electrophoretic particles and an insulating fluid for dispersing the electrophoretic particles, and a back electrode formed on the display area and a transparent gas shielding laminate film sealed to cover said display unit, the laminate film comprising a first transparent gas shielding film, a self-adhesive layer of 1–30 microns formed on the first transparent gas shielding film, and a second transparent gas shielding film formed on the self-adhesive layer.

More specifically, the present invention provides aspects of

(1) A transparent gas shielding laminate film characterized by comprising a first transparent gas shielding film, a pressure-sensitive binder layer formed on the first transparent gas shielding film of 1–30 microns, and a second transparent gas shielding film formed on said pressure-sensitive binder layer.

(2) The transparent gas shielding laminate film according to item 1 above, characterized in that at least one of the first and second transparent gas shielding films comprises a substrate and an inorganic oxide deposition layer deposited on the substrate.
(3) The transparent gas shielding laminate film according to item 1 above, characterized in that the first transparent gas shielding film comprises a substrate and an inorganic oxide deposit layer deposited on the substrate, and the second transparent gas shielding film comprises a gas shielding resin layer.
(4) The transparent gas shielding laminate film according to item 2 above, characterized in that said inorganic oxide comprises at least one selected from aluminum oxide, silicon oxide, and magnesium oxide.
(5) A transparent gas shielding laminate film according to item 2 above is characterized in addition by comprising: a gas shielding coating formed on the inorganic oxide deposit layer by applying a coating agent and drying the coating agent by heating: said coating agent comprising as its main component an aqueous solution or an aqueous/alcoholic solution mixture, said aqueous solution or aqueous/alcoholic solution mixture comprising a water-soluble polymer and at least one of (a) and (b), wherein (a) is one or more metal alcohol salts and/or hydrolysis products thereof, and (b) is tin chloride.
(6) The transparent gas shielding laminate film according to item 5 above, characterized in that said metal alcohol salt comprises at least one of tetraethoxysilane and triisopropoxyaluminum selected from among the above.
(7) The transparent gas shielding laminate film according to clause 5 above, characterized in that said water-soluble polymer is polyvinyl alcohol.
(8) A transparent gas shielding laminate film according to item 1 above, characterized in that it additionally comprises a heat-sealing layer formed on at least one of the first and second transparent gas shielding films.
(9) An EL element characterized by comprising a laminate structure comprising a transparent electrode layer, an EL layer formed on the transparent electrode layer, a dielectric layer and an insulating layer formed on the electroluminescent layer, and a back electrode layer formed on either the dielectric layer or the insulating layer; and a transparent gas shielding laminate film as described in any one of items 1–8 above, and said laminate film being laminated to a transparent gas shielding film according to item 1 above, and said laminate film is sealed to cover said laminate structure.
(10) An EL display device, characterized in that it comprises the EL element as described in item 9 above and a color filter layer formed on said transparent electrode layer between said transparent electrode layer and said transparent gas shielding laminate films.
(11) An electrophoretic display panel characterized in that it comprises a display unit, said display unit comprising a transparent electrode layer, a display area containing electrophoretic particles and an insulating fluid for dispersing the electrophoretic particles, and a back electrode formed on the display area; and a transparent gas shielding laminate film as described in any of items 1–8 above, and said laminate film being sealed to cover said display unit.

(12) The electrophoretic display panel according to item 11 above, characterized in that said display unit contains a capsule encapsulating an insulating fluid with electrophoretic particles dispersed therein.

9.1.11 BASF(Rolic)-Germany

BASF SE (Germany) is one of the world's largest chemical companies. The BASF Group has wholly owned or joint venture subsidiaries in more than 80 countries in Europe, Asia, North America, and South America, six integrated innovation sites, and 338 other product production sites, and employs approximately 112,000 people worldwide. Its main products are plastics, rubber, high-performance polymers, rubber additives, and BASF plastic additives. Its main business areas are chemicals, materials, industrial solutions, surface treatment technologies, nutrition and care, and agricultural solutions [21].

BASF pursues customers to bring high-value products and technologies that also contribute to the advancement of the LCD and OLED industries. In 2017, BASF completed the acquisition of Rolic AG. Rolic is a specialized company in the LCD/OLED industry, so in the flat panel display sector, BASF is able to provide the market with thin-film transistor arrays, color filters, polarizers, touch panels, and 3D Rolic develops barrier films for OLEDs with a WVTR as low as 10^{-6} g/m^2/day.

Some Patents

CN107405893 B

Barrier Film Laminates Including Submicron-Absorbing Particles and Electronic Devices Including Such Laminates [22].

A barrier film laminate (1) comprising an organic layer (4) disposed between two inorganic layers (2,3). The organic layer comprises submicron-absorbing particles between 0.01% and 0.9% by weight. The barrier film laminate can be used to package OEDs, such as OLEDs. The long-lasting uniform transparency makes the laminate particularly suitable for protecting the light-emitting side of OLEDs.

Contents of the invention

Deterioration of OEDs means a deterioration of the functional properties of the device. The luminous intensity of the OLED may decrease or the distribution of the light intensity along the emitting surface may become uneven. Black spots may appear in OLEDs due to local degradation of the electrodes, such as oxidation of the cathode. This is due to the use of commonly used low-function metals, such as calcium or barium, which can be unstable under the influence of air and moisture. Degradation of the OLED due to physical or chemical changes at the interface between different layers of the device, for example, can lead to changes in the IV characteristics of the OLED. Oxidation occurring at the interface of the device is a well-known source of degradation. The degradation of functional properties may eventually lead to a complete loss of any functionality, in which case the OLED will no longer emit light. In the case of organic transistors, degradation may be reflected in reduced, suppressed charge carrier mobility and/or reduced current output, resulting in reduced charge transport in the transistor. Well-known sources of such degradation in organic electronics are the oxidation of electrodes and charge carrier capture. Degradation of organic photovoltaic cells can lead to a reduction in the efficiency of light energy conversion.

To avoid water vapor diffusion into the electrode of an OLED, for example, a barrier film laminate including a stack of organic layers sandwiched between two inorganic layers can be applied. The acceptable lifetime can be further improved by dispersing absorbing particles in the organic layer that will at least partially absorb the incoming water. The problem with such a barrier film, however, is that the optical performance of the transparent layer is altered by the absorption of water. Although the present inventors do not intend to be limited by any theory, it is generally believed that the problem is caused by a change in the refractive index of the absorbing particles due to the absorption of water. As a result of such an effect, the difference between the refractive index of the absorbing particle and the organic matrix material surrounding the absorbing particle is changed during the lifetime of the barrier, and more particularly, this is due to the absorption of water. When referring to an absorbing particle or absorbing material, it means a particle or material suitable for absorbing water.

In order to protect a light-emitting device such as an OLED from electrode degradation or other degradation of the device by the environment, a barrier may be used to protect the light-emitting side of such a device, and the barrier must be transparent. During the lifetime of the device, the optical properties of the barrier, such as transparency and color, may change. In the case where the barrier film includes a transparent organic layer in which the absorbing particles are dispersed, the initially homogeneous barrier film can become inhomogeneous due to the absorption of water by the organic layer and, more particularly, due to the absorption of water by the absorbing particles.

It is the object of the present invention to overcome the problems mentioned above with respect to the encapsulation of existing OEDs. In particular, it is intended to provide a barrier film laminate that acts as a good moisture barrier.

The object is obtained by means of a barrier film laminate comprising a first inorganic layer, a second inorganic layer, and a first organic layer comprising submicron-absorbing particles, the organic layer being located between the first inorganic layer and the second inorganic layer, characterized in that the number of submicron-absorbing particles in the organic layer is between 0.01% and 0.9% by weight of the organic layer. The first organic layer may be the first and only organic layer of the barrier film laminate. The first organic layer may also be the first organic layer of two or more organic layers of the laminate.

The word "film" as used herein is for a product having a sheet geometry. One dimension of the film, i.e. the thickness, is substantially smaller than the dimension in either of the other two directions, typically by at least an order of magnitude. Film laminates are stacks of at least two films that adhere to each other and therefore form an adherent product, which is itself usually a film.

The surprising effect of the number of submicron-absorbing particles whose weight is calculated to lie between 0.01% and 0.9% is that, although the number is less than 1% by weight, the number still provides sufficient resistance to water vapor blocking during the time period corresponding to the service life of the OED.

In the preferred embodiment of the present invention, the first inorganic layer, the second inorganic layer, and the first organic layer are optically transparent. The advantage of the transparent layer is that the transparent barrier laminate can be used as a barrier on the side of the OED that must be transparent for proper operation. Therefore, especially for the OLED emission side or the OPV sunlight receiving side. The surprising effect is that the light transmission of transparent blocking film laminates that include submicron-absorbing particles between 0.01% and 0.9% by weight in the organic layer does not change significantly during the time period matching the service life of the electronic device based on such a laminate or including such a laminate (in particular, light-emitting diodes). Thus, such an optical laminate provides a barrier having stable optical properties. More particularly, such a laminate provides a transparent barrier having an appearance and transparency that remain uniform over a long period of time.

Here, the term "optically transparent" or simply the word "transparent" is used in connection with the layer or laminate to refer to the property of such a layer or laminate that allows light, particularly the visible portion of the electromagnetic spectrum, to pass through the layer or laminate. Thus, "transparent" in this context means not only the property of allowing a person to see through such a layer or laminate and see the part behind it but also translucent layers and laminates, which, for example, cause blurred vision due to light scattering.

In an embodiment of the present invention, the barrier film laminate includes a second organic layer that does not include an absorber (getter). Such a second organic layer may be an intermediate layer located between the environment and an inorganic layer. Such a layer may have a different function. It may protect the inorganic layer from damage, for example, mechanical damage. The second organic layer may also be a flattening layer between the substrate and the inorganic layer. The second organic layer may also be located between the organic layer comprising submicron-absorbing particles and the inorganic layer, for example, to provide a suitable interface.

In embodiments of the present invention, the barrier film laminate includes a third inorganic layer and a second organic layer that includes a quantity of submicron-absorbing particles between 0.01% and 0.9% by weight of the second organic layer. In the preferred embodiment, the third inorganic layer and the second organic layer are transparent. The second organic layer is disposed between the first inorganic layer and the third inorganic layer such that the barrier film laminate comprises an alternating stack of organic and inorganic layers. The advantage of such laminates is that improved barrier properties can be obtained, or the individual layers need not be of very high quality, as in the case of laminates including a single organic layer, but still provide the same barrier properties. For example, the density of pinholes or other defects in the inorganic layer may be high, or the number of submicron-absorbing particles in an organic layer or two organic layers may be at a lower limit, e.g. between 0.01% and 0.5% by weight, or between 0.01% and 0.1% by weight, and still provide blocking properties such that, in other cases, the blocking properties can only be achieved by having a higher number of absorbing organic layers of material or a high-quality inorganic layer.

In embodiments of the present invention, the number of submicron-absorbing particles in the sole organic layer, or in one or more organic layers, has an average particle size of 200 nm or less. Such small absorbent particles have the advantage of having a large surface area compared to their volume and therefore provide effective water absorption. When such small particles are uniformly dispersed in the organic layer and do not aggregate into large agglomerates (conglomerate), they are not visible. Another advantage of small absorbent particles is that the organic layer does not need to be thickened in order to encapsulate the absorbent particles.

In embodiments of the present invention, submicron absorber particles include calcium oxide, magnesium oxide, barium oxide, or strontium oxide. Such metal oxides have good water absorption. Preferably, one of the first two metal oxides, more preferably calcium oxide, is used as the absorber.

In embodiments of the present invention, the submicron absorbent particles are embedded in a radiation-cured organic material. Radiation curing is advantageous in that it enables the material surrounding the absorber particles to cure rapidly, thus enabling rapid solidification of the particles in the organic layer.

A further object of the present invention is to provide an encapsulated OED with long-term stability and good performance. This object can be obtained by an encapsulated OED comprising a barrier film laminate as described above. The use of such a barrier film laminate is useful in that it will significantly reduce the moisture penetrating into the device. Thus, the deterioration of the electronic device will be delayed, making it have good performance for an acceptable period of time.

In the embodiment of the present invention, the exposed electronic device is located between the substrate and the barrier film laminate, according to the present invention. The advantage of such an embodiment is that (especially for asymmetric devices), both the substrate and the barrier can be optimized for use on all sides of the electronics. For example, the transparency of one side of the OLED is optimized.

9.1.12 Vitex(Samsung)-The United States

Vitex Corporation has commercialized ultrahigh-barrier films using the PML process, which was developed from the process used by GE to manufacture capacitors in the United States. In the late 1970s and early 1980s, GE manufactured a multilayer polymer/aluminum structure for capacitors that had a very smooth surface, and some researchers realized that this structure could improve the performance of barrier materials.

The key to this process is to first deposit a liquid monomer film (polyacrylate) on the substrate by flash evaporation. Because this film is liquid, it fills in surface defects, resulting in a smooth, flat surface. The film is then UV-cured to form a film. This allows the surface roughness to be less than 1 nm, and then the barrier material is deposited by reactive sputtering on the smooth surface. The organic and inorganic films are deposited in several cycles, and the product obtained from depositing several pairs (one organic layer plus one inorganic layer) can achieve ultrahigh barrier performance. The organic layer in this structure allows the defects in the inorganic layer to be unrelated to each other, so that water and oxygen passing through one layer cannot pass directly through the entire structure.

The PML process can be used either as a separate barrier substrate (flexible substrate) or as an encapsulation material (BarixTM). Multilayer barrier films are made continuously on a roll-to-roll process (no printing) and can be up to 1.5 m wide with an upper-speed limit of 3 m/min. Production costs vary depending on process design, size, and production capacity, ranging from \$0.05 to \$0.10/in^2 for Barix encapsulation materials and \$5 to \$50/m^2 for flexible barrier films (depending on the substrate and required properties). The layer thickness varies by application, from 0.25 to a few microns for a single polymer layer and 20–100 nm for a single inorganic layer (Al_2O_3). Alumina layers are deposited using DC reactive sputtering, which has a low process temperature of 40 °C on average and only 80 °C at the highest temperature. The transparency of the product can reach 90% in the visible range. Barix package coated with an organic layer on top of the inorganic layer not only fills the pinholes of the inorganic layer but also enhances the bending resistance of the device. With this inorganic–organic multilayer package structure, the water permeability can be 10^{-4} to 10^{-6} g/m^2/day and the oxygen permeability can be 0.005 cc/cm^2/day (Tables 9.10 and 9.11).

Barix products are used as direct replacements for glass covers on OLED displays. In 2008, Samsung SDI used Barix products to create the world's thinnest display.

Table 9.10 Vitex product parameters.

Company	Product model	Product specification	Price (¥/m^2)	WVTR (g/m^2/day)	OTR (cc/m^2/day)
Vitex	Barix®	1520 m	500–1000	$<10^{-6}$	$<10^{-4}$

Table 9.11 Vitex product structure.

Product structure	WVTR (g/m^2/day)	Diagram	Method of production	
			Organic layer	Oxide layer
PEN/Organic layer/Oxide layer)/Organic layer	10^{-4}–10^{-7}	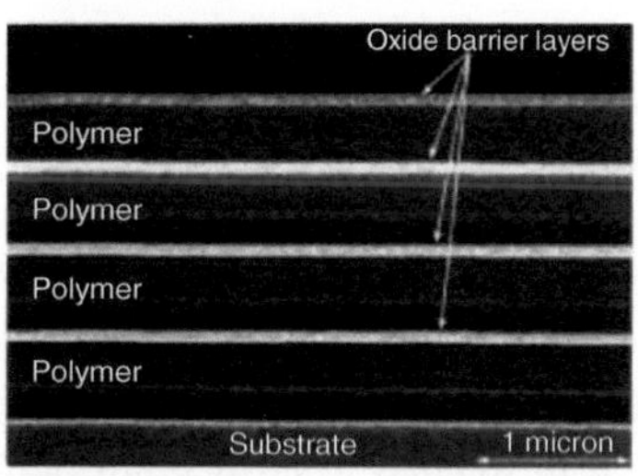	CVD	Sputter

Some Patents

CN101231974 B Multilayer Coating and Devices and Methods Using the Multilayer Coating [23].

A multilayer coating for environmentally sensitive devices, a method for using the multilayer encapsulated article, and an apparatus for depositing a multilayer coating on a substrate. The apparatus includes a chamber defining a vacuum chamber connected to a vacuum source, several deposition stations each configured to deposit layers of the multilayer coating on the substrate, a curing station, and a contamination reduction device. At least one deposition station is configured to deposit an inorganic layer, while at least one additional deposition station is configured to deposit an organic layer. This device is particularly suitable for depositing multilayer coatings onto flexible substrates and for encapsulating environmentally sensitive parts placed on such flexible substrates.

Content of the Invention

In accordance with an aspect of the present invention, an apparatus for inline deposition of multilayer coatings on discrete thin substrates is disclosed. In this context, the inline apparatus differs from the tape and reel apparatus because, first, the inline apparatus is configured to process discrete flakes, while the tape and reel apparatus processes continuous webs, and second, the deposition station also generally follows a linear planar path along the inline apparatus (which can move around unidirectional/once through or shuttle/multiple passes) so that during the coating process the apparatus does not encounter extreme bending of the substrate (and any of the devices to which it is mounted) extremely curved paths that would otherwise be detrimental to the coating or the devices encapsulated by the coating. For the purposes of this document, the deposition path is considered to be essentially linear. Inline devices differ from cluster devices in that for inline devices, the deposition of each layer of a multilayer coating is performed in a continuous, sequential path in a common environment, while for cluster devices, each coating is deposited in a separate chamber isolated from the surrounding environment and adjacent chambers. The apparatus includes a proximal end configured to receive the substrate, a distal end opposed to the proximal end, and at least one chamber (housing) disposed substantially between the proximal and distal ends. This housing defines a substantially linear deposition path to facilitate the transport of the substrate through the apparatus and comprises at least one organic layer deposition station, a curing station, an inorganic layer deposition master, and at least one contamination reduction apparatus to control the migration of material comprising said organic layer from said organic layer deposition station. Said material occurs at said organic layer deposition station. The chamber also defines a shared vacuum in which the entire organic, curing, and inorganic stations are placed in such a way that at least the inorganic layer deposition station and the organic layer deposition station are in vacuum communication with each other after the operation of the vacuum source connected to this shared vacuum. The inorganic layer deposition master is configured to deposit at least one inorganic layer of the multilayer coating, while the organic layer deposition station is configured to deposit at least one organic layer of the multilayer coating, and the curing station is configured to cure the organic layer deposited by the organic layer deposition station. Both the organic and inorganic layer deposition stations are configured so that they can deposit the respective coatings on the substrate. In this context, depositing a coating "onto" a substrate includes coating in direct contact with the underlying substrate and coating on one or more coatings previously deposited on the substrate as adjacent stacked portions. In this way, an organic or inorganic layer can be deposited first, and both coatings, even in the form of a multilayer structure, are considered to be deposited on the substrate. The essentially linear deposition path of this inline structure is not required to have a physical separation in a separate isolated station.

Optionally, the device may include additional devices, including one or more mask stations. These can be assembled from organic mask placement devices and inorganic mask placement devices, each configured to place an appropriately shaped and sized mask onto the substrate prior to entering the organic and inorganic stations, respectively. One or both proximal and distal ends can be configured as an accumulator so that it can contain a batch of substrate at least before, after, or between the multilayer coating deposition steps. In this context, a "batch" of substrates includes one or more substrate sheets placed in environmentally isolated and controlled equipment so that they are processed in a single equipment run. Thus, while a multisubstrate may comprise a continuous stream of such substrates added individually to the apparatus, a batch of such substrates is a subset of many multisubstrates whose formation is limited by collector capacity rather than continuous stream size (size). For devices with collectors at both ends, the device can handle multiple batches (preferably two) of substrates simultaneously. In this context, the term "simultaneous" refers not to subjecting batches of substrates from two batches to the same deposition in parallel (which would essentially be equivalent to the case of two batches combined), but rather to the ability of the equipment to sequence its deposition and nationalization steps so that all substrates in a batch can be shuttled through the corresponding station or reserve station, and the next step in the process can be performed on another batch in the equipment. equipment before proceeding to the next deposition step for another batch of material in the equipment and returning it to one of the collectors. Thus, more than one batch can be produced between the time the substrate batch is loaded and separated in the apparatus and the time it leaves the apparatus. This collector can further be configured to reverse the substrate along a substantially linear deposition path to deposit multiple layers of a multilayer coating. As the first device where the substrate intersects along the inline apparatus, the collector can be configured to at least partially isolate the surrounding external environment from one or more pieces of substrate. In this context, partial environmental isolation includes the ability of the collector to reduce at least one of the vacuums or temperatures in the area containing the substrate to a level necessary for normal operating conditions of the substrate prior to or during the multilayer coating deposition process. This collector may include a temperature regulation feature to reduce the temperature within the collector. As a result of this feature, the collector acts as a temperature control unit, suppressing the temperature rise of the substrate or device due to the deposition process. The collector may also include an environmental isolation valve so that the valve can be closed once the substrate (or substrates) is placed inside the collector, after which the temperature and vacuum level can be changed at will.

The apparatus may also include at least one surface treatment chamber configured to enhance the adhesion of each layer of the multilayer coating to the substrate or adjacent layers. The surface treatment chamber may be placed in a chamber, in a collector, or adjacent to one of the two. The apparatus may place the inorganic layer on top of the substrate prior to placing the first organic layer. While the use of sputtering facilitates the application of multiple impermeable layers to the device, other forms, including thermal evaporation, can be used to achieve the coating effect without exposing the environmentally sensitive devices being coated to harmful environments such as high temperatures and/or plasma. Special measures can be taken to avoid damage to environmentally sensitive devices (OLEDs) that might otherwise be caused by exposure to the plasma and/or temperature of the sputtering coating process. In one approach, the first deposited inorganic layer can be deposited by thermal evaporation rather than sputtering. For example, since thermal evaporation is the method currently used to form the metal top electrode of OLEDs, this inorganic layer deposition method can also be used as an enhanced coating method. Unlike oxides such as alumina (Al_2O_3), which are typically sputtered reactively, inorganic species such as fluoride (LiF) and fluoride (MgF_2), both of which are optically transparent, can also be coated by thermal evaporation to form a protective layer without exposing environmentally sensitive devices to the plasma. Again, this approach may utilize an inorganically transparent metal halide that has been thermally evaporated, a sputtered transparent inorganic or first-deposited organic, or a simpler method where thermal evaporation can be used for first-deposited inorganic). The latter may require a first-deposited inorganic material that can be coated by thermal evaporation and that provides a combination of adhesion and transparency.

In one form, the apparatus for reducing contamination is a thermal control device, which is placed close to at least one side of the organic layer deposition station, preferably close to the upstream and downstream sides of the organic layer deposition station. For example, the thermal control apparatus may be a chiller configured to mitigate atmospheric conditions in and around the organic layer deposition station, or it may be a heat accumulator. For example, a first chiller may be placed in a first migration control chamber at the inlet of an adjacent station and a second chiller in a second migration control chamber at the outlet of an adjacent station. Other cooler placements and configurations can also be envisioned depending on system requirements. For example, these coolers can be cold inert gas (e.g. nitrogen) type injection devices that are placed upstream and downstream of the organic deposition station. In addition to thermal control devices, the contamination reduction apparatus may include at least one baffle placed on at least one side of an adjacent organic deposition station to form a curved path that makes it more difficult for excess organic layer material to migrate out of the organic deposition station and into other stations.

The apparatus may be configured to shuttle the substrate through the chamber as many times as necessary to deposit multiple layers of coatings on the substrate. For reciprocal movement, one or more conveyor belts may be included that extend through one or more chambers to transport the substrate through at least a portion of the apparatus. The belts may be configured to move in both directions between the proximal and distal ends. Such equipment may also include an inspection chamber to facilitate testing the resistance of the multilayer coating to environmental attack. Examples of multilayer impermeable coatings configured to prevent environmental leaching include oxygen and water penetration. Therefore, a current method for testing permeation resistance is based on vacuum depositing a thin layer that is sensitive to oxygen or water (e.g. calcium) and then depositing a multilayer impermeable coating to form a sample that can be tested. One assay chamber that facilitates this approach consists of vacuum depositing a thin sensitive layer on an uncoated substrate at the main occupation point, forming a detection blank with similar sensitivity to an OLED. The ability to prepare test samples for the application of multiple impermeable coatings in the same environment (maintaining a vacuum throughout the process) increases the accuracy (validity) of the assay results and reduces turnaround time.

A control system can be included to determine the operability of the various equipment devices and process conditions within the chamber plug, as well as to be responsive to process parameters such as temperature, scan rate, and presence of contaminants. In the deposition of inorganic layers, the vacuum source provides a different vacuum level than in the deposition of organic layers. For example, the vacuum level during the deposition of the inorganic layer can be about 3 mbar, while it can be about 10 mbar during the deposition of the organic layer. In another option, the inorganic layer deposition station includes a rotating sputtering source, which may include a rotatable cathode.

Preferably, the inorganic layer is deposited on the substrate before the organic layer is placed. The present inventors have found that placing the inorganic (e.g. oxide) layer first results in improved adhesion between substrates and layers, and also improves barrier properties. The inventors also found that the "inorganic-first" approach resulted in strong adhesion and excellent impermeability in the case of a target (such as QLFD) envelope placed on the substrate. Thus, although the inclusion of an organic layer continues to make a valuable contribution to the total performance of the multilayer coating, the present inventors' research illustrates that one or more inorganic layer/organic layer pairs, guided by an inorganic layer, can be used to obtain a suitable base (or substrate) sufficiently to achieve effective separation of the impermeable layer from the underlying substrate ("or" device). By first placing the inorganic layer on the substrate (e.g. glass or plastic), the inventor achieves bonding to the substrate, bonding to the device placed on the substrate, and bonding between multiple layers of environmental impermeable layers, all of which withstand the harsh conditions of the physical and thermal environment to which they must be subjected for operation. In addition, when these layers are formed on the surface on which the device is placed, they withstand all the processing associated with the manufacture of the device. The present inventors believe that at least one explanation may be that the migration of organic material from the substrate to the first coating is reduced compared to the case where the first layer is organic, and that this reduced migration promotes and maintains enhanced adhesion between the substrate and the first coating. Furthermore, in cases involving the deposition of a device mounted on a substrate, the present inventors believe that for the first deposited organic layer, this layer does not sufficiently wet or uniformly cover the surface of the device. This may be due to the fact that the substance is derived from the organic layer of the device being coated, and that there is no suitable formulation for the first deposited organic layer relative to the device or a combination of the two. On the other hand, an "organic-first" approach (at least in the case of an envelope) may reduce or even eliminate possible damage to the device due to the plasma used in depositing the inorganic layer.

In accordance with another aspect of the invention, an apparatus is disclosed for coating a target between a multilayer coating and a flexible substrate. The apparatus comprises at least one chamber (which is substantially defined by a shared vacuum and a substantially linear deposition path therein), a vacuum pump connected to the vacuum chamber, a first collector placed upstream of the chamber, and a second collector placed downstream of the chamber. The first collector is configured to isolate at least a portion of the environment of the substrate from the external surroundings once placed on the substantially linear deposition path, and the second collector is configured to isolate at least a portion of the environment of the substrate from the external surroundings and to reverse the substrate along the substantially linear deposition path so that multiple layers of multilayer coatings can be deposited on the substrate. The chamber contains at least one organic layer deposition station, at least one nationalization station, at least one inorganic layer deposition station, a shielding station configured to place organic and inorganic shielding on the substrate, and at least one contamination reduction device to control migration of the organic layer formulation material. Optionally, the apparatus may also include a positioning device placed upstream of the first collector, while at least one collector may include a thermal control device. Alternatively, the first collector includes a substrate entry path and a substrate exit path, which are separated from each other by the substrate entry path.

In accordance with another aspect of the present invention, an envelope apparatus is disclosed for the inline deposition of a multilayer coating on a substrate to protect the target material placed thereon. The apparatus includes at least one chamber substantially defining a shared vacuum and a substantially linear deposition path therein, apparatus (means) for depositing the first material on the target in at least one chamber, apparatus for curing the first material on the target in the at least one chamber, apparatus for depositing a second material on the target in the at least one chamber, apparatus for depositing a first material on the target in the at least one chamber, apparatus for forming a vacuum in the at least one chamber so as to deposit the first material in the at least one chamber, and apparatus for depositing a second material in the at least one chamber. The apparatus for depositing the first material by creating a vacuum in at least one chamber, the apparatus for depositing the second material and the apparatus for nationalizing the first material both in vacuum communication with each other, and the apparatus for controlling migration of the first material from the apparatus for depositing the first material. Optionally, the wrapping apparatus is configured such that the first or second material is first coated adjacent to the substrate, and the chambers may be configured as multiple chambers sequentially connected to each other so that the multiple chambers all share a vacuum. The chambers may be configured as multiple chambers sequentially connected to each other so that the multiple chambers all share a vacuum. As in previous aspects, at least one collector may be included to at least partially isolate the substrate from the external surroundings. The McCollector may share a selective vacuum with the chamber and may include a device configured to reduce the temperature within the collector.

In accordance with another aspect of the invention, a method of depositing a multilayer coating on a substrate is disclosed. The structure of the apparatus is in accordance with at least one of the preceding aspects. The method comprises several steps: loading the substrate into the chamber, forming at least a partial vacuum within the chamber, introducing inorganic material into the inorganic layer deposition station, depositing at least a portion of the inorganic material as a component of the multilayer coating, introducing organic material into the organic layer deposition station, depositing at least a portion of the organic material as a component of the multilayer coating, curing the deposited organic material, and controlling the migration of excess organic material out of the organic layer deposition station. migration out of the organic layer deposition station. Optionally, the method includes this further step of treating at least one surface of the substrate prior to forming the first layer of the multilayer coating. In one form, the step of controlling the migration of excess organic material includes cooling at least a portion of the space defined by the organic layer deposition station to reduce some of the excess organic material remaining in the vapor phase in and around the organic layer deposition station. For example, the cooler can be placed in such a way that it can exchange heat with the space defined by the organic layer deposition station. Baffles can be used to reduce the transfer of vapor contaminants across adjacent stations by reducing the basin of permeable gas crossing between stations. Additional steps may include placing an inorganic shield over the substrate prior to the deposition of the inorganic material step and an organic shield over the substrate prior to the deposition of the organic material step. To reduce the incidence of leakage and related capillarity, the shield may be stacked to form an undercut shield, or the organic shield may be removed prior to the nationalization step. Removing this shield prior to curing can also increase the speed of curing by eliminating the organic material edge shielding. Controlling the migration of excess organic material involves cooling at least a portion of the space within the organic layer deposition station, thereby reducing the amount of excess organic material retained in the vapor phase at the organic layer deposition station. The cooler may be placed in heat exchange with the space defined by the organic layer deposition station for cooling, while additional steps may be to place a first collector upstream of the chamber, place a second collector downstream of the chamber, and include a device in at least one of the collectors configured to reduce the temperature rise on the substrate caused by deposition of organic material, solidification of organic material, and deposition of inorganic material.

This invention also includes a method for coating a target material placed on a substrate. The coating apparatus may be configured as described above. The process of coating the target includes the following steps: loading the substrate onto the device and placing it into the chamber, creating a partial vacuum in the chamber, introducing inorganic material into the inorganic layer deposition station, depositing at least a portion of the inorganic material, introducing organic material into the organic layer deposition station, depositing at least a portion of the decoupled organic material, and removing excess organic material from the organic layer deposition station to reduce contamination. The excess (undeposited) decoupled organic material is separated, and the deposited organic material is solidified. Optionally, the steps of depositing the organic and inorganic materials can be repeated at least once, and materials consistent with the two layers may be applied in any alternating order.

To facilitate the evaporation of the organic layer, the vaporized organic material can be drawn into the main organic layer deposition, using techniques such as vacuum flash evaporation. The undeposited organic material can be separated by cooling at least a portion of the organic layer deposition station, which causes the undeposited vapor organic material to condense. The inorganic material may be a ceramic, while the organic material may be a polymer precursor, such as a monomer. These materials can be selected using any of the previously discussed aspects.

9.1.13 General Electrics-The United States

GE USA employs a PECVD process that incorporates a flat plate capacitively coupled plasma source to deposit a barrier layer featuring a continuous gradient structure. The continuous gradient structure is primarily composed of silicon compounds, including silane, silicon nitride, or silicon nitride oxide, in the organic–inorganic layers, as shown in the figure.

The organic barrier layer consists mainly of silane, and the inorganic barrier layer consists of silicon nitride. Compared with the multilayer structure with abrupt changes between interfaces, this gradient structure can reduce the delamination between multilayers and enhance the adhesion effect. The gradient structure of the barrier film can prevent defect expansion in terms of thickness and also improve mechanical stability and stress relaxation. This film is thinner compared to barrier films produced by PML, which helps to reduce water transmission in the horizontal direction. GE's barrier layer is coated with high-temperature polycarbonate (Tg about 240 °C). The polycarbonate is coated on both sides with a chemically resistant epoxy resin. The epoxy resin coating improves the chemical resistance of the base film. High temperature can make the film deposition more uniform, with fewer defects and better barrier performance. The complete barrier film from GE is smooth, with an average roughness of <1 nm and peaks and valleys of about 20 nm. Its shrinkage rate at 200 °C is 25×10^6/h. The WVTR of the barrier film has reached 5×10^{-6} g/m^2/day.

Some Patent

US20090286010A1

High-Throughput Processes and Systems for Barrier Film Deposition and/or Encapsulation of Optoelectronic Devices [24].

Contents of the Invention

In brief, one aspect of the present technology provides an organic device package. This organic device package includes a flexible substrate (flexible substrate) having a top surface and a bottom surface. In addition, the organic device package includes an OED with a first side and a second side set on the top surface of the flexible substrate. In addition, the organic device package includes a first-blocking layer provided on the bottom surface of the flexible substrate.

According to another aspect of the present technology, a method of manufacturing an organic device package is provided. The method includes providing a flexible substrate having a top surface and a bottom surface. Further, the method includes providing an OED having a first side and a second side on the top side of the flexible substrate. The method also includes providing a first barrier layer on the bottom surface of the flexible substrate. Alternatively, the method includes providing a second barrier layer close to the first face of the OED such that the perimeter of the second barrier layer is adapted to surround the edge of the organic device package. In addition, the method includes connecting (coupling) the second barrier layer to the side of the first barrier layer opposite to the flexible substrate.

According to another aspect of the present technology, an organic device package is provided. The organic device package includes a flexible substrate having a top surface and a bottom surface. In addition, the organic device package includes an OED having a first surface and a second surface provided on the top surface of the flexible substrate. In addition, the organic device package includes a first barrier layer having a first inner surface and a first outer surface provided on the bottom surface of the organic substrate. In addition, the organic device package includes a second barrier layer having a second inner surface and a second outer surface provided close to the first surface of the OED. The organic device package further includes an edge sealing material disposed around the perimeter of the OED between the first and second barrier layers, wherein the edge sealing material is disposed to hermetically seal the perimeter edge of the organic device package.

[0011] According to a further aspect of the present technology, there is a method of manufacturing an organic device package. The method includes providing a flexible substrate having a top surface and a bottom surface. The method includes providing an OED having a first side and a second side on the top side of the flexible substrate. Further, the method includes providing a first barrier layer having an inner surface and an outer surface on the bottom surface of the flexible substrate. Alternatively, the method includes providing a second barrier layer having an inner surface and an outer surface close to the first surface of the OED. Alternatively, the method includes connecting the first and second barrier layers.

[0012] According to a further aspect of the present technology, an organic device package is provided. The organic device package includes a flexible substrate having a top surface and a bottom surface. Alternatively, the organic device package includes an OED having a first side and a second side disposed on the top surface of the flexible substrate. Further, the organic device package includes a first barrier layer having a first inner surface and a first outer surface provided on the bottom surface of the flexible substrate. The organic device package includes a second barrier layer having a second inner surface and a second outer surface provided close to the first surface of the OED, wherein the second barrier layer includes an edge adapted to surround the organic device package such that the periphery of the second barrier layer is connected to the side of the first barrier layer opposite to the flexible substrate. Further, the organic device package includes a functional layer having a first side and a second side provided on a portion of the first outer surface, wherein the functional layer is configured to enhance the output of the organic device package.

9.1.14 Mitsui Chem-Japan

Mitsui Chem is one of the largest chemical groups in Japan and a globally integrated chemical company with a long-standing history in the country. Its headquarter is based in Tokyo, Japan, and the company had 8046 employees at the time of the merger. Mitsui Chemicals' core business includes basic petrochemical and synthetic fiber raw materials, as well as functional products, fine chemicals, and licensed chemicals. Additionally, the company is involved in the production of synthetic resins and basic chemicals [25].

In recent years, in the field of general petrochemical products, the construction of cheap ethane plants in the Middle East and the rise of the petrochemical industries in China and Korea, especially the financial crisis in 2008, made the naphtha-based Japanese petrochemical companies face a huge business crisis. Mitsui Chemicals' sales revenue plummeted from 1786.7 billion yen in fiscal 2007 to 1487.6 billion yen in fiscal 2008, a drop of 16.7%, and operating profit and net profit were both substantial losses in the same period.

Mitsui Chemicals has maintained a strong commitment to research and development, with a focus on basic technology development and the creation of high-performance products such as functional resins, thin-film sheets, and functional chemicals. The company is dedicated to driving industrial transformation, as demonstrated by its efforts in the development of OLED packaging materials. In June 2017, Mitsui Chemicals announced its plan to expand production and sales of transparent encapsulation materials for OLED displays. Its Structbon XMF T sealing material offers high barrier, moisture resistance, and transparency. The material boasts excellent water resistance and barrier properties, with official trial samples having already been conducted. With Chinese panel makers aggressively increasing their production capacity for encapsulation materials, Mitsui Chemicals aims to commence mass production of the product in 2018, marking another milestone in its transformation journey.

Some Patents

CN103140940B Solar Cell Sealing Material and Solar Cell Module [26].

(a) The purpose of the present invention is to provide solar cell sealing materials with excellent characteristics such as transparency, flexibility, adhesion, heat resistance, appearance, crosslinking characteristics, electrical characteristics, and calendering properties. In order to achieve the above purpose, solar cell sealing materials containing ethylene-a-olefin copolymers satisfying the following elements a1) to a4) are made. a 1) The proportion of constituent units derived from ethylene is 80 to 90 mol%, and the proportion of constituent units derived from alpha-olefin with carbon atoms of 3 to 20 is 10 to 20 mol%. a2) According to ASTMD1238 a2) The MFR measured at 190°C and 2.16 kg load is 2 g/10 min or more and less than l0 g/10 min. a3) The density measured in accordance with ASTM D1505 is 0.865–0.884 g/cm3. a4) The Shore A hardness measured in accordance with ASTM D2240 is 60–85.

(b) Contents of the invention

(c) The subject matter is to be solved by the invention.

(d) According to the research conducted by the present inventors and others, the above polyolefin-based compositions have difficulties in meeting various properties such as transparency, adhesion resistance, and moldability during calendering at the same time. In addition, the polyolefin copolymers described in Patent 2 have insufficient crosslinking characteristics or have a high strain with crosslinking, which may lead to deformation and breakage of the glass substrate. In addition, the balance between the electrical properties (insulation) and crosslinking properties (inhibition of gelation) of the resin composition for solar cell sealing materials containing ethylene-alpha-olefin copolymer described in Patent Document 3 is not sufficient.

(e) The present invention is proposed in view of the problems of the prior art, and its subject is to provide a solar cell sealing material that is a solar cell sealing material containing an ethylene-alpha-olefin copolymer satisfying specific elements, which is used to obtain the desired physical properties with clear guidelines, and each property such as transparency, flexibility, adhesion, heat resistance, appearance, crosslinking property, electrical property, and calendering property It is excellent. The subject of the present invention is to provide a solar cell module using such a sealing material for solar cells. In addition, the subject of the present invention is to provide solar cell sealing materials that can suppress the occurrence of the abovementioned PID and to provide solar cell modules using such sealing materials for solar cells.

(f) Method for solving the subject matter.

(g) As a result, it was found that by using a specific ethylene-alpha-olefin copolymer whose ethylene unit content ratio, density, MFR, and Shore A hardness meet the specified requirements, solar cell sealing materials with excellent characteristics such as transparency, flexibility, adhesion, heat resistance, appearance, crosslinking properties, electrical properties, and calendering properties can be obtained. The present invention was thus completed. In addition, the present invention is completed by finding that the crosslinking characteristics (inhibition of gelation) and electrical characteristics (insulating properties) are superior by making the content of aluminum elements meet a specific range.

That is, according to the present invention, the solar cell sealing material shown below can be provided.

A solar cell sealing material comprising an ethylene-alpha-olefin copolymer satisfying the following elements: (a_1) to (a_4)

(a) The proportion of the constituent units derived from ethylene is 80–90 mol%, and the proportion of the constituent units derived from α-olefin with carbon atoms 3–20 is 10–20 mol%,
(b) MFR of 2 g/10 min or more and less than 10 g/10 min, measured at 190 °C and 2.16 kg load, according to ASTM D1238.
(c) Density of 0.865–0.884 g/cm^3 as determined by ASTM 01505.
(d) Shore A hardness of 60–85 as determined by ASTM 02240.

9.1.15 Mitsubishi Chem-Japan

Mitsubishi Chemical Corporation, the largest chemical company in Japan, offers a broad range of products through three main divisions: functional materials and plastic products (including information and electronic products, specialty chemicals, and pharmaceuticals); petrochemicals; and carbon and agricultural products. While functional materials such as aluminum, sheet plastic products, and plastic packaging account for more than 55% of sales, the company has set its sights on specialty chemicals and pharmaceuticals as its long-term goal. Mitsubishi Chemical is undertaking structural and management reforms to enhance its profitability. Its product portfolio includes industrial and petrochemical raw materials, polymer resins, functional and plastic chemicals, carbon materials, carbon fiber composites, information and electronic materials, display materials, battery and semiconductor materials, insulation and inorganic materials, and chemical fibers [27].

For film production, Mitsubishi has developed VIEW-BARRIER, a customized high-functional film/sheet for front-file films for solar modules. The product is designed to provide excellent gas barrier performance of 10^{-4} g/m^2/day and can be modified to meet the various performance requirements necessary for various solar module front-end materials, such as flexibility and lightweight.

TECHBARRIER is a transparent film with silicon plating on plastic films such as OPET and ONY. It has excellent properties such as a barrier to oxygen and water vapor, and also has excellent fragrance retention. It is widely used in the global market as a packaging material for medical, pharmaceutical, industrial, and food products. It is particularly good at printing and bag making, as well as maintaining the stability of gas barrier properties at high temperatures and high humidity.

Some Patents

CN102804440 B Organic Electronic Device and Manufacturing Method [28].

The present invention provides an OED and a manufacturing method thereof, wherein the device function of said OED does not deteriorate during long-term use, for example, in the case of a thin-film organic solar cell element, and wherein a reduction in power generation efficiency can be suppressed. Said OED has, in sequence, a stack of organic semiconductor elements (B) having at least one pair of electrodes, a layer (C) containing a trapping agent that absorbs at least one of water and oxygen, and a gas barrier film (D), having at least one corrosion-resistant layer (E) between at least one of said pair of electrodes and said layer (C) containing a trapping agent that absorbs at least one of water and oxygen, said corrosion-resistant layer (E) having a film thickness of 20 μm.) having a film thickness of 20 μm or more, said corrosion barrier (E) having a WVTR Pe ($g/m^2/day$) at 40 °C, 90% RH relative to the WVTR Pd of the gas barrier film (D) at 40 °C, 90% RH satisfying the condition of $15\,g/m^2/day > Pe > Pd$, and the WVTR Pd of the gas barrier film (D) satisfying the condition of $104 \leq Pd \leq 101\,g/m^2/day/m^2/day$.

9.1.16 Fujifilm-Japan

Since its founding in 1934, Fujifilm Corporation (hereinafter referred to as "Fujifilm") has moved forward with determination and has become one of the world's largest manufacturers and suppliers of comprehensive imaging, information, and document processing products and services.

Fujifilm has three major business areas: (1) the imaging business area, which includes two product groups: conventional and digital (film, cameras, photo paper, chemicals, development and expansion equipment, etc.); (2) the information business area, which includes printing systems, medical systems, liquid crystal materials, recording media, and other products; and (3) the document processing equipment (copiers, printers, multifunctional digital centers, consumables, etc.) manufactured and sold by Fujifilm's subsidiary Fuji Xerox Corporation. (4) The document business field consists of document processing equipment (copiers, printers, multifunctional digital print centers, consumables, etc.) manufactured and sold by Fujifilm's subsidiary, Fuji Xerox. Its main products are as follows:

(1) Photographic and imaging field: Various kinds of residential and professional photographic films, photo papers, film films, film with lenses, primary imaging films, etc. 35mm cameras, APS cameras, professional cameras, digital cameras, primary imaging cameras, etc. Digital photo printers, digital printing equipment, etc.
(2) medical image diagnosis field: digital X-ray imaging diagnosis system, dry examination system, etc.
(3) the field of printing: multi-beam laser B_1, B_2 laser phototypesetting machine, B_1 and B_2 format multi-beam direct plate making machine and related drugs, materials, etc.
(4) recording media: computer storage media, AV storage media, etc.
(5) office image information system: micro-scanning printer, information recording paper, etc.
(6) Industrial materials: wide-angle films, polymer porous films, color filters, etc.

In addition, Fujifilm has developed the world's highest water vapor resistance and transparency, "ultra-high airtight transparent film." This film can be used in flexible electronic components such as organic EL displays and solar cells. Fujifilm will be actively promoting the film to the market next.

Flexible organic EL displays, electronic paper, organic EL lighting, and thin-film solar cells are considered the next generation of electronic devices. These devices all require flexible, transparent, ultrathin, and lightweight high-performance films as the main component, replacing today's heavy glass and metal materials. However, the use of high-performance films as the main component must ensure that they are highly resistant to moisture and water, in order to protect the performance and service life of electronic devices.

Fujifilm has developed "ultra-high airtight transparent films" with water vapor tightness as high as 10^{-6} g/m^2/day and has realized a film-forming technology that prevents defects that affect airtightness by utilizing the company's organic material knowledge and film quality/thickness control technology accumulated through its research and development of film and flat panel display materials. In addition, we have established a roll-to-roll film-forming technology that is necessary for mass production.

This film-forming technology is capable of forming partitions on various base films such as polyethylene terephthalate (PET) and polyethylene naphthalene dicarboxylate (PEN), which are transparent and bend-resistant and can be used in a wide range of fields as high-performance films to meet the needs of flexible electronic devices.

Some Patents

CN101800241 B Organic Electroluminescent Display Device [29].

The present invention provides an organic EL display device comprising: an organic EL element, said organic EL element comprising a transparent electrode, a counter electrode and an organic compound layer positioned between said transparent electrode and said counter electrode, said organic compound layer comprising a luminescent layer; and a fines-containing layer, said fines-containing layer being arranged on the optical range of the light emitted from said luminescent layer and adjacent to said transparent electrode wherein said fines-containing layer contains an organic resin material and fines, said organic resin material having a refractive index equal to or lower than the refractive index of said transparent electrode, said fines having a refractive index higher than the refractive index of said organic resin material and a weight average particle size of 0.5–5 μm, and said fines-containing layer having a thickness of 2 μm to 10 μm.

Content of the Invention

The present invention is intended to solve the abovementioned problems in the related art and to achieve the following objectives: Specifically, the object of the present invention is to provide an organic EL display device capable of improving light extraction efficiency and reducing image bleed.

The means for solving the above problem are as follows:

<1> An organic EL display device, said organic EL display device comprising an organic EL element, said organic EL element comprising a transparent electrode, a counter electrode, and an organic compound layer disposed between said transparent electrode and said counter electrode, said organic compound layer comprising a light-emitting layer, and a fine-grain-containing layer, said fine-grain-containing layer being arranged on the light path of light emitted from said light-emitting layer

and adjacent to the said transparent electrode, wherein said fine-grain-containing layer contains an organic resin material and said organic resin material has a refractive index equal to or lower than the refractive index of said transparent electrode, said fines have a refractive index higher than the refractive index of said organic resin material and have a weight average particle size of 0.5–5μm, and said fines-containing layer has a thickness of 2–10 μm. In the organic EL display device according to <>, when light passes through the fines-containing layer, the light emitted from the light-emitting layer is The light is scattered.

<2> The organic EL display device according to <1>, wherein said fines in said fines-containing layer are primary particles.

<3> The organic EL display device according to any one of <1> and <2>, wherein said fines-containing layer further comprises a fines dispersant.

<4> The organic EL display device according to any one of <1> to <3>, further comprising: a reflector sheet, said reflector sheet being provided on a surface of said fines-containing layer opposite to a surface on which said transparent electrode is disposed: a sandwich insulating layer, said sandwich insulating layer being provided on a surface of said reflector sheet opposite to a surface adjacent to said fines-containing layer: and a substrate, said substrate being provided on a surface of said fines-containing layer opposite to a surface of said fines-containing layer: and a substrate, said substrate being provided on a surface of said fines-containing layer opposite to a surface of said fines-containing layer. said surface of said sandwich insulating layer opposite to a surface of said reflective sheet adjacent to said reflective sheet. [0018] The organic EL display device according to <4> employs a top-emitting type configuration in which light emitted from the light-emitting layer passes through a transparent electrode, light that has passed through the transparent electrode passes through the fine-grain-containing layer, light that has passed through the fine-grain-containing layer is reflected on the surface of the reflector sheet, reflected light on the surface of the reflector sheet then passes through the fine-grain-containing layer, and light that has passed through the fine-grain-containing layer then passes through the transparent electrode.

<5> The organic EL display device according to any one of <I> to <3>, which further comprises a substrate, said substrate being provided on a surface of said fine-grain-containing layer opposite to a surface on which said transparent electrode is disposed.

The organic EL display device according to <5> employs a bottom-emitting structure in which light emitted from the light-emitting layer passes through the transparent electrode, light that has passed through the transparent electrode passes through the fine-grain-containing layer, and light that has passed through the fine-grain-containing layer passes through the substrate.

<6> The organic EL display device according to any one of <4> to <5>, further comprising an organic thin-film transistor provided on a passivation layer.

<7> The organic EL display device according to <6>, wherein the driver system of said organic electroluminescent display device is a thin-film transistor driver.

<8> The organic EL display device according to one of <6> and <7>, wherein in said fine-grain-containing layer, contact holes are formed for connecting said transparent electrode and said organic thin-film transistor.

According to the present invention, it is possible to solve the problems in the related technology and to provide an organic EL display device capable of improving light extraction efficiency and reducing image bleeding.

9.1.17 Konica Minolta-Japan

Konica Minolta Corporation (hereinafter referred to as Konica Minolta) was established in 1873 as a major Japanese manufacturer and seller of optical films for TAC-centered polarizing panels for LCD monitors. Konica Minolta is a giant in TAC films for PVA protection. In 2017, Konica Minolta started to develop phase-difference films made of COP and PMMA materials other than TAC, and from 2018, Konica Minolta's 60um P-TAC business for TVs will be reduced, and in 2019, Konica Minolta's TAC film production capacity will reach 31,500 m^2/year.

Konica Minolta exhibited white light-emitting organic EL lighting panels with a luminous efficiency of 103 Lm/W and transparent conductive barrier films for the panels at the New Energy Development Organization (NEDO) exhibition area in Japan. The size of the organic EL lighting panel is approximately 45 × 35 mm, and the luminous efficiency is 103 lm/W (luminance of 1000 cd/m^2), which is the highest among Konica Minolta's published lighting panels.

As for the conductive barrier film, the light transmittance is as high as 90% in all visible areas, and the film resistance value is only 3 Ω/□. Konica Minolta made a transparent conductive film on a glass plate in 2012 and achieved a film resistance value of 1 Ω/□ and light transmittance of 84%. This time, although the film resistance value was slightly higher, a flexible substrate was used. The company's self-developed barrier film has a very high barrier performance, with a WVTR of about 10^{-6} g/m^2/day.

Konica Minolta does not use a single inorganic layer but a barrier film with an inorganic layer laminated with a polymer. If only a single inorganic layer is thickened, there will be openings for water vapor to pass through easily in case of defects. By overlapping the inorganic and organic layers in multiple layers, like a sandwich, even if there is a defect in one part, the other layers can still be blocked. Konica Minolta has improved the performance of barrier films by leaps and bounds by adopting a laminated structure [30].

Some Patents

CN103269851B Gas Barrier Film and Electronic Device [31].

The purpose of the present invention is to provide gas barrier films with excellent gas barrier performance and heat resistance and electronic devices with excellent durability. The present invention has a first gas barrier layer containing Si and N formed by physical vapor deposition or chemical vapor deposition on a substrate in the following order and a second gas barrier layer formed by coating a solution containing a polysilicon nitrogen base compound, which is modified by irradiating the second gas barrier layer with vacuum ultraviolet light, and when the composition of the respective layer is expressed in terms of SiO_xN_y, the thickness direction of the second gas barrier layer The distribution of the composition SiO_xN_y of the second gas barrier layer satisfies the conditions specified in (A) below.

(A) The second gas barrier layer has a region of $0.25 < x < 1.1$ and $0.4 < y < 0.75$ in the thickness direction of 50 nm or more.

The above purpose of the present invention can be achieved by the following composition:

1. A gas barrier film having a first gas barrier layer containing Si and N formed by physical vapor deposition or chemical vapor deposition on a substrate in the following order, and a second gas barrier layer formed by coating a solution containing a polysilicon nitrogen base compound adjacent to it, wherein the second gas barrier layer is a layer modified by irradiation with vacuum ultraviolet light, and the first gas barrier layer and the second gas barrier layer are composed of a layer of the same type as the first gas barrier layer. When the composition of the first gas barrier layer and the second gas barrier layer is expressed in terms of SiO_xN_y, the distribution of the composition SiO_xN_y in the thickness direction of the second gas barrier layer satisfies the conditions specified in (A) below.

(A) The second gas barrier layer has a region of $0.25 < x < 1.1$ and $0.4 < y < 0.75$ in the thickness direction of more than 50 nm.

2. The gas barrier film according to 1 above, wherein the distribution of the composition SiO_xNy in the thickness direction of the said 2nd gas barrier layer satisfies the condition specified by (B) below.

(B) The 2nd gas barrier layer has a region of $0.25 < x < 0.55$ and $0.55 < y < 0.75$ in the thickness direction of more than 50 nm.

3. The distribution of the composition SiO_xN_y in the thickness direction of the barrier layer satisfies the condition specified by (C) below.

(C) This 1st gas barrier layer has the maximum value of y in the 1st gas barrier layer and the 2nd gas barrier layer, as described above.

Effect of the Invention

The present invention provides a multilayer gas barrier layer that can be formed without using the vapor phase method. This results in a gas barrier film that has high productivity, exceptional gas barrier performance, and excellent heat resistance, thanks to its composite coating. The film can be used to create electronic devices with superior durability.

9.1.18 KDX-China

Without high-grade water vapor barrier film products, it will be difficult to start downstream flexible industries such as flexible display, flexible electronics, flexible lighting, flexible photovoltaic, and flexible sensing [2].

After 3M took the lead in developing this grade of encapsulation barrier film, Contec invested heavily, and the domestic and European R&D teams worked closely together for many years to develop it, and succeeded in reaching the top in one fell swoop. The world's first winding-type large-width high-performance packaging barrier film production line independently developed by Contec has been successfully put into production, and the water vapor barrier WVTR value can reach 10^{-6} g/m^2/day, which is the highest performance level of packaging barrier film in the industry today. It has a wide range of applications and huge market demand. Not only is the barrier grade of the product the highest in the world, but also the process route is unique and innovative, and the production line is completely self-designed and has independent intellectual property rights. The packaging barrier film produced by this route has many advantages, such as simple film structure, stable process, high yield, high capacity, and high cost-effective, which makes it the most ideal packaging solution in the industry.

Prior to the successful mass production of KDX, high-grade vapor barrier films were only available in small quantities from individual manufacturers in the United States and Japan, but the complex product structure, low yield and capacity, and high price were the main bottlenecks limiting the development of the above flexible market in China. The successful production of KDX's high-performance vapor barrier film has completely solved the core material bottleneck faced by the development of the domestic flexible industry and will drive China's flexible industry to achieve a comprehensive rise.

At present, the production capacity of KDX's high-performance vapor barrier film line has been fully booked by downstream enterprises, and the company will rapidly expand its production capacity to meet the huge demand for different flexible applications downstream. High-grade water vapor barrier film products, with their huge market space, diversification, high technical threshold, and high added value, will quickly become an important new profit growth point for the company and also provide a core material guarantee for the development and industrial upgrading of China's flexible industry, which is of great social significance.

The company has leveraged its self-production of top packaging barrier materials to integrate global industrial resources and expand into various areas of the flexible industry. As a result, it has successfully developed and mastered several key technologies in flexible devices, including multifunctional films, flexible substrates, flexible touch, low-resistance electrodes, and device processes. KDX has also collaborated with renowned universities such as Stanford, Cambridge, and Manchester to develop flexible sensors, which are part of an international science and technology cooperation project supported by the Ministry of Science and Technology. The research and development of these technologies are progressing smoothly.

Some Patents

CN107393987 A Packaging Materials and Optoelectronic Devices [32].

The present invention mainly provides an encapsulation material for the front-end encapsulation of flexible thin-film solar cells. The encapsulation material has a base material layer, a wavelength conversion layer, a barrier layer, and a weathering layer, which has a high water vapor barrier performance and has the feature of being used outdoors for a long time without aging the polymer material, in addition to the unique wavelength conversion layer in the encapsulation material, which enables the encapsulation material to convert sunlight in the ultraviolet band into visible light, such as the blue light part, so that it can be absorbed by the photovoltaic device in the lower layer and converted into

Content of the Invention

The technical problem to be solved by the present invention is to provide an encapsulation material and a photovoltaic device. The prepared encapsulation material can make full use of the energy of ultraviolet light and has a high power generation efficiency.

The present invention provides an encapsulation material including a substrate layer, a barrier layer set on any surface of said substrate layer, a weathering layer set on a surface of said barrier layer away from said substrate layer, and a wavelength conversion layer set on a surface of said substrate layer away from said barrier layer or between said substrate layer and said barrier layer or between said barrier layer and said weathering layer.

wherein said wavelength conversion layer is used to convert light in the wavelength range of 200–410 nm to light in the wavelength range of 450–700 nm.

Preferably, said base material layer is PET or PEN.

Preferably, said wavelength conversion layer comprises a photoconverter, said photoconverter being an inorganic oxide doped with rare earths.

Preferably, said rare earth-doped inorganic oxide is one or more of $Sr_5(PO_4)_3Cl$: Eu, YPO_4:Tb, and $LaP0_4$:Tm.

Preferably, said wavelength conversion layer has a thickness of 3–100 μm.

Preferably, said barrier layer is an inorganic oxide.

Preferably, said inorganic oxide is iron oxide or aluminum oxide.

Preferably, said weathering layer is fluoride.

Preferably, said weathering layer is a fluorocarbon coating or a composite layer of a bonding layer and a fluorine-containing film.

The present invention also provides an optoelectronic device, including an encapsulating material.

The invention provides an encapsulation material, mainly suitable for the front-end encapsulation of flexible thin-film solar cells. The encapsulation material has a base material layer, wavelength conversion layer, barrier layer, and weathering layer, has a high water vapor barrier performance, and has a long time in outdoor use without making the polymer material aging characteristics, in addition to the unique wavelength conversion layer in the encapsulation material, so that the encapsulation material can convert the ultraviolet band of sunlight into visible light, such as blue light part, so that the lower layer of photovoltaic devices can be absorbed and converted into electrical energy, the ultraviolet light energy will be converted into electricity. This can be absorbed by the lower layer of the PV device and converted into electrical energy, making full use of the energy of UV light to improve the power generation efficiency of the PV device.

9.1.19 Wanshun-China

Founded in 1998, Shantou Wansun New Material Group Co., Ltd. is a national high-tech enterprise based in the Shantou Free Trade Zone in Guangdong Province.

The company's three major businesses include environmental protection packaging materials, high-precision aluminum foil, and functional films. Its diverse range of products includes intelligent light-controlled thermal insulation film, ITO conductive film, high-barrier film, intelligent electronically controlled liquid crystal film, paper packaging materials, aluminum foil, and aluminized film. With a focus on research and development, Wanshun New Material Group Co., Ltd. is dedicated to functional films, intelligent light-controlled thermal insulation films, conductive films, barrier films, double-zero aluminum foil, and new environmental protection materials. The company is currently constructing a high-barrier film material production base project, which will produce high-barrier film materials for new display components in the field of optoelectronics and high-end packaging. The production capacity of high-barrier film material for new display components is expected to reach 6 million square meters per year [4].

Some Patents

CN105408104 B Encapsulation Barrier Stack Containing Nanoparticles Encapsulated in A Dendritic Polymer [33].

The present invention discloses an encapsulated barrier stack capable of encapsulating a moisture and/or oxygen-sensitive article and comprising a multilayer membrane layer, wherein said multilayer membrane layer comprises: one or more barrier layers having low moisture and/or oxygen permeability, and - one or more sealing layers arranged to contact the surface of at least one barrier layer so as to cover defects in the barrier stack, wherein said one or more sealing layers comprise a plurality of dendrimer-encapsulated nanoparticles, said nanoparticles being reactive in that they are capable of interacting with moisture and/or oxygen to prevent penetration of moisture and/or oxygen through the defects present in the barrier layer.

The present invention provides a barrier stack with improved flexibility, gas barrier properties, weather resistance, optical properties, mechanical properties, and reliability, and also provides a cost-effective solution.

According to the first aspect, the present invention provides an encapsulation barrier stack. Said encapsulation barrier laminate is capable of encapsulating articles that are sensitive to moisture and/or oxygen. Said encapsulation barrier laminate comprises a multilayer membrane. Said multilayer membrane includes one or more barrier layers and one or more sealing layers comprising nanoparticles encapsulated in a dendritic polymer providing low moisture and/or oxygen permeability. Said multilayer membrane further comprises one or more sealing layers. Said one or more sealing layers are arranged to be in contact with a surface of said at least one barrier layer. Said one or more sealing layers thereby cover defects present in the said barrier layer. Said one or more sealing layers comprise a plurality of dendritic polymers and/or dendrimers and organic substances (e.g. nanoparticles encapsulated in dendritic polymers). Said nanoparticles are reactive in that they are capable of interacting with moisture and/or oxygen to prevent penetration of moisture and/or oxygen through defects present in said barrier layer.

According to a second aspect, the present invention provides an electronic device. Said electronic device comprises an active component that is sensitive to moisture and/or oxygen. Said active component is arranged within an encapsulated barrier stack according to the first aspect.

According to a third aspect, the invention provides a method of manufacturing an encapsulation barrier stack according to the first aspect. Said method comprises providing one or more barrier layers. Said method further comprises forming one or more sealing layers. In a first embodiment of a third aspect, forming said one or more sealing layers comprises mixing an encapsulating material according to the present invention with a plurality of nanoparticles or functionalized nanoparticles, said encapsulating material comprising or comprising: a dendrimer or precursor thereof containing a plurality of nanoparticles or functionalized nanoparticles, a dendrimer or precursor thereof, a dendrimer/polymerizable compound mixture, a dendrimer/polymerizable compound mixture, a dendrimer/polymerizable compound mixture, dendrimer/polymerizable compound blends, dendrimer/cross-linkable compound blends, and dendrimer/cross-linkable compound blends. Polymerizable or cross-linkable substances include monomers, polymers, and/or oligomers or combinations thereof.

Optionally, in a second embodiment of the third aspect, forming said one or more sealing layers comprises, optionally, mixing an encapsulating material with a polymerizable or cross-linkable compound, said encapsulating material comprising or comprising nanoparticles encapsulated by a dendrimer polymer. The polymerizable or cross-linkable material comprises monomers, polymers, and/or oligomers or combinations thereof.

According to a fourth aspect, the present invention relates to the use of dendrimer-encapsulated reactive nanoparticles for the preparation of sealing layers for barrier stacks. Said nanoparticles are reactive in that they are capable of interacting with moisture and/or oxygen to prevent penetration of moisture and/or oxygen through defects present in said barrier layer.

According to a fifth aspect, the present invention relates to the use of reactive nanoparticles encapsulated in dendritic polymers for encapsulating electronic devices or for food packaging, pharmaceutical packaging, or medical packaging.

9.1.20 Lucky-China

Lucky Ltd. is a leading state-owned high-tech enterprise that caters to the printing industry with its production capacity of offset printing plates, printing films, and flexible resin plates. The company has played a significant role in the evolution of China's printing industry, from traditional "lead and fire" to modern "light and electricity" and digitalization, while simultaneously promoting green environmental protection. The product range includes PS plate series, printing film series, CTP plate series, flexible resin plate series, Huaguang brand polyester film, supporting chemicals, SGS-certified products, Huaguang UV ink series, Huaguang UV curing products, and Huaguang intelligent interactive all-in-one machines. The company has recently completed the installation of a transparent high-barrier film production line in November 2018, a high-barrier film coating production line in December 2018, and a high-barrier project coating line in February 2019.

These lines manufacture high-barrier film materials for new display components, particularly for quantum dot film packaging and OLED device packaging [4].

Some Patents

CN104962875B A Preparation Method of High-Barrier Film [34].

The invention relates to a method of preparing high-barrier film, the technical solution of which is to first place the substrate on the substrate table or wind it around the electrode series, and evacuate the reaction chamber, then surface active treatment of the substrate, then pass the film forming gas such as silicone compound vapor and oxygen-containing gas and adjust the air pressure, then turn on the power of the dust particle removal device and adjust the voltage applied to the trapping electrode, then turn on the RF power switch and adjust the feed power, start the silicon oxide film deposition, until the flexible and transparent ultrahigh barrier silicon oxide film sample is obtained, finally turn off the RF power supply, stop the film-forming gas, then turn off the power supply of the dust particle removal device, stop the vacuum pumping, hold the vacuum chamber after the completion of deflation, and remove the sample. The invention can realize the preparation of silicon carbon oxide thin film without pollution, has the advantages of a simple preparation process, a high film deposition rate, and is conducive to the realization of large-scale industrial production.

Invention Content

The purpose of the invention is to provide a method for the preparation of high-barrier films that can remove the dust particles generated in the plasma during the plasma-enhanced chemical vapor deposition process in real time so that even under the plasma chemical conditions where dust particles are generated, the barrier layer can still be prepared without pollution, which greatly expands the scope of the preparation of high-barrier films using the plasma-enhanced chemical vapor deposition technology and the adjustment of the experimental parameters. The adjustment range of each experimental parameter makes the production process easier to control, and at the same time can improve the deposition rate of the barrier layer, improve the yield, reduce costs, and get better barrier performance of flexible transparent high-resistance films.

The technical problem described in the present invention is solved by the following technical solution:

A method for preparing a high-barrier film, comprises the following process steps.

(1) Placing the substrate on a substrate table or wrapping it around an electrode series and evacuating the reaction chamber to below 1×10^{-2} Pa.
(2) Reactive treatment of the substrate surface using Ar plasma.
(3) chemical vapor deposition of the film-forming gas using plasma to deposit a high barrier layer on the substrate to obtain a flexible, transparent high-barrier film.

Said high-barrier film has a light transmission rate of more than 80%.

Said plasma chemical vapor deposition using a dust particle removal device in a vacuum chamber and turning on the dust particle removal device after the film-forming gas is introduced and before the RF power is turned on.

Said dust particle removal device is a device comprising at least one electrode for trapping dust particles, a power supply and circuitry for controlling the potential of the trapping electrode, dust particle storage, and exclusion of trapped dust particles from the vacuum chamber.

In the above method of preparing a high-barrier film, the surface of said electrode for trapping dust particles is covered by an insulating material.

The method of preparing the abovementioned high-barrier film, wherein said substrate is a film or sheet with a light transmission rate of 85% or more.

The method of preparing the abovementioned high-barrier film, said film-forming gas is one or more containing silicon as well as carbon, oxygen, and nitrogen.

The method of preparing the abovementioned high-barrier film, said film-forming gas containing silicon, oxygen, and carbon.

The high barrier layer of the high-barrier film of the method of preparing the above high-barrier film is a 50–3000 nm thick silicon carbon oxide film.

The thickness of the silicon oxide film is 100–2000 nm.

The preparation method of the high-barrier film mentioned above involves using various resin films such as polyethylene terephthalate (PET), polyethylene naphthalate (PEN), polycarbonate (PC), polyvinyl chloride (PVC), polyethylene (PE), polypropylene (PP), polystyrene (PS), polyimide (PI), and heat-resistant transparent films with organic-inorganic hybrid structure based on bisphenol-A siloxane with a thickness ranging from 5.0 to 500μm as the substrate. It also includes a resin film composed of two or more layers of the aforementioned resins. A type of resin film.

Beneficial Effects

1. The present invention uses a dust particle removal device in the plasma-enhanced chemical vapor deposition process to capture and remove dust particles in the plasma-enhanced chemical vapor deposition process in real-time, eliminating contamination of the deposited barrier film by large dust particles, making it possible to prepare a high-barrier film without contamination even under the plasma chemical conditions where large dust particles are generated, thereby This greatly expands the adjustment range of experimental parameters in the preparation of silicon oxide films by plasma-enhanced chemical vapor deposition, and makes it easier to optimize the experimental conditions for the preparation of high quality flexible and transparent silicon oxide films with high barrier.
2. Since the contamination of the deposited films by dust is not considered, high-quality flexible transparent ultrahigh-barrier films can be prepared under higher power and air pressure conditions, and higher film deposition rates can be obtained, thus increasing production efficiency, improving yield, and reducing production costs.

9.2 Analysis of Flexible Electronic Packaging Enterprises

Several countries, including the United States, Japan, and South Korea, have strategically invested in flexible electronics projects, resulting in significant progress in commercialization. High-precision technology in this field is expected to maintain a long-term, high-speed growth trend. Domestic companies are also actively investing and positioning themselves to seize this historic opportunity. According to IDTechEX's forecast, the global flexible electronics market was worth $46.94 billion in 2018 and is expected to grow to $301 billion by 2028. The high-barrier film market, which is a necessary component for flexible electronics, is also expanding. Water and oxygen can degrade or corrode the electrodes and active layers of flexible electronic materials made from organic matter, which negatively affects the performance of flexible devices. As a result, the encapsulation of devices with high-barrier film is necessary even in laboratory testing phases, and various flexible device research companies have an urgent demand for ultrahigh-barrier film [6] (Figures 9.11 and 9.12).

Thin-film packaging is currently the most widely used technology for flexible electronic packaging. There are three main types of TFE materials: inorganic, organic, and inorganic–organic composite. Inorganic–organic composite encapsulation materials have the advantages of both inorganic and organic encapsulation materials, providing good water and oxygen barrier and good film formation characteristics.

However, the production technology of thin-film packaging materials has been monopolized by foreign countries since the mass production of flexible screens. Development is being driven towards thinner packaging with better film-forming and printing characteristics, and the sensitive nature of the materials requires demanding production environments and quality control management. This presents a challenge for domestic companies to compete in the market and develop their own proprietary technology.

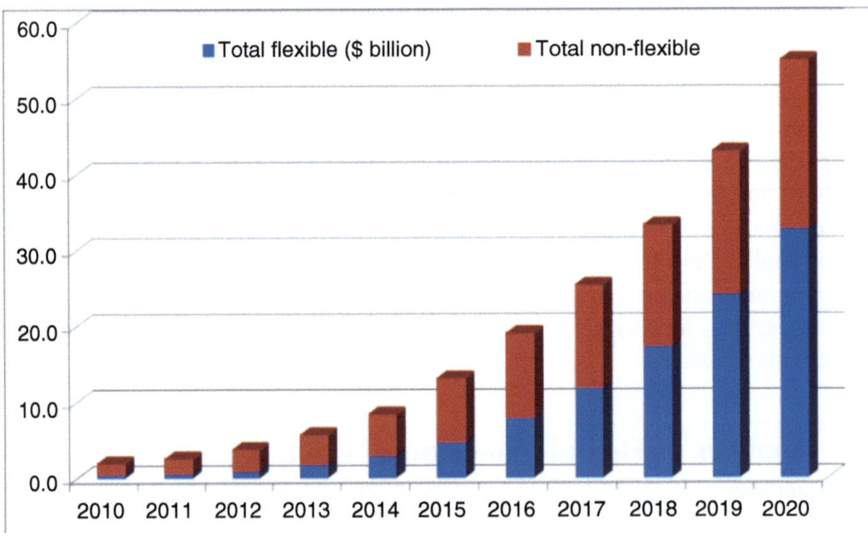

Figure 9.11 Flexible and non-flexible electronic devices market 2010–2020. Source: Reproduced with permission from https://www.grandviewresearch.com.

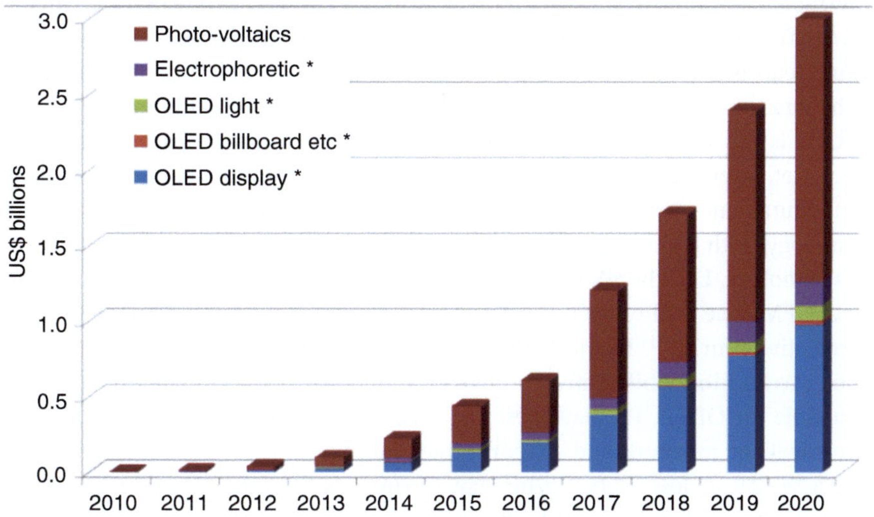

Figure 9.12 Flexible electronic component packaging high-barrier film market 2010–2020. Source: Reproduced with permission from https://www.grandviewresearch.com.

The TFE materials market was valued at USD 13.1 million in 2017 and is projected to reach USD 101.1 million by 2023, growing at a CAGR of 26.1% from 2018 to 2023. The growth of this market is primarily driven by the increasing popularity of flexible OLED displays in smartphones and smart wearable devices, where thin-film protective layers are necessary for the longevity and durability of organic devices.

The thin-film packaging materials market can be segmented based on its applications, including OLED displays, OLED lighting, thin-film photovoltaic cells, electrochromic devices, and others. In 2017, the market for thin-film packaging materials was primarily driven by the demand for OLED displays. Samsung SDI and LG Chem dominated the market by developing TFE materials for flexible OLED displays and working with thin-film packaging equipment suppliers such as Kateeva and Applied Materials. During the forecast period, the market for TFE materials for OLED lighting applications is expected to register the highest CAGR of 31.8%.

The thin-film packaging market can be segmented into North America, Europe, Asia Pacific (APAC), and the Rest of the World based on regional segmentation. Many market players are optimistic about the market outlook for thin-film packaging in APAC due to industry leaders such as Samsung and LG Display. In Europe, OLED lighting manufacturers dominate the market, while North America relies heavily on various solar cell developers.

In 2017, Asia Pacific dominated the thin-film packaging materials market due to its concentration of major display panel and brand manufacturers in countries such as China, Japan, Taiwan, South Korea, and Hong Kong. South Korea, with industry leaders like Samsung and LG Display, accounted for 74.2% of the TFE materials market share in Asia Pacific. Additionally, the demand for TFE materials is expected to grow rapidly in China from 2018 to 2023, as Chinese companies like BOE Technology and CSOT plan to build production lines for flexible OLED panels.

European market growth is driven by the adoption of OLED lighting solutions in the automotive industry and government funding for OLED lighting research. OSRAM is the leading company in the region. In the thin-film packaging materials market, foreign companies have developed earlier with more mature technology, occupying a major position. Samsung, LG, and 3M are leading in thin-film packaging technology, with Samsung and LG at the forefront of industrialization and packaging technology. LG Chem's main products include POLED encapsulation material BFL, FSPM Face Seal type OLED encapsulation material, OLED components, and a protective film TPF. In the United States, 3M produces high-end display water vapor barrier film, while Universal barrier film encapsulation material from UDC is suitable for OLED, PV, batteries, and other fields. Vitriflex's ultra barrier films have excellent moisture and oxygen protection, while BASF's Rolic barrier films have a WVTR as low as 10^{-6} g/m^2/day. Germany's Rfaunhofer produces ORMOCER barrier materials with a WVTR of 10^{-5} g/m^2/day and OTR of 10^{-5} cm^3/m^2/day. Japan's Mitsui Chem produces Structbon XMFT sealing material with high barrier, moisture resistance, and transparency, and Mitsubishi Chem's Techbarrier transparent vapor deposition high-performance gas barrier film has excellent oxygen and water vapor barrier performance. Fujifilm's transparent high-gas barrier film has an organic–inorganic nano-multilayer structure for superior barrier function.

Domestic enterprises engaged in high-barrier materials mainly focus on food and pharmaceutical packaging, while few enterprises are dedicated to the research and development and production of flexible displays of water vapor barrier film, and the materials mainly rely on imports. Domestic enterprises mainly focus on Kangde Xin, Wanshun New Material, and Lekai Guanghua. Their products are basically in the early stage of industrialization or still in the research and development stage and have not yet taken a significant share in the field of flexible electronic packaging, but the huge investment of these companies and the country's attention to this field, will certainly quickly catch up with the leading foreign enterprises and take a place in the field.

References

1 https://www.samsungsdi.com.cn, 10 March 2023.
2 Ton-Soo, C., Daho, L. and Sun, L.J. CN108249764 A, Encapsulates the composition of organic light-emitting diodes (OLED) and a display which prepared by it [P]. 6 June 2018.
3 https://www.lgchem.com, 10 March 2023.
4 Hyun-Suk, K., Hyun-Ji, R. and Yu, W. J. CN106463649 B, An enveloping film and a method of using the enveloping film to encapsulate an organic electronic device [P]. 3 November 2018.
5 https://www.3m.com.cn, 10 March 2023.

6 Vigor, M.D. and Nahtigarh, A. CN103079816 B, A blocking module having an enveloping agent and a photovoltaic cell [P]. 2 January 2018.
7 https://oled.com/, 10 March 2023.
8 Hosoda, H. and Wuqing, H.G. CN101901825 B, Organic electroluminescent display devices [P]. 1 June 2016.
9 https://www.amcor.com, 10 March 2023.
10 Ettridge, P., Christofferson, R. and Birkhoff, I. CN107969124 A, Flexible multilayer packaging film with super high barrier properties [P]. 27 April 2018.
11 https://www.oled-info.com/vitriflex.
12 Brassard, N. CN105647324 A, New barrier layer reactor and its method [P]. 8 June 2016.
13 http://www.tera-barrier.com, 10 March 2023.
14 Ramadas, S.K. and Shanmugavel S. US20190027414A1, An encapsulation barrier stack consisting of nanoparticles encapsulated in a tree molecule [P]. 24 January 2019.
15 https://www.polo.fraunhofer.de/en, 10 March 2023.
16 Doris, T. B., Hanselmann, Gellermann, C. and Probst J. EP2424655A2, Method of encapsulating liquid or paste material in a crosslinked packaging material [P]. 23 August 2017.
17 https://www.sigmalabs.com/, 10 March 2023.
18 Mihael, M.G. and Janrichs, A. CN1960811 A large-area electroluminescent light-emitting device [P]. 24 November 2006.
19 https://www.toppan.com/en/, 10 March 2023.
20 Yamamoto, A. CN1791510B, Transparent gas shielding laminated film, electroluminescent element, electroluminescent display device and electrophoretic display panel [P]. 13 October 2010.
21 https://www.basf.com/ch/de/who-we-are/BASF-in-Switzerland/group-companies/Rolic-Technologies-AG.htm, 10 March 2023.
22 Franz, R., Maria, A., Wanmore, B. and Seiler, B. CN107405893 B A, Barrier thin film laminate including submicron absorbing particles and an electronic device including such laminate [P]. 16 March 2021.
23 Pagano, J.C., Nelson, K.J., Burrows, P.E., Gross, M.E., Zumhoff, M.R., Martin, P.M., Bonham, C.C. and Graff G.L. CN101231974 B, Multilayer coating and a device and a method for using the multilayer coating [P]. 2 June 2010.
24 Erlat, A.G., Duggal, A.R. and Yan, M. US20090286010A1, High throughput processes and systems for barrier film deposition and/or encapsulation of optoelectronic devices [P]. 12 July 2011.
25 https://jp.mitsuichemicals.com/, 10 March 2023.
26 Ikenaga, N., Takeuchi, S., Watanabe, K., Tokuhiro, A., Murobushi, T., Sendomizu, K., Ito, T. and Maruko, S. CN103140940B, Encapsulating material for solar cell and solar cell module [P]. 3 April 2014.
27 https://www.m-chemical.co.jp/, 10 March 2023.
28 Oizumi, S., Funayama, K., Fujiwara, T., Yoneyama, T. and Handa T. CN102804440B, Organic Electronic Devices and Manufacturing Methods [P]. 20 January 2016.

29 Mori, H., Hosoda, H. and Hiroda M. CN101800241B, Organic Electroluminescent Display Devices [P]. 13 April 2016.
30 https://www.konicaminolta.com/al/index.html, 10 March 2023.
31 Mori, T. CN103269851B, Gas Barrier Films and Electronic Devices [P]. 1 April 2015.
32 Shuo, L., Haili, C., Jinku, X. and Zhen, Y. CN107393987A, Packaging Materials and Photoelectric Devices [P]. 24 November 2017.
33 Ramadas, S.K. CN105408104B, Encapsulating Barrier Layers Containing Nanoparticles Encapsulated in Dendrimers [P]. 25 June 2019.
34 Wanbing, L., Xianhao, L., Li, Wei, Y. and Shanghong, H. CN104962875B, A preparation method of high barrier film [P]. 12 October 2018.

10

Flexible Electronics Packaging Development Trends

Mingqiang Liu

School of Advanced Materials, Peking University Shenzhen Graduate School, Lishui Road, Shenzhen, 518055, China

10.1 Flexible Electronics Packaging Trends Overview

Thin-film packaging is a key technology for the commercialization of flexible organic electronics. The development of wearable and IoT device applications is increasing the demand for flexible and stretchable electronic device technologies. In the coming years, it will become even more important to finally integrate and function in a monolithic structure through reliable packaging technologies for various device components, etc. Effective encapsulation to prevent the penetration of water vapor and oxygen, which would degrade organic devices, is the key to achieve the required lifetime for most applications [1–3]. The main research contribution of this book is a systematic study of the parameters affecting multilayer and monolayer encapsulation films and their integration with organic electronic devices. Through the efforts of this work, we were able to determine the importance of defects in determining the overall film barrier performance and, as a result, to create a new hybrid layer barrier film. This new film effectively provides an ultrahigh barrier performance film with a simplified structure by addressing defects commonly found in cryogenic vacuum-deposited inorganic materials. Analytical simulations of permeation through the encapsulated film provide a fundamental understanding of the water vapor permeation mechanism and key guidelines for designing an effective barrier layer. Comparison with water vapor transmission rate (WVTR) and hysteresis time measurements in the films elucidates the role of diffusion coefficient, solubility, and defects in barrier layer performance. Overall, this analytical modeling provides a tool to understand how the operation of the barrier film structure affects the overall performance of the package. The additional study of the mechanical behavior of the film enables the development of highly flexible package structures by controlling the position of the neutral axis within the device architecture [4]. Such a concept is important to extend the flexibility of the device while maintaining good barrier performance. The following are the key findings of this book and the insights gained from these findings. In addition,

Flexible Electronic Packaging and Encapsulation Technology, First Edition.
Edited by Hong Meng and Wei Huang.

based on the current results of this book, future work is proposed to address the salient research questions. Experimental results of multilayer encapsulated structures are given, including a comprehensive study of water vapor transmission through various films. SiN_x/poly(parylene) and SiO_x/poly(parylene) dimers were prepared, and the barrier properties were investigated to correlate them with the number of dimers. This study is one of the first to show a strong correlation between the number of binary elements and barrier performance and a diminishing effect of barrier performance improvement after quaternary elements. This diminishing effect suggests that there should be limitations in improving the barrier performance of multilayer structures by simply adding binary [5]. The development of high-quality films is essential to improve the overall potential barrier properties of multilayer structures, rather than increasing the number of binary films. In addition, the results of improving the barrier properties by annealing the poly(parylene) layer are presented. The poly(parylene) annealing provided a smoother surface, which helped to produce subsequent inorganic layers with higher quality. This finding also supports the idea that the quality of the inorganic layer is the main factor determining the barrier performance of multilayer structures. In addition, it is worthwhile to try to find other organic layers that can reduce the processing time and provide a smoother surface for subsequent inorganic layer deposition and to investigate their effect on the overall barrier performance.

Analytical simulation studies were also performed in this paper to investigate how the individual properties of the constituent films affect the overall barrier performance. By comparing the analytical simulation results with experimental results from calcium corrosion tests, expected ranges of permeation properties, such as diffusion coefficients and solubility coefficients of individual membranes, were determined. The permeation properties of the inorganic membranes play a dominant role in determining the permeation rate in the steady-state region as well as the lag time. In addition, the results show that the hysteresis time can be further increased to several years, which can be longer than the required lifetime of organic devices. This hysteresis time depends on the permeation characteristics of the constituent films as well as on the effective length of the permeation path through the organic layer, which is directly related to the defect spacing in the case of multilayer package structures. Therefore, it was concluded that reducing the steady-state permeation rate as well as increasing the hysteresis time is essential to extend the shelf life of encapsulated organic devices. However, little attention has been paid to the engineering of the hysteresis time effect. Future work should consider the use of desiccant layers to improve the hysteresis time effect. Layers such as TiO_x or other materials can be easily integrated into the film to give this behavior. Although monolayers fabricated by atomic layer deposition (ALD) as a barrier layer have the advantage of simple process steps and structure and high barrier performance, the long process time makes them difficult to be widely used. A hybrid encapsulation structure, which combines the advantages of both monolayer and multilayer structures, was proposed, developed, and tested. The structure showed good results due to its high-quality barrier performance and simpler structure compared to multilayer packages. The ultrathin layers deposited by ALD can passivate well the defects in the films deposited rapidly by Plasma enhanced chemical vapor deposition (PECVD). The synergistic effect of combining PECVD and ALD was found by applying lamination theory and verified by comparing the activation energy with the current results [6].

Although this contribution of creating hybrid layer packages is considered a successful approach, more work must be done to fully capture the effects of the hybrid approach. First, detailed experiments should be performed to study the cross-section of the interlayer between the ALD-deposited film and the PECVD-deposited film using TEM. In addition, the defect-filling method using ALD can be extended to other techniques, such as pulsed PECVD. It is known that pulsed PECVD can deposit higher-quality films and would allow the method to be a single-chamber deposition method, thus further simplifying the process. In addition, techniques such as plasma ALD that are being developed should be considered, as they increase the deposition of ALD while maintaining excellent film quality. Often, additional processing methods that can take advantage of rapid defect filling must be investigated.

The mechanical response of the films was studied using nanoindentation and wafer curvature methods to measure modulus and residual stress. The cracking behavior of various films under tensile stress was investigated, and cracks were found at a radius of curvature of 6.4 mm, which corresponds to a strain of 0.8% in SiO_x, SiN_x, and mixed layers. In order to improve the mechanical limitations of the brittle inorganic layer, which is applied to meet the stringent requirements for barrier properties, a new approach for highly flexible encapsulation structures is proposed. By placing the brittle barrier layer close to the neutral axis using an additional epoxy resin layer with a suitable thickness, the encapsulated structure can maintain its barrier properties in bending with a smaller radius of curvature due to the reduction of strain on the film. For the practical application of this structure, long-term stability and fatigue tests should be performed on the proposed structure under repeated bending.

The compatibility of the encapsulation process with organic devices is investigated by comparing the device performance parameters before and after encapsulation. The encapsulated organic devices were stored under constant conditions, and the performance parameters were measured periodically to compare with the initial values. Both multilayer packages and hybrid packages with three SiN_x/poly(parylene) dimers provided a shelf life in excess of 7000 hours. Therefore, it was concluded that 10^{-5} g/m^2/day is the minimum requirement for barrier performance of at least 7000 hours of shelf life in a PVA/C60-based organic solar cell. In addition, the shelf life of encapsulated organic devices can be further improved by applying in-line processing. Organic devices begin to degrade during encapsulation due to exposure to air during handling and transfer from one process to another. This degradation can be prevented by in-line processing without exposure to air. The process can also reduce barrier failure due to particle contamination that can occur during the transfer of organic devices. Additional work should also consider barrier performance at high temperatures and high humidity (65 °C and 85% RH). In addition, barrier performance under UV or AM 1.5 illumination must be addressed for solar cell applications.

The main advantage of plastic organic electronics is their ability to produce lightweight, mechanically flexible, and large-size electronic devices at a low price and low manufacturing cost. However, there are still many challenges to fully realize the paradigm shift from traditional silicon-based to plastic-organic electronics, including developing manufacturing processes, improving device performance, and improving reliability. In terms of reliability, additional thin-film packaging technologies must be developed to provide sufficient barrier performance while being compatible with the unique properties of plastic electronics. More fundamental research behind barrier layers, such as water vapor diffusion, which shows non-ideal gas behavior through barrier layers. Showing non-ideal gas behavior through various films and the interaction with water vapor leading to changes in barrier film properties, its impact on the performance of organic devices should be extensively investigated. In addition to applications in organic solar cells, the developed thin-film barrier layers can also be used in the packaging of other organic electronic devices, such as organic light emitting diode (OLED) and organic thin film transistor (OTFT). Therefore, further investigation of the compatibility and effectiveness of barrier layers with these devices should be an area of future research. In addition, it is very important to consider the optical effects produced by the barrier layer. The transparency of the barrier layer is a major consideration for display, lighting, and solar cell applications, as it directly affects the efficiency of these devices.

Yoshida et al. used silicon oxide (SiON) to encapsulate the OELD and changed the oxygen density ratio (O/O + N) to improve the transparency (>90%) (Quantum Electron. 2004, 10, 107). Therefore, these efforts should be combined with improvements in barrier properties and mechanical flexibility to fully exploit the advantages of thin-film packaging. With regard to the commercialization of organic electronics, manufacturers require increased line speed and reduced material consumption for thin-film barrier processing. Therefore, based on the available results in this book using widely adapted deposition techniques, further research should be conducted to reduce processing times and steps. Spin-coating processes that reduce annealing time, inkjet printing, roll-to-roll (R2R) processing, and spray methods to deposit thin films could be a few of the candidates. However, it should be noted that the film quality of these processes does not exceed that of the processes used in our study. processes used in our study so far. Therefore, efforts should be made to improve the quality of the films before adopting these techniques. Therefore, efforts should be made to improve the quality of the films before adopting these techniques. As an alternative, based on the barrier structures developed in this study, optimizing the size of individual films or replacing some of the constituent films with films deposited by different processes could be considered. can be considered. In addition to reducing processing steps and time, research on the reliability and continuous yield of the encapsulation process and efforts to reduce failure rates should be conducted in addition to reducing processing steps and time. This is because an increase in the failure rate during processing can lead to an increase in manufacturing costs. The ability to produce uniform films over a large area for cost-efficient package manufacturing. When the formed package barrier layer can be processed with efficient organic electronics, the realization of plastic electronics is expected to become a reality in our daily lives.

Organic electronics is a promising technology for the future, especially for flexible devices. In order to improve stability, more advanced technologies are needed to package devices than traditional methods. Previous studies have shown the potential of thin-film packaging technology for packaging performance. The operating time of organic devices was significantly increased after the deposition of thin films by PECVD and ALD. Water vapor transfer rates were significantly reduced by optimizing materials, barrier layer structures, and deposition processes. Recently, ALD/MLD technology had offered great application potential for flexible device packaging. However, there are still problems in fabrication, and in fact, thin-film packaging is problematic for large-area/flexible applications. This book discusses current issues and potential technologies for improving the potential performance of organic electronic devices. With the development of thin-film packaging methods, flexible devices will be a promising technology for the future.

10.2 Introduction of Three Packaging Technologies for Flexible Electronic Devices

This book focuses on flexible electronics packaging technology, describes the development trend and research progress of flexible electronics packaging technology, reviews the special processes in flexible electronics manufacturing, and looks forward to the development trend and outlook of packaging technology in the fields including OLED, biomaterials, and calcium titanite solar cell materials.

10.2.1 Application of Electronic Packaging Technology in the OLED Field

OLEDs have attracted much attention because of their high luminescence efficiency, fast response time, wide viewing angle, and compatibility with flexible electronics. However, OLEDs are susceptible to environmental moisture/oxygen. Therefore, how to protect OLEDs from environmental moisture/oxygen is of great importance to ensure their stability and reliability. ALD-based thin-film encapsulation (TFE) is a potential method to meet the requirements of OLED encapsulation due to its unique advantages. This paper first introduces the problems faced by TFEs, including pinholes, crystallization, cracking, and overheating. Researchers developed ALD-based monolayers, composite structures, and hybrid laminates to improve barrier properties, flexibility, and thermal conductivity. In addition, ALD reactors and processes for the preparation of TFE are reviewed. Finally, the problems and future directions of OLED stabilization using ALD are discussed [7].

As shown in Figure 10.1, Bulusu et al. demonstrated their contribution to mitigating stress in the Al_2O_3/HfO_2 nanolayer plate by incorporating a SiN_x layer with compressive stress. This innovative approach resulted in a remarkable reduction in the probability of cracking [8]. Figure 10.2 presents the findings of George and

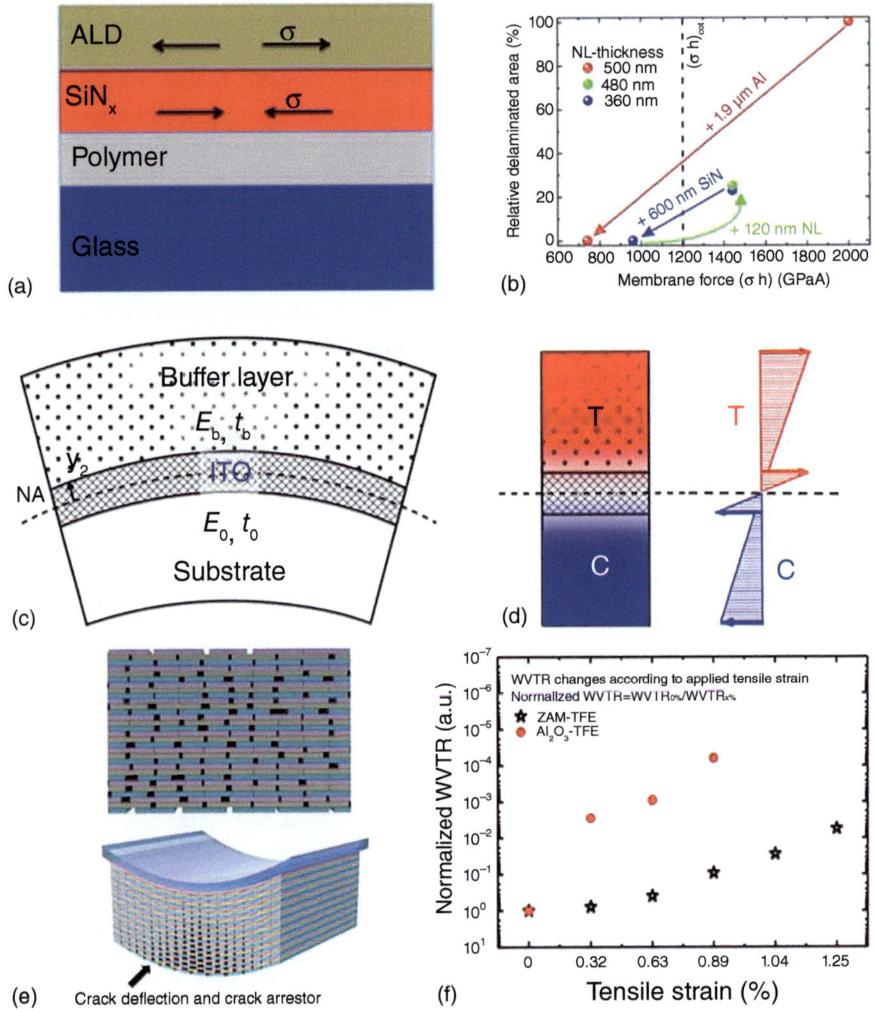

Figure 10.1 Showcases several key aspects related to stress management and barrier properties in OLEDs. (a) Stress management by inserting SiN_x layers with compressive stress; (b) relative layered area NL GDBs of OLEDs composed of various Al_2O_3/TiO_2 with or without stress management layer (SiN_x and Ag); (c) NA location of flexible OLEDs including ITO with buffer layer; (d) stress distribution (E: modulus of elasticity, T: thickness); (e) structural schematic of bead layer cross-section with energy dissipation system and bent ZAM film for flexure and crack arrest; (f) effect of tensile strain on the barrier properties of Al_2O_3-TFE and ZAM-TFE. Source: Reproduced with permission from Li et al. [7]; © 2020, Springer Nature.

his colleagues, who introduced MLD (molecular layer deposition) interlayers to alleviate stress in the inorganic layer. Additionally, they implemented an organic interlayer with minimal crosslinking between polymer chains to alleviate the compressive stress on the inorganic film, thereby decreasing the likelihood of cracking [9]. Kwon et al. reported a dielectric organometallic encapsulation structure to ensure the reliability of flexible OLEDs [10], as shown in Figure 10.3.

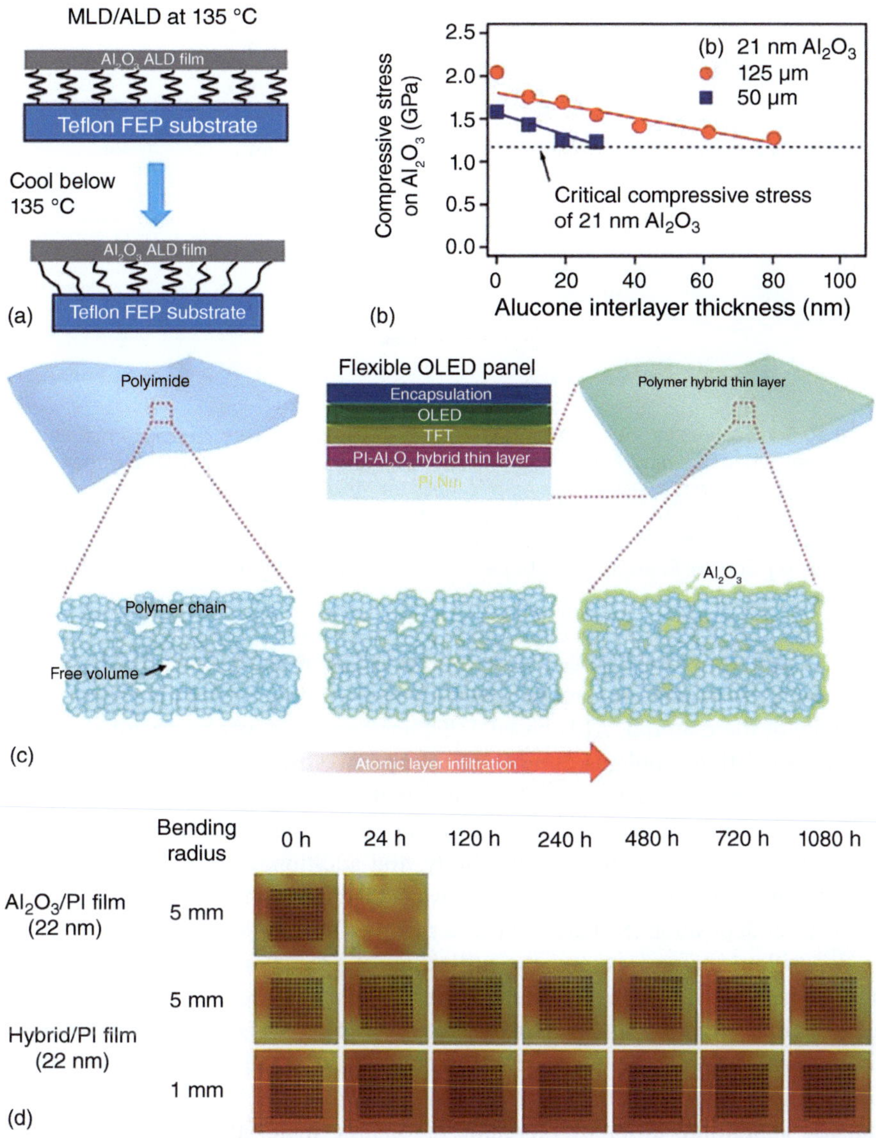

Figure 10.2 Illustrates several aspects related to the nature of the aluminol salt interlayer and the preparation of polymer heterogeneous thin layers. (a) Schematic depicting the "spring-like" nature of the aluminol salt interlayer between the Al_2O_3 ALD film and the Teflon FEP substrate; (b) compressive stress of 21 nm ALD-based Al_2O_3 on Teflon fluorinated ethylene propylene (FEP) substrates with thicknesses of 50 and 125 μm versus aluminol salt interlayer thickness; (c) schematic diagram of the ALI method for the preparation of polymer heterogeneous thin layers; (d) 10,000-year bending cycle of PI films after corrosion by Ca pitting corrosion test. Source: Reproduced with permission from Li et al. [7]; © 2020, Springer Nature.

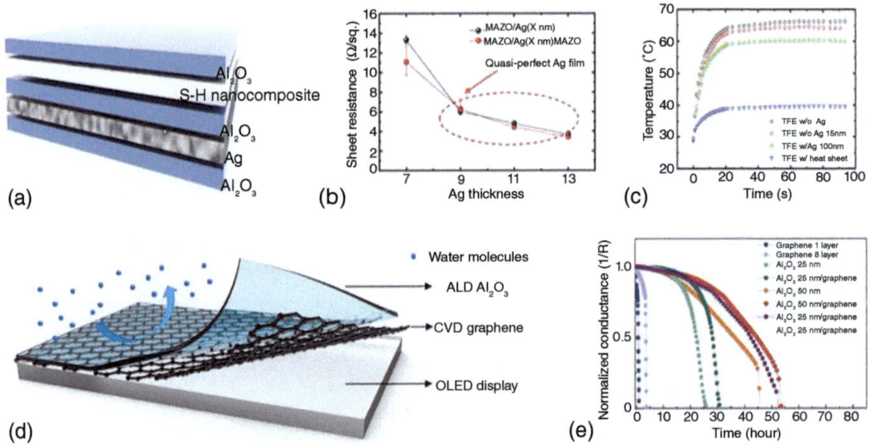

Figure 10.3 Presents various aspects related to the barrier structure and performance of FOLEDs. (a) Schematic diagram of Al_2O_3/Ag/Al_2O_3/S-H nanocomposite/Al_2O_3 barrier structure; (b) sheet resistance versus Ag thickness for MAZO/Ag and MAM structures; (c) real-time temperature distribution of FOLEDs with time for an input power of 80 mW; (d) Al_2O_3/graphene composite barrier layer; and (e) different electrocalcium tests with barrier structures [7]. Source: Reproduced with permission from Li et al. [7]; © 2020, Springer Nature.

OLED is attractive for flexible displays; however, overcoming obstacle layers such as pinholes, crystallization, cracks, and overheating poses a great challenge to the reliability of the material. ALD-based TFEs have emerged as a powerful approach to address existing problems. Single and composite structures fabricated by ALD can decouple pinhole defects, and nanolaminates have excellent water corrosion resistance and inhibit crystallization. In addition, hybrid laminate structures can reduce internal stresses, improve flexibility, and additionally embed heat dissipation systems to transfer heat away and avoid thermal aging. In addition, spatial atomic layer deposition (SALD) and peard methods have been proposed to improve the deposition rate and expand the utilization of precursors. However, there are some remaining challenges that need to be addressed in order to further improve the flexibility and potential stretchability of the displays. As the film thickness continues to decrease, the package structure needs to be further optimized while considering optical emissivity, barrier resistance, and mechanical robustness. A database containing different inorganic or organic materials based on different experimental processes is needed, which covers the mechanical, optical, thermal, and potential barrier parameters corresponding to the different experimental processes. The interface between multilayer hybrid films requires more accurate theoretical simulations considering the interface. In addition, the barrier structure is expected to replace the original capping layer of OLEDs, allowing for a simplified device structure and an effective barrier to moisture. Another current obstacle is how to fabricate large-scale organic layers with precise thickness control and good throughput since organic layers play an important role in the design of ultra-flexible barriers. Inspired by the SALD approach, spatial molecular layer deposition (SMLD) can be achieved by replacing the oxide precursors with organic materials. However, the separation between different precursors in SMLD will be more complicated due to the high viscosity of organic precursors. A more reliable method for organic layer preparation is inkjet printing, which can integrate multiple nozzles to meet the requirements of large-area substrates. In addition, TFE will also be needed for future stretchable electronics, requiring the development of organic–inorganic hybrid layers with good stretching properties. Another pressing issue currently facing ALDs is self-cleaning. Current ALD components rely heavily on alumina and other difficult-to-etch materials. Research on SiO_x and large-scale low-temperature growth is of great importance. The development of ALD precursors also needs to be addressed. On the other hand, masks are needed to prevent electrodes from being covered by dielectric films during the preparation of potential barrier structures. Also, they play an important role in the fabrication of patterned structures for OLEDs, especially for microLEDs in electronic devices. Therefore, to avoid particle contamination, masks need to be cleaned frequently, or thin-film deposition processes that enable in-situ self-cleaning. In addition, it is important to improve the accuracy of linewidth resolution in the edge region. Area-selective ALD is expected to simplify the process and resolve broken inorganic layers with a small folding radius, and TFE is deposited only where it is needed. Designed flexible package structure OLED is systematic and complex engineering; ALD method will play a more important role in the future because of its unique advantages.

10.2.2 Advances in Packaging Research for Flexible Bioelectronic Implants

Bioelectronic implant systems for biomedical and clinical research should combine long-term performance with in vivo biointegration. TFEs are a necessary component for next-generation bioelectronic implants with diagnostic or therapeutic functions in human healthcare and precision medicine. Considerable effort has been invested in TFE, particularly organic–inorganic multilayer structures, as a flexible and conformal alternative to conventional rigid canisters. TFEs are in direct contact with biological media and must therefore exhibit not only biocompatibility, inertness, and hermeticity but also mechanical robustness, consistency, and compatibility with micromachined device fabrication [11] (Figure 10.4).

Many research efforts are underway to develop long-life TFE biocompatible materials with ultrahigh barrier properties, which can guarantee decades of device lifetime and in vivo reliability (2–3 years in vivo). Table 10.1 lists the key quantitative requirements and guidelines for TFE fabrication for next-generation bioelectronic implants. They need to be ultrathin ($\approx 10^{-1}$–10 μm), have a WVTR below 10^{-6} g/m^2/day^1 (at 37 °C, 85% RH), and guarantee an in vivo barrier at 85 and 37 °C/device lifetimes of at least 80 days and 2–3 years. For inorganic barriers, the COS should be greater than 3% to ensure optimal mechanical integrity and the critical radius of curvature should be less than 1 mm. The required uniformity can be described by an equivalent aspect ratio of at least 102 to 103. Optical transparency, biocompatibility, CMOS compatibility, and functionalization are other required aspects.

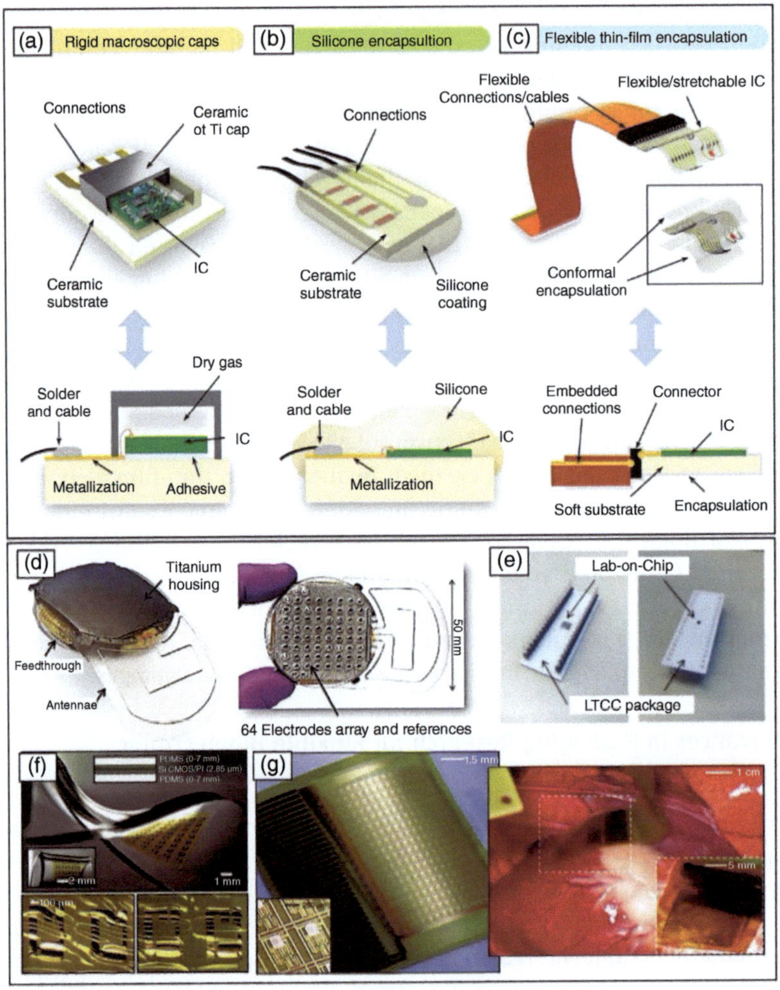

Figure 10.4 Provides an overview of different packaging and encapsulation solutions for implantable bioelectronic devices. (a) The diagram displays external macro capsules made of ceramic or titanium, which are used to encapsulate integrated circuits (ICs) in a dry gas atmosphere on rigid substrates. (b) It depicts conformal soft packaging using silicone on rigid substrates. (c) The figure showcases flexible, conformal, and shape-adaptive thin-film encapsulations (TFEs) designed for ultra-flexible next-generation devices. (d) IC stands for integrated circuit, and Ti represents titanium [12]. (e) It shows a laboratory on-chip sensor "flip chip bonded" to the bottom of an LTCC (Low-Temperature Cofired Ceramic) substrate [13]. (f) The image portrays a twisted and bent wavy Si-CMOS (Complementary Metal-Oxide- Semiconductor) circuit utilizing a dual neutral plane design. An optical microscope image of the sample inverter is provided at both the center and edge of the twisted configuration. (g) The photograph captures a complete flexible electro-physiological multiplexing mapping device in the microbending state. The inset offers a magnified view of a pair of unit cells, while the photograph on the right showcases the device placed on a pig heart. Sources: (d) Reproduced with permission from Mestais et al. [12]; © 2015, IEEE Transactions on Neural Systems and Rehabilitation Engineering. (e) Adapted from Halonen et al. [13]; © 2016, Beilstein-Institut Zur Forderung der Chemischen Wissenschaften. (f) Reproduced with permission from Kim et al. [14]; © 2008, American Association for the Advancement of Science. (g) Reproduced with permission from Viventi et al. [15]; © 2010, American Association for the Advancement of Science.

Table 10.1 Required properties for next-generation bioimplantable TFEs.

Property	Requirement
Thickness	$t_f = 10^{-0}$ to 10^1 μm
Barrier	WVTR < 10^{-0} g m^{-0} day^{-1} (at 37 °C, 84%RH)
	$LT_b \geq 80$ days (at 85 °C in PBS)
	$LT_b \geq$ 2–3 years (at 37 °C in vivo)
Mechanical integrity	COS ≥ 3%
Mechanical flexibility	$R_c = t_f/2$ COS $\leq 10^0 - 10^1$ mm
Conformality	$EAR = LpjAA \geq 10^2$–10^3
	• L depth of coated asperity/hole
	• p, A perimeter and cross section of coated asperity/hole)
Optical	Transparency (Optical transmittance $\mathcal{T} \geq 90\%$)
Biocompatibility	In vitro and in vivo non-inflammatory/non-toxic response
CMOS compatibility	Resistance to solvents and temperature, compatibility with photoresists, suitability for wet/dry-etching, flatness, patternability
Functionalization	Compatibility with surface treatments, applicability with hydrogels, possibility of surface decoration or embedding of NPs

Source: Reproduced with with permission from Ref. [11]; © 2022, Wiley-Blackwell."/John Wiley & Sons/CC BY 4.0.

Multilayer hybrid packages are one of the best strategies for combining high flexibility, optimal uniformity, and ultralow permeability because they rely on layered structural designs that can be produced without harsh or extensive manipulation of the substrate. The principle of defect decoupling and the tortuous diffusion pathways are inherent advantages that guarantee high barrier properties of composite coatings. The production of high-quality barrier packages at low temperatures (<100 °C) remains an unsolved challenge. Process-induced defect distribution within thc inorganic film is the main cause of residual permeation in multilayer packages. The generation of these defects is associated with out-of-equilibrium intrinsic or growth stresses and thermal stresses, all of which depend on the deposition rate and temperature. Therefore, the possibility of creating defect-free, highly dense, and high-dot-density encapsulation layers at temperatures below 100 °C would represent a significant advantage in reducing the risk of cracking and failure due to increased toughness and reduced thermal stress in the deposited material. As an alternative to defect-free encapsulation, new functionalized methods or materials can be considered for the production of smart/active encapsulation, such as conductive polymer coatings or drug-eluting matrices, in the context of electron deformation. Progress requires bioinspired compliance strategies, device/functional tolerance of defects, diffusion, and damage processes, including moisture removal and self-healing mechanisms. Efforts to create stretchable diffusion barrier materials or soft materials for TFE with COS >10% have also emerged in the last few years.

A leading approach is based on various "nanobrick" concepts, often biologically inspired by the microstructure of pearl layers, where 2D nanoparticles, such as clay, graphene, or ultrathin coatings processed with elastomeric materials, form densely arranged arrays of platelets that greatly increase molecular diffusion pathways. The stretch-permeability balance can be improved by considering the 2D structure of the film, e.g. 2D wrinkling. However, a reliable solution to ensure multiaxial deformation and stable sealing is still missing.

Finally, to reliably and effectively test ultrahigh barrier encapsulation, novel water vapor permeation methods with sensitivity limits below 10^{-6}–10^{-7} g/m^2/day need to be designed, as well as in situ investigation methods, especially regarding moisture entry, which would provide real-time monitoring of barrier performance and overall mechanical/functional integrity. Accelerated in vitro testing is the most convenient way to measure barrier and device lifetime, while in vivo testing of encapsulated implanted devices is necessary to assess barrier performance in real-world applications. Corrosive biofluids and micromotion can easily alter TFE integrity and barrier performance in a physiological environment, so in vivo experiments are critical to validate the actual performance of TFEs. To accurately predict durability, reliable time-temperature equivalence relationships for accelerated aging are needed, particularly to verify common empirical or semi-empirical relationships (e.g. Baker's rule and Arrhenius' law). Attention should also be paid to the occurrence of phase transitions between aging temperature and body temperature and the presence of specific energy state interfaces, both of which affect the diffusivity of water. Interfacial surfaces are characterized by interfacial energies and interfacial defects, which affect the transport of water molecules, the formation of water clusters, swelling, and plasticization phenomena through polar interactions. Various coupling processes may also occur between water transport and the physical properties of the encapsulated material, such as changes in the glass transition temperature of the organic phase, resulting in a variation of the water diffusivity with increasing concentration.

Research in materials science and microsystems technology will support further advances in this research area, as well as insights into device needs and requirements gained from end-use applications such as neuroengineering, biomedical engineering, human health care, drug targeting, and precision medicine. Finding optimal ultrahigh barrier packaging for flexible active implants requires an interdisciplinary effort and is an important goal for the next generation of the IoB revolution.

10.2.3 Advances in Packaging Research of Flexible Chalcogenide and Organic Photovoltaics

Chalcogenide solar cells (PSCs) and organic photovoltaics (OPVs) have been rapidly developed over the past decade, exhibiting exciting properties such as high efficiency, flexibility, and potential for large-scale manufacturing via R2R processing. Nevertheless, operational stability is considered an ongoing challenge, as extended equipment life has hardly been observed. This instability can be narrowed down to "intrinsic degradation" and "extrinsic degradation," where exposure to moisture and oxygen has a detrimental effect on device performance. One way to delay the degradation of flexible PSC and OPV devices is through barrier packaging. While glass packages exhibit ideal barrier performance, the potential for flexible devices and high-throughput R2R fabrication requires the development of flexible barrier materials and packaging strategies. These barriers must exhibit excellent resistance to moisture permeation, high transparency, chemical and thermal stability, and must be able to withstand repeated mechanical deformation [16]. This section provides an overview of the latest flexible barrier materials and encapsulation strategies, presents stability studies of flexible PSCs and OPVs, and suggests future research directions.

Methods to compensate for these ionic defects include the introduction of multiple barrier coatings that create more tortuous diffusion paths for the penetrating moisture, as shown in Figure 10.5.

As shown in Figure 10.6, thin-film packaging does not require edge sealing or barrier adhesive materials because the films are typically deposited directly onto the device [17, 18]. High-precision thin-film deposition techniques such as sputtering,

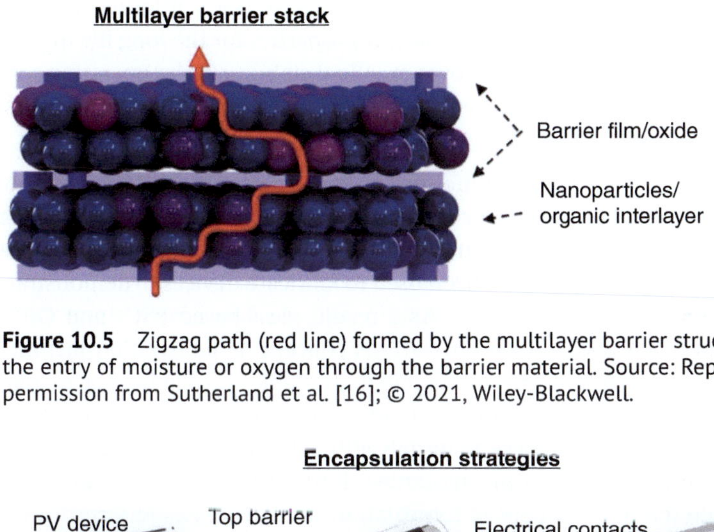

Figure 10.5 Zigzag path (red line) formed by the multilayer barrier structure to prevent the entry of moisture or oxygen through the barrier material. Source: Reproduced with permission from Sutherland et al. [16]; © 2021, Wiley-Blackwell.

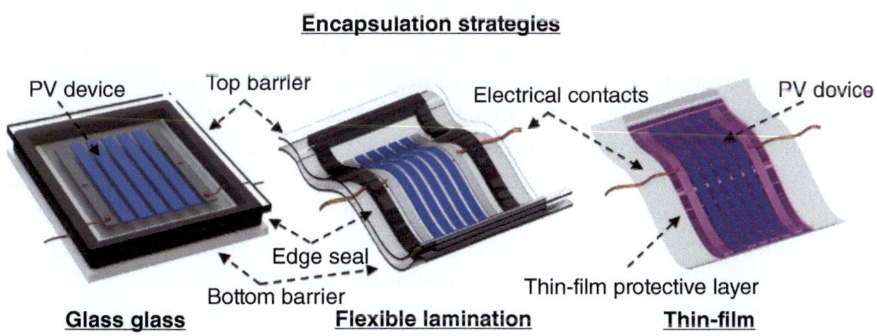

Figure 10.6 presents a schematic diagram showcasing three distinct packaging technologies for PV devices. The leftmost illustration portrays rigid glass–glass packaging, while the middle image depicts a flexible polymer lamination method. Lastly, the rightmost depiction illustrates the employment of flexible thin-film technology. Source: Reproduced with permission from Singh et al. [17]; © 2020, Elsevier.

ALD, and CVD are often used to create dense, pinhole-free layers with the ability to protect the device from moisture and oxygen entry.

OPV and PSC devices have made amazing advances in the last decade of development. The latest OPV and PSC devices have demonstrated competitive efficiencies in laboratory environments compared to widely commercialized silicon photovoltaics. An important development path to bring these flexible electronics into mass production is to improve long-term operational stability so that the devices can demonstrate a longer lifetime. In this perspective, the sensitivity of the BHJ active layer and the chalcogenide absorber layer is emphasized. In addition, the intrinsic and extrinsic stability of OPV and 0OPV was examined for PSC devices, and the penetration of oxygen and water molecules was shown to be a key external factor leading to device degradation. As a means to prevent oxygen and water penetration, various barrier materials and encapsulants have been developed. These not only prevent molecular penetration but also harmful UV light, thermal stress, and mechanical degradation. The widely adopted glass-to-glass encapsulation shows excellent resistance to moisture permeation, however, cannot be applied to flexible electronic design concepts. Therefore, various flexible packaging technologies such as polymer rod grid materials, nanoparticle polymer matrices, and thin films have been introduced. These strategies have proven to have impressive barrier properties for the long life of OPV and PSC devices, some of which have shown excellent stability under operating conditions for several years. Here we also point out what we believe to be the strongest directions for further research in this area [16]:

(1) Barrier properties: To extend the lifetime of PSCs and OPVs, it is essential to continue developing flexible barrier materials with low oxygen and water vapor permeation properties. The literature contains many examples of glass–glass encapsulation protocols that have been used to fabricate devices to demonstrate outstanding environmental stability. As a result, glass-based PSCs and OPVs are being fabricated and installed in various concept demonstration facilities around the world. While some commercially available flexible barrier materials have been shown to significantly extend the lifetime of encapsulated PSCs under ambient conditions, these materials still exhibit WVTR values that are not suitable for a 25-year lifetime guarantee. Further research on flexible barriers could see the development of formulations and processes to improve the performance of flexible substrate barriers. We also expect the continued emergence of organic/inorganic multilayer barrier films that, for example, exhibit the impermeability of organic layers while also allowing for improved mechanical corrosion resistance due to the presence of organic interlayers. The current technologies used to fabricate these barriers are highly specific and costly, thus requiring further development of high-throughput processes.

(2) High-throughput manufacturing: The development of scale-up manufacturing processes for flexible barrier films will significantly reduce the costs associated with the manufacture of flexible PSCs and OPVs. With the advent of R2R solution processing technologies for fabricating PSC and OPV functional layers, the final stage in realizing the manufacturing potential of this technology is to incorporate high-throughput packaging techniques. While this

has been demonstrated in commercial barrier lamination devices, for OPV, full reel-to-reel manufacturing devices showing high reproducibility and high operational stability have not yet been fully realized. We anticipate that further developments will see the continued emergence of R2R film deposition strategies. This can be either direct deposition to seal the functional layers or high-throughput deposition on a flexible substrate and then gluing them together to make a sensitive device with sand.

(3) Room temperature air handling: Even though flexible barrier materials can exhibit appropriate WVTR and OTR properties and high-throughput fabrication processes can be developed, the susceptibility of OPVs and PSCs to moisture and oxygen still puts them at risk for off-gassing effects after encapsulation. As we have discussed, barrier materials need to be adequately pretreated to avoid rapid degradation of the device during operation due to moisture and oxygen expulsion. This would require large-scale material pretreatment and inert-environment fabrication, which is both expensive and unrealistic for commercial production. Therefore, processes need to be developed to mitigate the effects of this exhaust and allow the fabrication and assembly of these devices in ambient air. Further studies are also needed to fully understand the impact of oxygen and moisture exhaust on the newly developed high-efficiency PSCs and non-fullerene-based OPV devices.

Meeting these three conditions will hopefully lead to the production of highly stable OPV and PSC devices entirely by R2R processes at room temperature.

10.3 Flexible Electronics Packaging Development Trend Summary

The development of future flexible electronic packaging technology will certainly insist on improving the cost and efficiency of packaging technology, taking user experience as the starting point of design and achieving more humane purposes, such as flexible sensors, flexible electronic eyes, wearable electronic clothing, flexible electronic paper, flexible circuit boards, artificial muscles, flexible heart monitoring clothing, flexible keyboards, and flexible electronic displays. Compared with traditional electronic devices, their unique flexibility and ductility make malleable flexible electronic devices have very wide and good application prospects in the fields of communication and information, biomedicine, machinery manufacturing, aerospace and defense security, etc.

References

1 Yuan, J. (2017). *Research Progress and Prospect of Flexible Electronic Packaging Technology*. Hangchow: Zhejiang Yusee Technology Co., LTD.

2 Wang, G.Z.Q., He, X.Y., Tang, H.W. et al. (2016). *Research Progress of Flexible Electronic Packaging Technology Outlook*. Wuhan, Hubei: School of Materials Science and Engineering, Huazhong University of Science and Technology.

3 Yin, Z.P., Huang, Y.A., Bu, N.B. et al. (2010). Flexible electronic printing manufacturing: materials, processes and equipment. *Chinese Science* 55 (25): 2487–2509.
4 Xu, W. and Lu, T.J. (2008). Flexible electronic systems and their mechanical properties. *Progress in Mechanics* 38 (2): 137–150.
5 Liu, G., Yan, W. High-power, high-density microwave film multilayer interconnection (HDI) packaging technology. Nanjing 1313 P.O. Box 111 divided.
6 Yu, D., Yang, Y., Chen, Z. et al. (2016). Recent progress on thin-film encapsulation technologies for organic electronic devices. *Optics Communication* 362: 43–49.
7 Li, Y., Xiong, Y., Yang, H. et al. (2020). Thin film encapsulation for the organic light-emitting diodes display via atomic layer deposition. *Journal of Materials Research* 35: 681–700.
8 Bulusu, A., Singh, A., Wang, C.Y. et al. (2015). Engineering the mechanical properties of ultrabarrier films grown by atomic layer deposition for the encapsulation of printed electronics. *Journal of Applied Physics* 118: 085501.
9 Jen, S.-H., George, S.M., McLean, R.S., and Carcia, P.F. (2013). Alucone interlayers to minimize stress caused by thermal expansion mismatch between Al_2O_3 films and Teflon substrates. *ACS Applied Materials & Interfaces* 5: 1165.
10 Kwon, J.H., Choi, S., Jeon, Y. et al. (2017). Functional design of dielectric–metal–dielectric-based thin-film encapsulation with heat transfer and flexibility for flexible displays. *ACS Applied Materials & Interfaces* 9: 27062.
11 Mariello, M., Kim, K., Wu, K. et al. (2022). Recent advances in encapsulation of flexible bioelectronic implants: materials, Technologies, and characterization methods. *Advanced Materials* 34: 2201129.
12 Mestais, C.S., Charvet, G., Sauter-Starace, F. et al. (2015). WIMAGINE: wireless 64-channel ECoG recording implant for long term clinical applications. *IEEE Transactions on Neural Systems and Rehabilitation Engineering* 23: 10.
13 Halonen, N., Kilpijärvi, J., Sobocinski, M. et al. (2016). Low temperature co-fired ceramic packaging of CMOS capacitive sensor chip towards cell viability monitoring. *Beilstein Journal of Nanotechnology* 7: 1871.
14 Kim, D.-H., Ahn, J.-H., Choi, W.M. et al. (2008). Stretchable and foldable silicon integrated circuits. *Science* 320: 507.
15 Viventi, J., Kim, D.-H., Moss, J.D. et al. (2010). A conformal, bio-interfaced class of silicon electronics for mapping cardiac electrophysiology. *Science Translational Medicine* 2: 24ra22.
16 Sutherland, L.J., Weerasinghe, H.C., and Simon, G.P. (2021). A review on emerging barrier materials and encapsulation strategies for flexible perovskite and organic photovoltaics. *Advanced Energy Materials* 11: 2101383.
17 Singh, R., Ghosh, S., Subbiah, A.S. et al. (2020). ALD Al_2O_3 on hybrid perovskite solar cells: unveiling the growth mechanism and long-term stability. *Solar Energy Materials & Solar Cells* 205: 110289.
18 Wang, H., Zhao, Y., Wang, Z. et al. (2020). Hermetic seal for perovskite solar cells: an improved plasma enhanced atomic layer deposition encapsulation. *Nano Energy* 69: 104375.

Index

k

l

m

u

v

w